Soft Computing in Industrial Electronics

Studies in Fuzziness and Soft Computing

Editor-in-chief
Prof. Janusz Kacprzyk
Systems Research Institute
Polish Academy of Sciences
ul. Newelska 6
01-447 Warsaw, Poland
E-mail: kacprzyk@ibspan.waw.pl
http://www.springer.de/cgi-bin/search_book.pl?series=2941

Further volumes of this series can be found at our homepage.

Vol. 80. P. Wong, F. Aminzadeh and M. Nikravesh (Eds.)
Soft Computing for Reservoir Characterization and Modeling, 2002
ISBN 3-7908-1421-0

Vol. 81. V. Dimitrov and V. Korotkich (Eds.)
Fuzzy Logic, 2002
ISBN 3-7908-1425-3

Vol. 82. Ch. Carlsson and R. Fullér
Fuzzy Reasoning in Decision Making and Optimization, 2002
ISBN 3-7908-1428-8

Vol. 83. S. Barro and R. Marín (Eds.)
Fuzzy Logic in Medicine, 2002
ISBN 3-7908-1429-6

Vol. 84. L.C. Jain and J. Kacprzyk (Eds.)
New Learning Paradigms in Soft Computing, 2002
ISBN 3-7908-1436-9

Vol. 85. D. Rutkowska
Neuro-Fuzzy Architectures and Hybrid Learning, 2002
ISBN 3-7908-1438-5

Vol. 86. M.B. Gorzałczany
Computational Intelligence Systems and Applications, 2002
ISBN 3-7908-1439-3

Vol. 87. C. Bertoluzza, M.Á. Gil and D.A. Ralescu (Eds.)
Statistical Modeling, Analysis and Management of Fuzzy Data, 2002
ISBN 3-7908-1440-7

Vol. 88. R.P. Srivastava and T.J. Mock (Eds.)
Belief Functions in Business Decisions, 2002
ISBN 3-7908-1451-2

Vol. 89. B. Bouchon-Meunier, J. Gutiérrez-Ríos, L. Magdalena and R.R. Yager (Eds.)
Technologies for Constructing Intelligent Systems 1, 2002
ISBN 3-7908-1454-7

Vol. 90. B. Bouchon-Meunier, J. Gutiérrez-Ríos, L. Magdalena and R.R. Yager (Eds.)
Technologies for Constructing Intelligent Systems 2, 2002
ISBN 3-7908-1455-5

Vol. 91. J.J. Buckley, E. Eslami and T. Feuring
Fuzzy Mathematics in Economics and Engineering, 2002
ISBN 3-7908-1456-3

Vol. 92. P.P. Angelov
Evolving Rule-Based Models, 2002
ISBN 3-7908-1457-1

Vol. 93. V.V. Cross and T.A. Sudkamp
Similarity and Compatibility in Fuzzy Set Theory, 2002
ISBN 3-7908-1458-X

Vol. 94. M. MacCrimmon and P. Tillers (Eds.)
The Dynamics of Judicial Proof, 2002
ISBN 3-7908-1459-8

Vol. 95. T.Y. Lin, Y.Y. Yao and L.A. Zadeh (Eds.)
Data Mining, Rough Sets and Granular Computing, 2002
ISBN 3-7908-1461-X

Vol. 96. M. Schmitt, H.-N. Teodorescu, A. Jain, A. Jain, S. Jain and L.C. Jain (Eds.)
Computational Intelligence Processing in Medical Diagnosis, 2002
ISBN 3-7908-1463-6

Vol. 97. T. Calvo, G. Mayor and R. Mesiar (Eds.)
Aggregation Operators, 2002
ISBN 3-7908-1468-7

Vol. 98. L.C. Jain, Z. Chen, N. Ichalkaranje (Eds.)
Intelligent Agents and Their Applications, 2002
ISBN 3-7908-1469-5

Vol. 99. C. Huang and Y. Shi
Towards Efficient Fuzzy Information Processing, 2002
ISBN 3-7908-1475-X

Vol. 100. S.-H. Chen (Ed.)
Evolutionary Computation in Economics and Finance, 2002
ISBN 3-7908-1476-8

Seppo J. Ovaska
Les M. Sztandera
Editors

Soft Computing in Industrial Electronics

With 164 Figures
and 21 Tables

Physica-Verlag
A Springer-Verlag Company

Professor Seppo J. Ovaska
Helsinki University of Technology
Institute of Intelligent Power Electronics
Otakaari 5 A
02150 Espoo
Finland
ovaska@ieee.org

Professor Les M. Sztandera
Philadelphia University
School of Science and Health
Computer Science Department
Philadelphia, PA 19144
USA
sztanderal@philau.edu

ISSN 1434-9922
ISBN 3-7908-1477-6 Physica-Verlag Heidelberg New York

Ovaska, Seppo J.: Soft computing in industrial electronics / Seppo J. Ovaska; Les M. Sztandera. – Heidelberg; New York: Physica-Verl., 2002
(Studies in fuzziness and soft computing; Vol. 101)
ISBN 3-7908-1477-6

Physica-Verlag Heidelberg New York
a member of BertelsmannSpringer Science+Business Media GmbH

Printed in Germany

Hardcover Design: Erich Kirchner, Heidelberg

SPIN 10866953 88/2202-5 4 3 2 1 0 – Printed on acid-free paper

To

my sons Sami and Samu

– S.J.O. –

To

my daughter Claudia

– L.M.S. –

"All one can really leave one's children is what's inside their heads. Education, in other words, and not earthly possessions, is the ultimate legacy, the only thing that cannot be taken away."

– Wernher von Braun –

Foreword

Soon after the development of the modern computer, researchers began to consider the possibility of improving upon the von Neumann architecture by employing computer architectures modeled on the human brain. This led to the development of the three methodologies, which form the core of Soft Computing: Neural Networks, Fuzzy Logic, and Evolutionary Computation; each of which takes a different approach to implementing computer hardware or software using a "brain-like" architecture.

Motivated by the layers of neurons and synapses in the brain, neural networks replace the complex single (or multi-) processor von Neumann architecture with a massively parallel network of extremely simple processors which are trained by repeated exposure to desirable responses, rather than programmed to implement a prescribed algorithm. Properly designed their massively parallel architecture can be highly redundant and fault tolerant, while given an appropriate database one can train a neural network to implement a function for which no obvious algorithm exists.

Fuzzy logic was motivated by the observation that human decisions are based on qualitative rather than quantitative information; big, small, fast, slow etc., rather than "equal to 3" or "greater than 5.193." This led to a new logic for dealing with qualitative processing and methods for "quantifying the qualitative" and "qualifying the quantitative," to facilitate the interaction between the fuzzy and crisp worlds. Fuzzy logic is ideally suited for applications where one does not have quantitative data and/or does not need a precise response, yielding systems which are insensitive to noise and other extraneous inputs, robust in the face of modeling errors, and also highly reliable.

Evolutionary computation is a methodology for discrete optimization, motivated by the manner in which nature optimizes biological systems. Each option is coded as a binary or floating-point sequence with the survival of a sequence determined by its fitness to achieve the desired goal. Components from the surviving sequences are then combined and perturbed by mutations to define a new generation of options, with the fittest of the new generation surviving to define the next generation until a satisfactory sequence that optimizes the fitness criteria is obtained. Though computationally intensive, evolutionary computation is perfectly suited for difficult problems where there is no derivative information available to aid the optimization process, and one would otherwise use an exhaustive search technique.

Although sometimes identified just as research, over the past decade soft computing has infiltrated into a myriad of industrial applications, typically within the infrastructure where the user is not even aware of the soft computing component in the application. Probably best known is the use of fuzzy logic in consumer electronics/appliances; algorithms for focusing cameras or controlling a washing machine, where the user simply specifies near or far, hot or cold without a specific quantitative input. Similarly neural networks are used for both acoustic and electrical echo cancellation within the infrastructure of a variety of communication systems, while evolutionary computation is used by system designers already in several engineering applications.

Although Professors Ovaska and Sztandera have chosen to emphasize electric power applications in this interesting book, soft computing applications can be found throughout the industry. In pattern recognition where crisp algorithms often do not exist, soft computing techniques are used to recognize and classify Kanji characters, postal codes, finger prints, and anomalous vibrational modes in machinery. In control engineering, soft computing techniques are used to design and implement intelligent systems where one does not have an accurate model of the process one desires to control, and/or where the plant may change in time due to environmental changes or system failures. Indeed, in my own field of *Aeronautics*, neural and fuzzy control, health monitoring, and sensor systems have been successfully flight tested, while evolutionary computation is frequently used to select between multiple vehicle design options.

The soft computing field is well along the way in making the transition from research laboratories to industrial applications, while Professors Ovaska and Sztandera have made a quantum jump in documenting that exciting transition. Their advanced book, *Soft Computing in Industrial Electronics*, is a must reading for those who are in the forefront of industrial electronics and soft computing (or computational intelligence) applications.

Richard E. Saeks
Accurate Automation Corporation
Chattanooga, Tennessee, U.S.A.

"But what we have not learned is how to mimic the remarkable human capability to perform a wide variety of physical and mental tasks without any measurements and any computations."

– Lotfi A. Zadeh –

Editors' Preface

There are nine chapters in this edited volume discussing various applications of soft computing (SC) in the field of industrial electronics. The applications are divided in two categories: *Electric Power Applications* and *Emerging Applications*. Originally, our idea was to concentrate fully on the category of electric power applications, but as the field of industrial electronics is continuously evolving to new frontiers, we decided to include three chapters discussing illustrative examples of emerging applications as well.

The early roots of soft computing could be traced back to Dr. Lotfi A. Zadeh's book chapter on *soft data analysis* [1] published in 1981. Nevertheless, the actual concept of 'soft computing' was launched about 10 years later, when also the Berkeley Initiative in Soft Computing (BISC), an industrial liaison program, was established at the University of California – Berkeley.

Soft computing is a collection of four core methodologies: evolutionary computation (EC), fuzzy logic (FL), neural networks (NN), and probabilistic computing (PC). In this book, however, only the first three methodologies – EC, FL, and NN – are considered [2, 3, 4]. The main characteristics of soft computing are listed below:

- Capability to approximate various kinds of real-world systems;
- Tolerance for imprecision, partial truth, and uncertainty; and
- Learning from the environment.

These characteristics are commonly leading to better rapport with reality, low solution cost, robustness, and tractability. Dr. Zadeh, the father of soft computing, has emphasized that soft computing provides a solid foundation for the conception, design, and application of intelligent systems employing its member methodologies *symbiotically* rather than in isolation.

From the practicing engineers point of view, this innovative symbiosis thinking should be extended to contain both soft computing and hard computing (HC). The term 'hard computing' is not nearly as widely accepted or known as 'soft computing' – many people use just the word 'computing' when they mean HC and 'computational intelligence' instead of SC. Hard computing solutions are typically straightforward to analyze; their overall behavior and stability can be predictable; and, the computational burden of practical algorithms is usually low to moderate. The advantages and disadvantages of particular HC algorithms and methods are often well known and widely understood in research and development organizations. Therefore, the majority of HC-oriented engineers may have considerable internal inertia to take the emerging SC techniques into active use;

this is definitely the case in the United States and Europe, whereas in Japan the use of soft computing methodologies in various industrial and consumer products has already been active for more than a decade.

In general, it is not relevant to question whether somebody should prefer SC to HC or vice-versa, just as it is not meaningful to ask a similar question concerning, for example, FL and NN. As the soft computing community is not looking for perfect solutions but competitive ones instead, there exists an implicit commitment to benefit from the fusion of various methodologies. Such a fusion can lead to cooperative and complementary combinations of the individual methodologies. Constructive fusion thinking has already been extended beyond the individual SC technologies [5]. There exist a large number of applications in which SC is used together with HC; those approaches aim at technically and economically competitive systems, products, and services. It should be pointed out that a successful methodological fusion can be made, not only at the lower algorithm level, but also at a higher system level.

Soft computing is a major area of research; nearly 50,000 soft computing-related journal and conference papers were published between 1990 and 2000 by the IEEE alone. However, the concept is still evolving, and new methodologies, e.g., chaos computing and immune networks, are nowadays considered to belong to SC. While this methodological evolution is taking place, the number of successful soft computing-based industrial products is increasing concurrently [6]. In the majority of such products, SC is hidden inside systems or sub-systems, and the end user does not necessarily know that soft computing methods are used in control, fault diagnosis, pattern recognition, signal processing, etc. This is the case when SC is mainly utilized for improving the performance of conventional hard computing algorithms or even replacing them. Another class of products uses soft computing for implementing novel intelligent and user-friendly features that could not be realized competitively by hard computing.

The idea of this book is to put together under one cover original contributions by authors who have, in most of the cases, contributed considerably in the field of hard computing, and have just recently pioneered in the application of soft computing to the field of industrial electronics. This makes the chapter contents highly relevant for practicing engineers, who have no specific preference to use SC but solve demanding real-world problems instead. It is hoped that the book will attract researchers and engineers both in the fields of industrial electronics (IE) and computational intelligence (CI). These two research and development communities could benefit each other through constructive cooperation; people in the field of IE are usually electrical engineers, while the CI people have typically a computer science or engineering background. Thus, their viewpoints are different and they could definitely complement each other. It is expected that the use of soft computing will increase greatly in industrial applications, because the conceptual structure of hard computing is much too precise in relation to the great imprecision of the world around us.

Chapter 1, by Gao and Ovaska, offers first a brief review of soft computing applications in motor fault detection and diagnosis. This discussion forms a solid basis for two case studies related to bearing fault diagnosis and gear fault detection. In the first case study, a modified adaptive neuro-fuzzy inference system (ANFIS) is proposed for detecting the severity of bearing faults. This neuro-fuzzy system provides linguistic values as its output. The center of the output membership function corresponds to the *severity* of the bearing fault, while the width indicates the *confidence level* of the diagnosis result. The second case study is an application where an Elman neural network (ENN) is used for detecting gear faults in an automatic transmission application. Genetic algorithm (GA) is used for finding optimal initial values for the internal states of the locally recurrent ENN. The GA-aided back propagation training offers accurate time-series prediction capabilities with practical learning time.

In Chapter 2, Kazmierkowski and Orlowska-Kowalska discuss neural network-based estimation and control methods in converter-fed induction motor drives. They propose the use of neural networks in *sensorless* motor control applications, and analyze the behavior of NN-based velocity estimators and field-oriented control strategies. Besides, they introduce a neuro-fuzzy direct torque control scheme, and provide comprehensive simulations and laboratory measurements to illustrate their interesting findings and conclusions. Various NN topologies and the ANFIS are used for estimation and control. The main motivation of using soft computing in this application is to improve the *low-speed* behavior of sensorless induction motor drives.

Cheok proposes and analyzes the robust operation of a fuzzy logic-based angle estimation algorithm for switched reluctance motors in Chapter 3. It is shown both analytically and experimentally that the FL-based angle estimation scheme offers a high level of operational *robustness* and *reliability* on a wide range of angular velocities. Thus, the proposed estimator is suitable for demanding real-world applications in noisy operating environments. An interesting discussion is also provided on the general concept of robustness.

Chapter 4, by Tomescu and VanLandingham, proposes an adaptive fuzzy logic controller for the master-slave current sharing loop of a paralleled DC-DC converter system. Their design approach is to use a "PID expert" to derive the fuzzy inference rules. Simulations show excellent, *parameter insensitive transient behavior*. The complexity of current sharing power converters justifies the advanced control methods of soft computing, whereas individual DC-DC converters can efficiently be compensated with pure hard computing techniques.

In order to minimize fossil fuel consumption for cost saving as well as carbon dioxide reduction, optimal operation of power energy systems is demanded. Implementation of such an advanced system involves three principal problems to be solved: the short-term load forecasting problem, the unit commitment problem, and the power plant start-up scheduling problem. All these are *highly nonlinear*. In

Chapter 5, Kamiya, Kato, Shimada, and Kobayashi discuss the applications of neural networks and genetic algorithms for achieving optimal load dispatching, and developing an automatic power plant start-up control system.

In Chapter 6, Halinka, Sowa, Szewczyk, and Sztandera propose a neural network-based approach to the protective systems problem in complex power generating units. An efficient *operation mode identification* system, which is one of the crucial components of an intelligent power management system, is proposed. This identification system consists of three NN modules dedicated to particular components of an individual power generating unit, as well as of the main unit responsible for global decision making. The other essential component of the power management system is a sub-system that identifies the *location of disturbances*. It is also based on feed-forward neural networks.

Fuzzy logic can be used successfully to model situations in which *people* make decisions in a *complex* environment. Therefore, fuzzy control and approximate reasoning form an attractive concept in dealing with traffic signal control. In addition, the *uncertainty* of input information to signal controller favors for approximate reasoning. In Chapter 7, Niittymäki, Nevala, and Mäenpää discuss the operation of a fuzzy logic-based traffic controller. Several control strategies in different isolated control environments were tested with encouraging results. This chapter is the first one belonging to the category Emerging Applications.

Chapter 8, by Gordon and Bentley, is concerned with an emerging field of evolvable hardware. In that new field, hardware abstracts are evolved using the methods of evolutionary computation. Three strands of current research and development are reviewed: *innovation*, *generalization*, and *evolvability*. During this, the advantages and challenges of evolving at low levels of abstraction are explored. Finally, two illustrative case studies are presented – one in simulation and the other using a commercial field programmable gate array (FPGA). The field of evolvable hardware is a truly new frontier that may become a future design technique for industrial electronics systems.

In Chapter 9, Fujinaka, Yoshioka, and Omatu apply a category classification method to remote sensing data. This is a supervised and non-parametric method, and employs both a self-organizing neural network and a *k*-nearest neighbor method. From experimental results, it can be seen that the proposed method yields superior classification results compared with other methods. However, the results depend largely on the weather condition, which is a major disadvantage. To overcome this problem, a classification method for radar data analysis is developed using a competitive neural network. It is shown that the proposed method outperforms other methods in average accuracy. Although the presented application is from the specific field of remote sensing, the developed image classification methods are applicable for *industrial image processing* as well.

The intended audience of this book includes industrial electronics professionals, researchers and developers of software/hardware tools for the design of soft computing-based systems, and the entire computational intelligence community. It is expected that the reader is a graduate of electrical engineering, computer engineering, or computer science study program with a modest mathematical background. Our book forms also a good basis for Ph.D. level seminars on soft computing applications.

The editors are grateful to Matti Pöllä, an undergraduate student at the Helsinki University of Technology, for his skillful and enthusiastic assistance in preparing the camera-ready manuscript.

Seppo J. Ovaska
Helsinki University of Technology
Espoo, Finland

Les M. Sztandera
Philadelphia University
Philadelphia, Pennsylvania, U.S.A.

References

1. Zadeh L.A. (1981), Possibility theory and soft data analysis, in *Mathematical Frontiers of the Social and Policy Sciences*, Cobb L. and Thrall R.M. (Eds.), Westview Press, Boulder, CO, U.S.A., pp. 69–129.

2. Fogel D.B. (1999), *Evolutionary Computation: Toward a New Philosophy of Machine Intelligence*, IEEE Press, Piscataway, NJ, U.S.A.

3. Kosko B. (1996), *Fuzzy Engineering*, Prentice-Hall, Upper Saddle River, NJ, U.S.A.

4. Haykin S. (1999), *Neural Networks: A Comprehensive Foundation*, Prentice-Hall, Upper Saddle River, NJ, U.S.A.

5. Ovaska S.J., Dote Y., Furuhashi T., Kamiya A., and VanLandingham H.F. (1999), Fusion of soft computing and hard computing techniques: A review of applications, *Proceedings of the IEEE International Conference on Systems, Man, and Cybernetics*, Tokyo, Japan, pp. I-370 – I-375.

6. Dote Y. and Ovaska S.J. (2001), Industrial applications of soft computing: A review, *Proceedings of the IEEE*, **89**, no. 9, pp. 1243–1265.

*"Sometimes in life we are called to do great things.
But as a saint of our times has said, every day we
are called to do small things with great love."*

– George W. Bush –

Table of Contents

Emerging Applications

Part 1: Electric Power Applications

Motor Fault Detection and Diagnosis Using Soft Computing

Xiao-Zhi Gao and Seppo J. Ovaska

Helsinki University of Technology

Keywords: *soft computing, motor, fault detection and diagnosis, neural networks, fuzzy logic, genetic algorithms, fuzzy-neural networks, optimization, identification, motor control.*

1 Introduction

AC and DC motors are intensively applied in various industrial applications [1]. Changing working environment and dynamical loading always strain and wear motors and cause incipient faults such as shorted turns, broken bearings, and damaged rotor bars [2]. These faults can result in serious performance degradation and eventual system failures, if they are not properly detected and handled. Improved safety and reliability can be achieved with appropriate early fault diagnosis strategies leading to the concept of preventive maintenance. Furthermore, great maintenance costs are saved by applying advanced detection methods to find those developing failures. Motor drive monitoring, fault detection and diagnosis are, therefore, very important and challenging topics in the electrical engineering field [3].

Soft computing is considered as an emerging approach to intelligent computing, which parallels the remarkable ability of the human mind to reason and learn in circumstances with uncertainty and imprecision [4]. In contrast with hard computing methods that only deal with precision, certainty, and rigor, it is effective in acquiring imprecise or sub-optimal, but economical and competitive solutions to real-world problems. The motivation of utilizing the human intuition is that a large number of real-world problems cannot be solved with hard computing methods, due to the fact that either they are too complex to handle or they cannot be described or catalogued by analytical and exact models. However, human experts are sometimes marvelously successful at dealing with these difficulties, e.g., speech recognition against a noisy background. Dr. L. A. Zadeh stresses that precise measurement and control are not always effective in such difficult cases, but *perception* can often help [5]. In a word, soft computing is capable of achieving simple, reliable, and low-cost solutions to challenging problems in modern industry by exploiting the imprecision and uncertainty of human decision making procedure. As we know, *qualitative* information from practicing operators may play an important role in accurate and robust diagnosis of motor faults at early stages. Therefore, introduction of soft computing to this area can provide us with the unique features of adaptation, flexibility, and embedded linguistic knowledge over conventional schemes [6]-[9].

This chapter is organized as follows. First, we give a concise introduction to the conventional motor fault diagnosis in Section 2. Soft computing-based approaches, including operating principles, system structures, and computational algorithms, are then discussed in the following sections. We present a few interesting motor fault diagnosis schemes using soft computing methods, such as neural networks, fuzzy logic, fuzzy-neural, and genetic algorithms in Sections 3, 4, 5, and 6, respectively. Their advantages and disadvantages are also briefly reviewed and compared. Particularly, we discuss two application examples in more details. Some conclusions are finally drawn at end of the chapter.

2 Conventional Motor Fault Diagnosis Methods

There are numerous conventional approaches for motor fault detection and diagnosis [10]. The most straightforward method is the *direct* inspection. It requires careful check-over of the condition of individual motor components to find defective faults. A similar procedure is named particle analysis of lubricate oil of the motor. The oil is first sampled and then taken for laboratory check, which detects the possible faults. This will, however, result in a time consuming and costly examination. The above two approaches are more suitable, on the other hand, for routine maintenance [11] [12].

Classical parameter estimation methods can also be reasonably applied for motor fault detection and diagnosis problems [13]. The underlying idea is that based on some measurement signals from the actual motor, we use parameter identification techniques to estimate relevant information of the motor working condition. Figure 1 illustrates such kind of fault diagnosis process. The parameter estimation strategy is well suited for real-time cases. Nevertheless, it requires a deep understanding of the operating principle of the motor as well as an exact mathematical model. In addition, with the aging of the motor, the original model becomes less accurate.

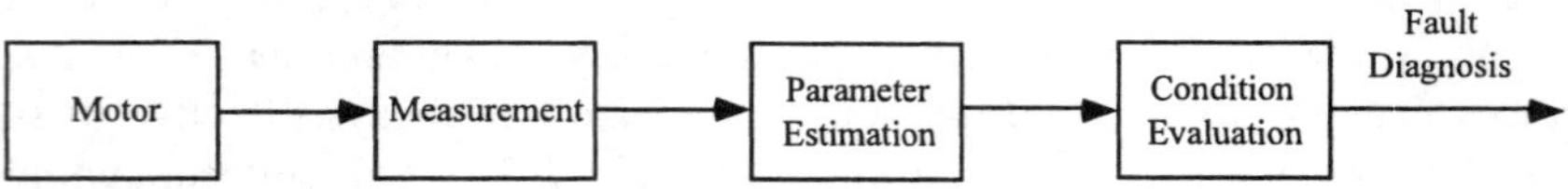

Fig. 1. Motor fault diagnosis using parameter estimation scheme.

During the past few years, soft computing has been employed to overcome the aforementioned difficulties that conventional diagnosis strategies are facing. In general, soft computing methods consist of three essential paradigms: neural networks, fuzzy logic, and genetic algorithms (evolutionary computation) [14] [15]. In this chapter, we discuss the recent progresses of soft computing methods-based motor fault diagnosis. The applications of neural networks, fuzzy logic, and genetic algorithms together with their fusion, e.g., fuzzy-neural networks, in this motor fault detection and diagnosis area will be presented in the following sections, respectively.

3 Neural Networks-Based Motor Fault Diagnosis

Due to their powerful nonlinear function approximation and adaptive learning capabilities, neural networks have drawn great attention in the motor fault detection and diagnosis field. Chow and his colleagues have carried out comprehensive investigation on various neural networks-based fault detection schemes [16]-[19]. They proposed a typical Back-Propagation (BP) neural network structure for incipient motor faults diagnosis, as illustrated in Fig. 2 [19].

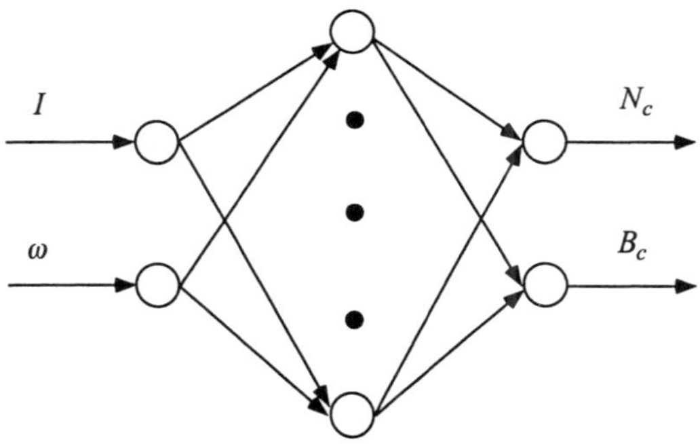

Fig. 2. BP neural network for incipient motor fault diagnosis.

The incipient faults here refer to the turn-to-turn insulation and bearing wear in a split-phase squirrel-cage induction motor. In Fig. 2, I is the steady-state current of the stator, and ω is the rotor speed. N_c and B_c are the conditions of the motor winding insulation and bearing. From the characteristic equations of induction motors, we know that the relationships between inputs (I, ω) and outputs (N_c, B_c) are highly nonlinear. Thus, a BP neural network is applied to approximate this relationship. The training structure is shown in Fig. 3. I and ω can be obtained easily from the on-line measurement data. In fact, the inputs of the BP neural network in Fig. 2 could be further expanded to include higher orders of I and ω, e.g., I^2 and ω^2, which would increase the convergence speed [19].

N_c and B_c, on the other hand, should be evaluated by a human expert, as Fig. 3 shows. More precisely, based on the observation of the working condition and qualitative fault diagnosis knowledge of a *training* motor, the values of N_c and B_c, which quantitatively describe the motor, are classified into three condition levels, {*good*, *fair*, *bad*}, to yield N_c and B_c, respectively.

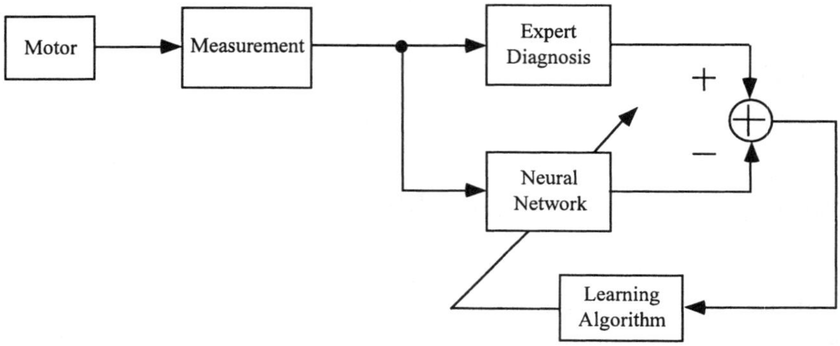

Fig. 3. Training phase for neural network-based motor fault diagnosis.

After the neural network has been trained to learn diagnosis experience from the expert, it is employed on-line, as illustrated in Fig. 4. Judging from the motor operating condition, stator current and rotor speed, the neural network can indicate incipient faults according to the above three fault levels. Filippetti *et al.* proposed a similar BP neural network-based motor fault diagnosis scheme to detect the number of broken rotor bars in [20]. The training data for the neural network is acquired from healthy as well as simulated faulty machines. Their promising scheme has the diagnosis accuracy of 100% in simulations.

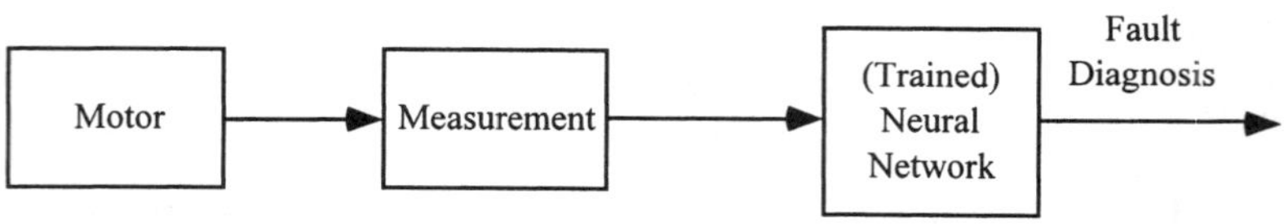

Fig. 4. Neural network-based motor fault diagnosis.

Besides BP neural network, unsupervised learning neural networks are also used in motor fault diagnosis [21] [22]. Penman and Jin applied the Kohonen's Self-Organizing Map (SOM) to classify *multiple* faults in a 3kW, 4-pole, 3-phase, cage induction motor [21]. It is well known that different faults such as stator voltage unbalance, stator interturn short-circuit, and misalignment can occur simultaneously. They are separated and detected from their representing clusters in the feature map. This approach has the practical advantage of learning and producing fault classifications without supervision. In many cases, however, fault data for training the neural network is not easy to collect, because motors have already been repaired during the routine maintenance before those faults really appear. To solve this problem, Tanaka *et al.* introduced a modified SOM-based motor fault diagnosis scheme in [22]. Their SOM-based diagnosis algorithm is constructed to classify unknown signal into *possible* fault and normal data. Real fault data will be identified by an expert and applied for SOM training afterwards. Interesting performance comparison is made with the diagnosis approach using BP neural network. Simulations show that in the case of a relatively small range of input signal, SOM can correctly detect fault conditions, while BP neural network fails.

From the discussions above, it is concluded that the motivation of employing neural networks for motor fault diagnosis is due to their self-adaptation and nonlinear approximation abilities, which can set up the relationship between the indication of faults and available measurement signals. We emphasize, nevertheless, that the learning procedure is usually guided by human experts. Basically, neural networks-based fault diagnosis is a general-purpose solution. No prior knowledge about the motor diagnosis model is needed. Only the training data should be obtained in advance. The inherent drawbacks of neural network learning, such as problem dependent selection of neural network structure, over-training and under-training, and slow convergence speed, may also affect the diagnosis performance. However, the critical shortcoming of neural networks-based motor fault diagnosis is that *qualitative* and *linguistic* information from the operator of motors cannot be directly utilized or embedded in the neural networks because of their numerically oriented '*black-box*' structures. Additionally, it is difficult if not impossible to interpret the input and output mapping of a trained neural network into meaningful fault detection and diagnosis rules.

4 Fuzzy Logic-Based Motor Fault Diagnosis

To take advantage of linguistic fault detection and diagnosis knowledge explicitly, numerous motor fault diagnosis methods using fuzzy logic have been studied [23]-[25]. In [23], Nejjari and Benbouzid applied fuzzy logic to the diagnosis of induction motor stator and phase conditions. Their diagnosis structure, whose kernel is just a representative fuzzy inference system, is illustrated in Fig. 5. Two faults are considered here, i.e., stator voltage unbalance and open phase. The diagnosis procedure is carried out based on the analysis of the amplitude characteristics of stator currents. Three stator current amplitudes I_a, I_b, and I_c are selected as the inputs to the fuzzy fault diagnosis system. Four fuzzy triangular and trapezoidal membership functions, namely *zero*, *small*, *medium*, and *big*, are assigned to each input. The conditions of the stator and phases are represented with three rectangular membership functions, i.e., *good*, *damaged*, and *seriously damaged*. Totally, there are 12 heuristic IF-THEN fuzzy inference rules applied to detect the two aforementioned faults. For example,

IF I_b is *small* THEN the stator is *damaged*,

IF I_c is *medium* THEN the stator is in *good* condition.

Note that the final diagnosis conclusions are based on the outputs from all these 12 rules. This scheme is verified with experimental data in computer simulations, which reveal that it has a good capability of fault diagnosis. The reported diagnosis accuracy rates for good, damaged, and seriously damaged conditions are 100%, 100%, and 94%, respectively.

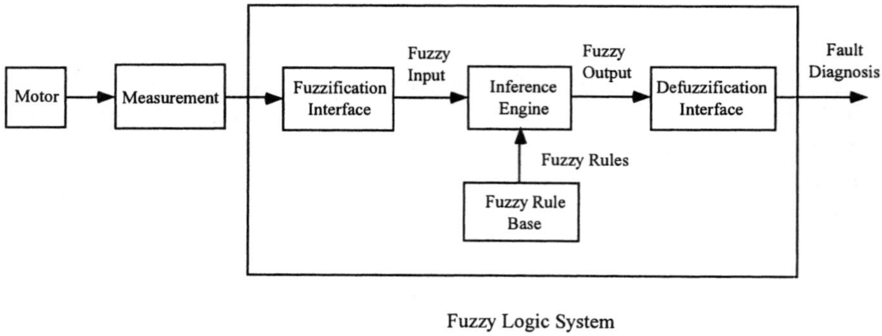

Fig. 5. Fuzzy logic-based motor fault diagnosis.

In [24], Filippetti *et al.* described a broken rotor bar diagnosis system using fuzzy logic in a squirrel cage induction motor. Not only the occurrence but also number of broken bars can be detected in their system. More precisely, there are five kinds of conditions for the rotor bars: no broken bar, an incipient fault, one broken bar, one or two broken bars, and two broken bars. It is known that in an induction motor with rotor asymmetry caused by broken rotor bars, the spectrum lines of harmonics are present at $(1 \pm 2s)f_1$, where s is the slip and f_1 is the stator excitation frequency. The amplitudes of the harmonic components at these two frequencies are denoted here by A_1 and A_2 in dB. The broken bar fault, therefore, can be detected based on these two variables. A typical fuzzy rule used is as follows:

IF A_1 is *large* and A_2 is *small* THEN one broken bar exists,

where *large* and *small* are predefined fuzzy terms. Note that the five possible faults, such as one broken bar, are described by fuzzy membership functions instead of numerical values as in the neural network-based fault diagnosis methods. These membership functions are shown in Fig. 6. The above fault diagnosis method has been successfully employed on a 5.5 kW 2-pole induction motor.

Goddu *et al.* proposed a similar fuzzy logic-based motor bearing fault (bearing looseness, defects of the inner raceway, and defects on the rolling element) diagnosis scheme in [25]. The principle is to analyze the frequency spectrum of the bearing vibration signal using fuzzy reasoning rules. This diagnosis approach achieves 91.7% accuracy in detecting 'severe' conditions and 100% accuracy at both 'good' and 'bad' conditions of the bearing. In the next section, we will revisit this bearing fault detection problem using the fuzzy-neural technique in a case study example.

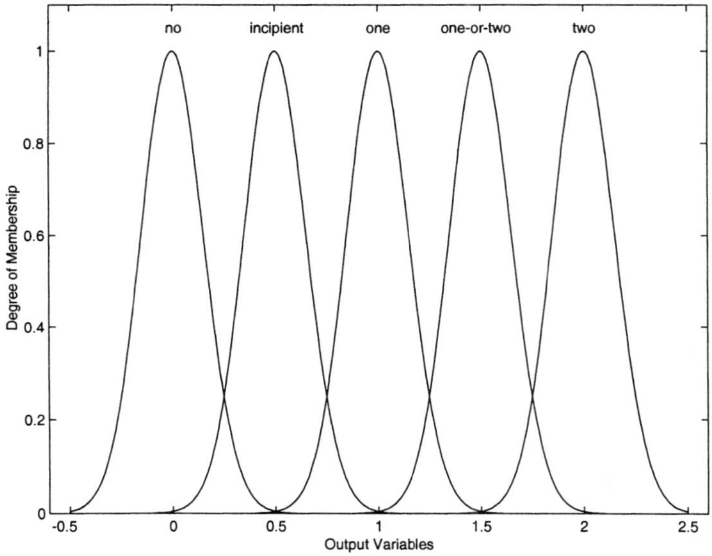

Fig. 6. Membership functions for the broken bar faults

(no = no broken bars, incipient = incipient fault, one = one broken bar, one-or-two = one or two broken bars, two = two broken bars).

Fuzzy logic-based motor fault diagnosis methods have the advantages of embedded linguistic knowledge and approximate reasoning capability. However, design of such a system heavily depends on the intuitive experience acquired from practicing operators. The fuzzy membership functions and fuzzy rules cannot be guaranteed to be optimal in any sense. Furthermore, fuzzy logic systems lack the ability of *self-learning*, which is compulsory in highly demanding real-time fault detection and diagnosis cases. The above two drawbacks can be partly overcome by the fusion of neural networks and fuzzy logic—fuzzy-neural technique.

5 Motor Fault Diagnosis Using Fuzzy-Neural Technique

5.1 Fuzzy-Neural Networks-Based Motor Fault Diagnosis

As we know, both neural networks and fuzzy logic have their own advantages and disadvantages. The major drawbacks of BP neural network are its 'black box' data processing structure and slow convergence speed in training. On the other hand, fuzzy logic has a similar inference mechanism to the human reasoning, while it lacks an effective learning capability. Auto-tuning the fuzzy rules and membership functions may be difficult in a classical fuzzy logic system. In a word, neural networks are regarded as *model free* numerical approaches, and fuzzy logic deals with rules and inference on a linguistic level. Therefore, it is natural to merge neural networks and fuzzy logic into a hybrid system—fuzzy-neural, so that both of them can overcome their individual drawbacks as well as benefit from each other's merits. In fact, fuzzy-neural technique has found many promising applications in the field of motor fault diagnosis [26]-[29].

In [28], Filippetti *et al.* investigated an Adaptive Network-based Fuzzy Inference System (ANFIS)-based motor fault diagnosis scheme, as shown in Fig. 7. Actually, ANFIS is an implementation of a fuzzy logic inference system with the architecture of a five-layer feedforward network [30]. Detailed discussions of the structure and learning algorithm of ANFIS are beyond the scope of this chapter. The explored diagnosis problem here is exactly the same as discussed above [24]. The selection of input and output variables is also similar. However, nine Sugeno-type instead of Mamdani-type fuzzy rules are used in ANFIS. For example,

$$\text{IF } A_1 \text{ is } \textit{large} \text{ and } A_2 \text{ is } \textit{small} \text{ THEN } f = pA_1 + qA_2 + r\,.$$

f is the output of the current fuzzy rule, and p, q, and r are adaptive consequent parameters, which can be adjusted with ANFIS learning algorithm. Numerical output that indicates the final diagnosis decision, i.e., the number of broken rotor bars, is calculated from individual outputs of these nine rules. Simulations demonstrate that this ANFIS-based motor fault diagnosis method can achieve the equivalent diagnosis performance as the pure fuzzy logic system in [24]. Nevertheless, in ANFIS, fuzzy membership functions and fuzzy rules are obtained 'automatically' from the training data instead of by *trial and error*. This alternative saves considerable efforts, and thus speeds up the initial design procedure.

Altug *et al.* discussed another fuzzy-neural network—Fuzzy Adaptive Learning Control/Decision Network (FALCON)-based three-phase induction motor friction fault detection method [29]. Their scheme can provide good diagnosis/detection under varying load torques. Besides the desired diagnosis results, consistent heuristic knowledge in terms of IF-THEN fuzzy rules can be extracted from the trained FALCON. As a matter of fact, rule acquisition is one of the distinguished features of fuzzy-neural networks. Comparison has also been made between FALCON and ANFIS in [29]. It is illustrated that diagnosis/detection acquired using ANFIS is more accurate than FALCON. Furthermore, the convergence speed of fault detection with ANFIS is faster than with FALCON. However, the extraction of diagnosis knowledge in ANFIS is not as straightforward as in FALCON. As a conclusion, fuzzy-neural networks can provide better motor fault detection performance than pure neural networks and fuzzy logic, and it is expected that an efficient fuzzy-neural-based fault diagnosis method should combine the fuzzy rule extraction capability of FALCON with the relatively fast training speed of ANFIS.

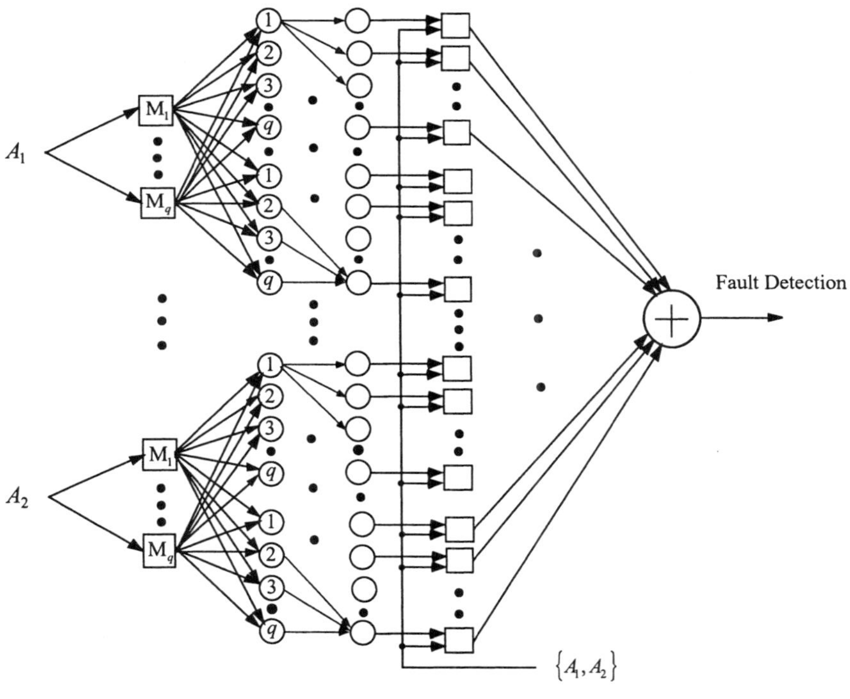

Fig. 7. Motor fault diagnosis using ANFIS.

5.2 Intelligent Motor Fault Diagnosis Using a Modified ANFIS: An Example

In this section, we show an example of intelligent motor fault diagnosis [31]. First, we introduce a new fuzzy logic system approximation method using neural networks. Based on the 'grade' presentation of fuzzy membership functions, any fuzzy relationship can be approximated by a neural network mapping. With the aid of this approach, a modified ANFIS is then presented, which is proven suitable to handle with fuzzy output information. The modified ANFIS-based motor fault diagnosis scheme is finally proposed. Our fault diagnosis scheme has the advantage of learning *directly* from supervising experts. In addition, the diagnosis results acquired can be easily interpreted in a *linguistic* way. The bearing fault diagnosis problem is employed here as a testbed for this approach. Computer simulations are carried out to illustrate its effectiveness.

5.2.1 Approximation of Fuzzy Logic Systems with Neural Networks

During the last decade, fuzzy-neural networks have found extensive applications in a number of engineering fields [14] [15]. Basically, the fusion of neural networks and fuzzy logic can overcome their individual drawbacks while benefiting from each other's merits. However, current fuzzy-neural networks often have complex structures and training algorithms [32] [33]. Besides, some of them still cannot directly deal with fuzzy information [30]. Inspired by the α-sets cut representation of fuzzy numbers [14], we propose a simple neural networks-based approach to processing linguistic information in this section. Our scheme has proven capable of coping with fuzzy input and output without introducing a new network topology or learning algorithm.

It is known that although possessing powerful nonlinear approximation capability, conventional neural networks cannot manipulate linguistic knowledge *explicitly*. On the other hand, in fuzzy logic systems, vague concepts such as 'small' and 'large' are defined in the form of fuzzy sets characterized by membership functions. Hence, in order to enable neural networks to process fuzzy information, we have to find some efficient way to represent membership functions with *numerical* values. The well-known resolution principle indicates that a fuzzy set can be expanded in terms of its α-cuts that are indeed crisp sets [14]. Similar to this idea, a new 'grade' representation scheme for fuzzy membership functions is illustrated in Fig. 8. To put it to more details, the output of a fuzzy membership function, $\mu_A(x)$, is first quantized to a certain number of discrete levels, for example, $\{0.1, 0.2, \cdots, 1.0\}$ in Fig. 8. We can, therefore, obtain the corresponding function input vector denoted as $\{x_1, x_2, \cdots, x_n\}$, where n is the number of quantized membership function grades ($n = 10$ here). It is obvious that if these discrete grade levels are fixed, this fuzzy membership function $\mu_A(x)$ is approximately but *uniquely* determined by $\{x_1, x_2, \cdots, x_n\}$ only. In fact, $\mu_A(x)$ does not necessarily need to be quantizied evenly as shown in Fig. 8. Since most fuzzy membership functions applied in practice are symmetrical, only the left-half part has to be decomposed by the above procedure. Briefly, given a set of discretion grades in general, all the membership functions defined on the same universe of discourse can be represented by their crisp 'input vectors'.

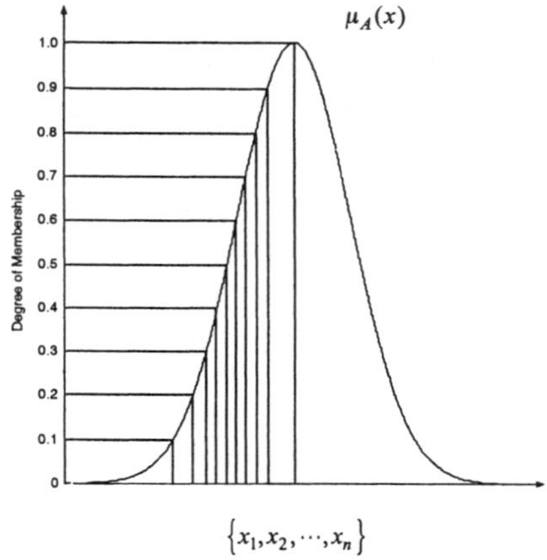

Fig. 8. Representation of a fuzzy membership function with discrete values.

In fuzzy logic theory, a fuzzy inference rule associates output membership functions with input membership functions. To simplify our presentation, we only discuss those Single-Input and Single-Output (SISO) fuzzy rules here. A typical SISO fuzzy rule is conceptually expressed as follows:

$$\text{IF } X \text{ is } \mu_A(x) \text{ THEN } Y \text{ is } \mu_B(y).$$

$\mu_A(x)$ and $\mu_B(y)$ are the membership functions for input and output variables X and Y, respectively. Thus, if $\mu_A(x)$ and $\mu_B(y)$ are represented by $\{x_1, x_2, \cdots, x_n\}$ and $\{y_1, y_2, \cdots, y_m\}$ as discussed above, neural networks can be applied to *approximate* this fuzzy rule by setting relationship between these two data sets, as depicted in Fig. 9. In Fig. 9, m is the number of quantized membership function grades for $\mu_B(y)$. We point out that usually $n \neq m$, because $\mu_A(x)$ and $\mu_B(y)$ are defined on two different universes of discourse.

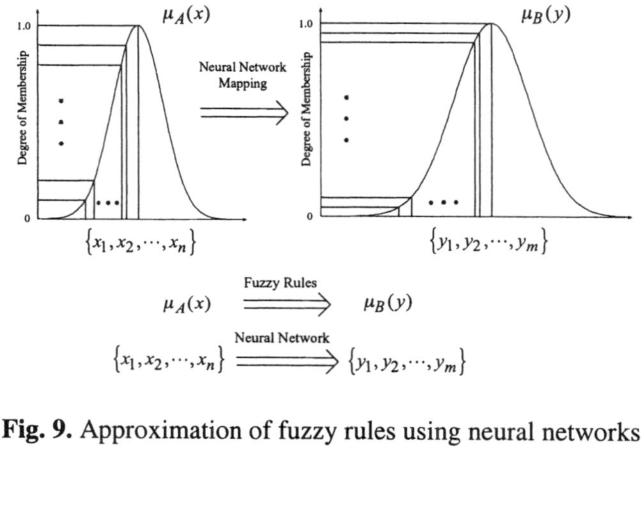

Fig. 9. Approximation of fuzzy rules using neural networks.

Known as a representative of neural networks, BP neural network is used here to implement our idea. Figure 10 reveals the concise structure of such a BP neural network. It has n input nodes and m output nodes, which correspond to $\{x_1, x_2, \cdots, x_n\}$ as well as $\{y_1, y_2, \cdots, y_m\}$, respectively. In this way, the construction of a fuzzy reasoning rule has been converted to a BP neural network-based nonlinear function approximation problem. Consequently, no compulsory variant has to be made on the topology of the neural network. Regular BP learning algorithm can still be employed for training. In addition, it is easy to generalize this scheme to the multi-input and multi-output fuzzy rules case. Comparing with other fuzzy-neural networks, our fuzzy information processing method has the advantages of simple structure and learning algorithm. We demonstrated by simulations that it can efficiently approximate existing fuzzy rules while also *implicitly* generating new rules [34]. In conclusion, with fuzzy membership functions represented in their discrete values, BP neural network is capable of tackling with both fuzzy input and output information. However, the critical shortcoming of this approach is that during the design procedure, prior fuzzy rules cannot be embedded into BP neural network in a straightforward way due to its '*black-box*' structure. We will introduce a modified ANFIS to overcome this drawback in the next section. Like most fuzzy-neural networks, human knowledge can be incorporated in the design phase of ANFIS, e.g., selection of the number and type of input membership functions.

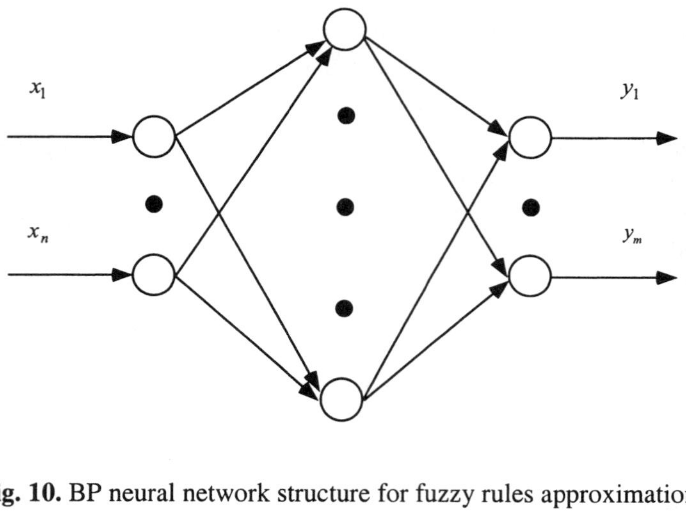

Fig. 10. BP neural network structure for fuzzy rules approximation.

5.2.2 Modified ANFIS with Fuzzy Output

As aforementioned, ANFIS has the features of intrinsic fuzzy inference rules and adaptive learning capability. Nevertheless, it cannot cope with fuzzy input/output information directly. In this section, based on the fuzzy system approximation strategy presented above, we introduce a modified ANFIS network model, as illustrated in Fig. 11, which can effectively tackle *numerical* input as well as *fuzzy* output.

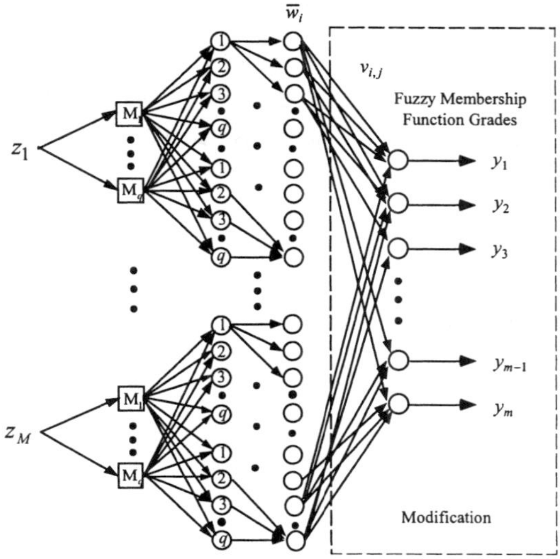

Fig. 11. Modified ANFIS structure with fuzzy output.

Sugeno-type fuzzy inference rules are used in the consequent part of the basic ANFIS [30]. Therefore, if $\overline{w}_i$ is denoted to be the normalized firing strength of rule i in the premise part, the output of ANFIS (single-output case), y, can be represented as:

$$y = \sum_{i=1}^{N} \left[\overline{w}_i \left(\sum_{j=1}^{M} p_j^i z_j + p_0^i \right) \right], \tag{1}$$

where $p_0^i, p_1^i, \cdots, p_M^i$ $(i = 1,2,\cdots,N)$ are the consequent parameters of ANFIS, $z_1, z_2, \cdots, z_M$ are the ANFIS inputs, M is the number of these inputs, and N is the total number of fuzzy rules in ANFIS. N is calculated as $N = q^M$, where q is the number of fuzzy membership functions for each input; refer to Fig. 11. As for our modified ANFIS, we replace the original consequent inference part with a linear *feedforward* network consisting of m output nodes that comply with the 'representation grades' of output fuzzy membership functions. Adjustable weights, v, connect the normalized fuzzy rule outputs, $\overline{w}$, with these nodes. Thus, desired output membership functions can be approximated with the node outputs, $y_1, y_2, \cdots, y_m$. In an original ANFIS with m output nodes, i.e., the number of 'grades' of output fuzzy membership functions is m, there are $(n+1) \times q^M \times m$ consequent parameters to be adjusted. On the other hand, concerning our modified

ANFIS, the number of adaptive weights is only $q^M \times m$. It is apparent that for the same targeted fuzzy membership function approximation problem, computation burden is substantially reduced in the modified ANFIS.

The learning algorithm for weights v is derived based on the gradient descent principle. We first denote $v_{i,j}$ as the connection between firing strength $\overline{w}_i$ and network output y_j. For node j, the output is calculated:

$$y_j^{(k)} = \sum_{i=1}^{m} v_{i,j}^{(k)} \overline{w}_i^{(k)}, \tag{2}$$

where k is the iteration step. $v_{i,j}^{(k+1)}$ can be updated as

$$v_{i,j}^{(k+1)} = v_{i,j}^{(k)} + \eta \Delta v_{i,j}^{(k)}. \tag{3}$$

Here, η is the learning rate. The fuzzy membership function approximation error E is defined:

$$\mathrm{E}^{(k)} = \frac{1}{2} \sum_{j=1}^{m} (d_j^{(k)} - y_j^{(k)})^2, \tag{4}$$

where d_j is the expected output for node j. Applying the derivative chain rule, we get

$$\Delta v_{i,j}^{(k)} = -\frac{\partial \mathrm{E}^{(k)}}{\partial v_{i,j}^{(k)}} = -\frac{\partial \mathrm{E}^{(k)}}{\partial y_j^{(k)}} \frac{\partial y_j^{(k)}}{\partial v_{i,j}^{(k)}} = (d_j^{(k)} - y_j^{(k)}) \overline{w}_i^{(k)}. \tag{5}$$

In fact, adaptation of these extra weights v is just the same as the training of a single-layer Adaline [35]. Additionally, like in the original ANFIS, approximation error at the output of our modified ANFIS can be further back-propagated through each hidden layer to adapt the parameters of the premise fuzzy membership functions. This training approach is proven to significantly accelerate the convergence speed [30].

As a conclusion, our modified ANFIS retains all the advantages of the original ANFIS, such as embedded prior fuzzy knowledge. With simple learning algorithm, it also has the unique feature of dealing with fuzzy output. Therefore, this modified ANFIS can be applied to learn linguistic teaching signals from human instructors. The drawback of our modified ANFIS is that the number of weights in its consequent part, $q^M \times m$, increases drastically with the scale of the fuzzy rule base and desired membership function approximation accuracy. This shortcoming could be relieved by using some adaptive methods to expunge redundant fuzzy rules. We will explore the application of the modified ANFIS in the motor fault diagnosis systems in Section 5.2.3.

5.2.3 Modified ANFIS-Based Motor Fault Diagnosis

As discussed in the previous sections, soft computing methods have shown great potential in motor fault diagnosis. The motivation of utilizing soft computing in this area is to take advantage of the intuitive knowledge from practicing experts. In a practical expert-based diagnosis system, the expert evaluates the working condition of the motor in operation using fault *feature signals*. The feature signals basically consist of real-time measurement signals and their frequency-domain transformations. Moreover, it is obvious that fault diagnosis results obtained in this way are often qualitative, such as *good*, *fair*, and *bad*. That is to say, the supervising diagnosis given by expert is in terms of fuzzy membership functions, which cannot be applied directly to train neural networks. One straightforward solution to this problem is converting these linguistic diagnosis results into appropriate discrete values [16]-[18]. Unfortunately, such an approach is always application dependent, and can easily result in information loss during the conversion.

Since the modified ANFIS is capable of coping with fuzzy output, it is well suited to be applied to mimic the expert's diagnosis behavior in this case. More precisely, the modified ANFIS is used here to approximate the decision making procedure of the expert, i.e., learning the diagnosis experience. The training phase of our modified ANFIS-based motor fault diagnosis scheme is shown in Fig. 12. Note that the modified ANFIS receives the same sensory input (feature signals) as the expert, and the qualitative teaching signal of diagnosis is *directly* provided for training. After the modified ANFIS has been trained to acquire the expert's knowledge, it is employed on-line as illustrated in Fig. 13. From available feature signals, the modified ANFIS can detect incipient motor faults.

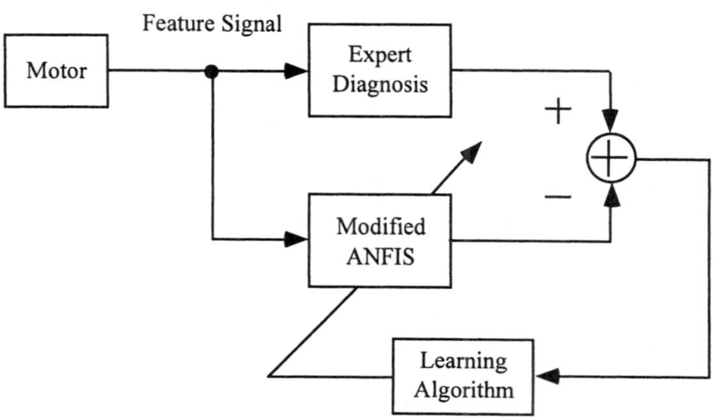

Fig. 12. Training phase of the modified ANFIS-based motor fault diagnosis.

There are several advantages of our modified ANFIS-based motor fault diagnosis method. First, linguistic diagnosis knowledge and rules from practicing operators can be naturally utilized and embedded in ANFIS at the initialization stage. This leads to a tremendously increased convergence speed. Second, by featuring the generalization capability of ANFIS, we are able to interpolate among existing fuzzy diagnosis rules, and thus achieve an enhanced diagnosis performance. Third, learning of the modified ANFIS is directly based on supervising diagnosis results from the expert with no information conversion. Finally, linguistic fault diagnosis output obtained from the trained ANFIS can help us not only detect the occurrence of motor faults but also evaluate their severity degrees. In the following sections, we will demonstrate the effectiveness of our scheme in a bearing damage fault diagnosis problem.

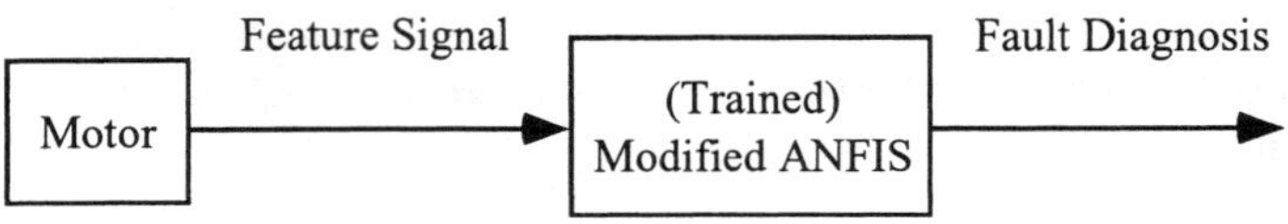

Fig. 13. Modified ANFIS-based motor fault diagnosis.

5.2.4 Bearing Fault Diagnosis in Motor Systems

Bearings are essential components applied in the motor systems [36]. However, stress and friction resulting from severe operation environments always wear aging bearings. In fact, even under normal operation conditions with balanced load and proper alignment, faults caused by bearing damage still take up a considerable proportion (over 40%) of all the motor defects. Therefore, appropriate detection of incipient bearing faults plays a crucial role in improving the reliability and safety of motor systems. Generally, there are three major bearing faults, i.e., bearing looseness, inner raceway damage, and ball damage. To keep our illustrative presentation simple, we only consider one of them, ball damage, in this section.

It is well known that the working condition of bearings is implicitly reflected in the bearing vibration signals [37]. In other words, bearing faults can be detected from five basic frequency responses generated by the motions of bearings. They are f_S, shaft rotational frequency, f_{FCF}, fundamental cage frequency, f_{BPOF}, ball pass outer raceway frequency, f_{BPIF}, ball pass inner raceway frequency, and f_B, ball rotational frequency. These frequencies are actually determined by several parameters of the ball bearing, such as the rotating frequencies of the inner raceway (f_i), ball center (f_c) as well as outer raceway (f_o), cage diameter of the bearing (D_c), number of bearing balls (N_B), diameter of the ball (D_b), and contact angle of the bearing (θ). Nevertheless, detailed calculation of f_S, f_{FCF},

f_{BPOF}, f_{BPIF}, and f_B does not belong to the coverage of our discussion. In practice, frequency analysis of the vibration signals from operating bearings is one of the most commonly employed bearing fault diagnosis methods [38] [39]. To put it to more details, during the diagnosis procedure, the vibration signals in time-domain are first acquired from real-time measurement of bearings under inspection. Fourier transform is used to transform them from the time domain to frequency domain. Both of these two signal types are then delivered to an expert. Based on the vibration signals and fault diagnosis experience gained from practical operations, the expert gives his diagnosis of possible bearing defects. The conceptual structure is illustrated in Fig. 14. As discussed in Section 5.2.3, the goal of our fault diagnosis scheme is to mimic the role of the expert with a modified ANFIS. Computer simulations are made in the next section.

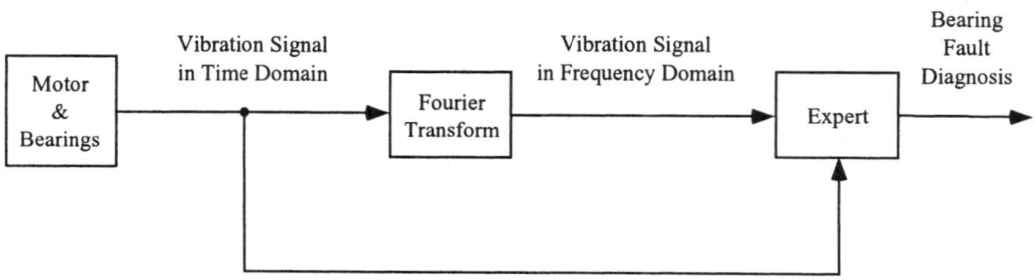

Fig. 14. Expert-based bearing fault diagnosis using vibration signals in both time and frequency domains.

5.2.5 Simulations

In this section, the proposed approach is applied to solve the aforementioned bearing fault (ball damage) diagnosis problem in a motor system. We first summarize the parameters of the ball bearing used in our simulations as follows: $D_c = 6.5\,\text{cm}$, $D_b = 1.27\,\text{cm}$, $\theta = 0$, $N_B = 9$, $f_S = 25\,\text{Hz}$, $f_{FCF} = 10.056\,\text{Hz}$, $f_{BPOF} = 90.85\,\text{Hz}$, $f_{BPIF} = 134.15\,\text{Hz}$, and $f_B = 61.61\,\text{Hz}$. It should be pointed out that all the data given here is quoted from [25], where bearing vibration signals were generated from a MATLAB/SIMULINK-based motor simulation program, namely MotorSim. By applying the Discrete Fourier Transform (DFT) method, we can obtain the power spectrum of these vibration signals. In fact, the vibration *feature vector* that combines time-domain vibration signals with the power spectrums of different basic frequencies (f_S, f_{FCF}, f_{BPOF}, f_{BPIF}, and f_B) is employed for our bearing fault diagnosis scheme. We denote the feature vector as: $\{x_{LOOSE}, x_{DAMAGE}, x_{TIME}\}$. Goddu *et al.* demonstrated in [25] that this feature vector actually consists of amplitudes of the vibration signals in both time and frequency domains. For example, x_{TIME} is calculated as follows:

$$x_{TIME} = \max_{0 \le k \le N-1} |d(k)| , \tag{6}$$

where $N = 2048$, and $d(k)$ denotes the time-domain data generated by MotorSim. More details on how the simulation data is collected and calculated can be found in the same reference paper.

Based on the vibration feature vector, we construct the structure of the modified ANFIS. There are three inputs to our modified ANFIS according to the three elements, x_{LOOSE}, x_{DAMAGE}, and x_{TIME}, in the feature vector. We assign two Gaussian type fuzzy membership functions, labeled *Low* and *High*, to each of the three inputs. In other words, the feature vector elements are linguistically described by these two vague concepts in the modified ANFIS. It is known that any Gaussian-shaped fuzzy membership function can be written as

$$\mu_F(x, \sigma, c) = e^{-\left(\frac{x-c}{\sigma}\right)^2} , \tag{7}$$

where c represents the membership function's center, and σ determines its width. Let $c_1, c_2, \sigma_1, \sigma_2$ be the parameters of the above two membership functions, *Low* and *High*, for each input. The membership functions together with their parameter values for all the three inputs of the modified ANFIS are illustrated in Figs. 15, 16, and 17, respectively. We emphasize that such design of fuzzy membership functions, i.e., choice of number, type, and parameters, is highly *tentative*, which depends on the designer's experience. Generally, the coefficients for the membership functions are initially chosen by *trial and error*, and further fine-tuned using the hybrid ANFIS learning algorithm [30]. In principle, more membership functions could be used to improve the diagnosis accuracy in our scheme. However, this would lead to a heavy inference computation burden. With three inputs and two membership functions for every input, there are totally eight inference rules in our modified ANFIS, which implies that all the possible fuzzy rules have been included here. Simplification of such a *fully-covered* rule base in ANFIS can be implemented by introducing some learning algorithm to delete possible redundant and irrelevant rules [41].

Intuitively, severity degrees of ball damage fault are classified into three *qualitative* categories, i.e., {*None*, *Some*, *Severe*}. Instead of numerical discrete levels, three fuzzy membership functions are utilized to characterize this fault, which are shown in Fig. 18. These fuzzy membership functions are placed *overlapped* with each other instead of *isolated.* The coefficients of the three Gaussian-shaped membership functions are $c_1 = 1$, $c_2 = 2$, $c_3 = 3$, $\sigma_1 = \sigma_2 = \sigma_3 = 0.25$. Again, selection of the above parameters is both problem and expert dependent.

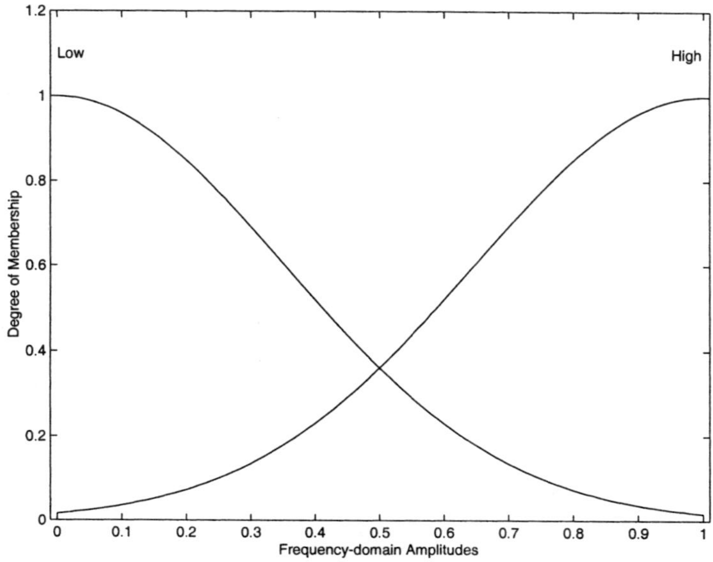

Fig. 15. Fuzzy membership functions for x_{LOOSE} ($c_1 = 0, c_2 = 1, \sigma_1 = \sigma_2 = 0.5$).

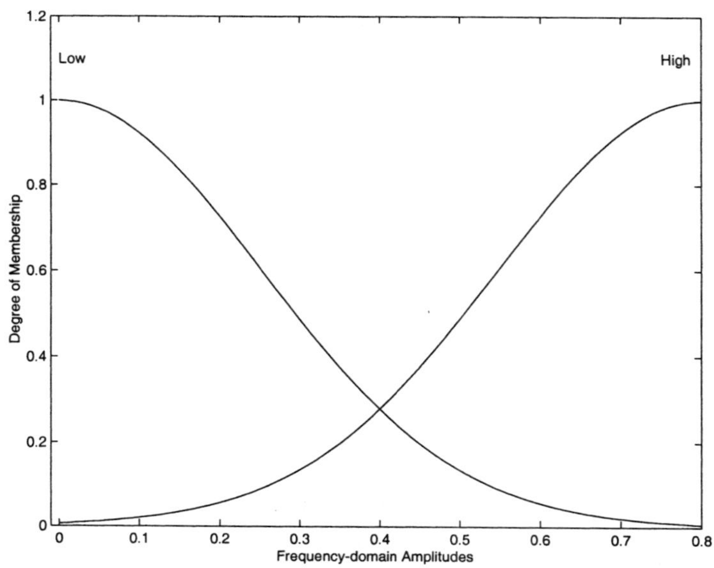

Fig. 16. Fuzzy membership functions for x_{DAMAGE} ($c_1 = 0, c_2 = 0.8, \sigma_1 = \sigma_2 = 0.4$).

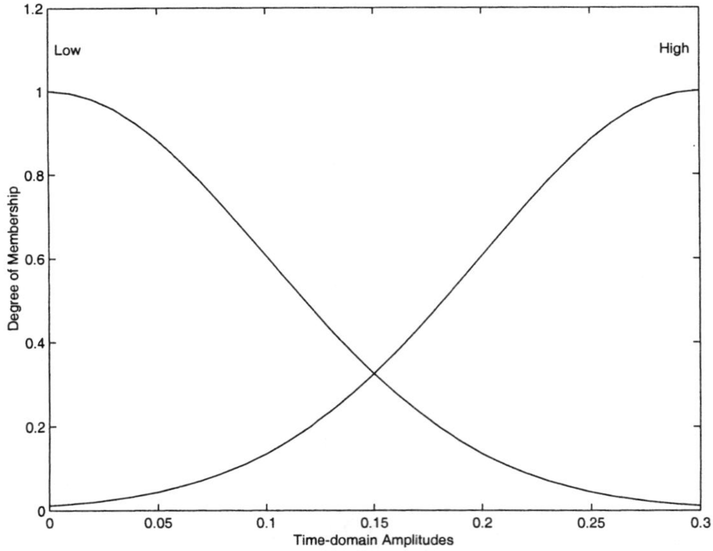

Fig. 17. Fuzzy membership functions for x_{TIME} ($c_1 = 0, c_2 = 0.3, \sigma_1 = \sigma_2 = 0.1$).

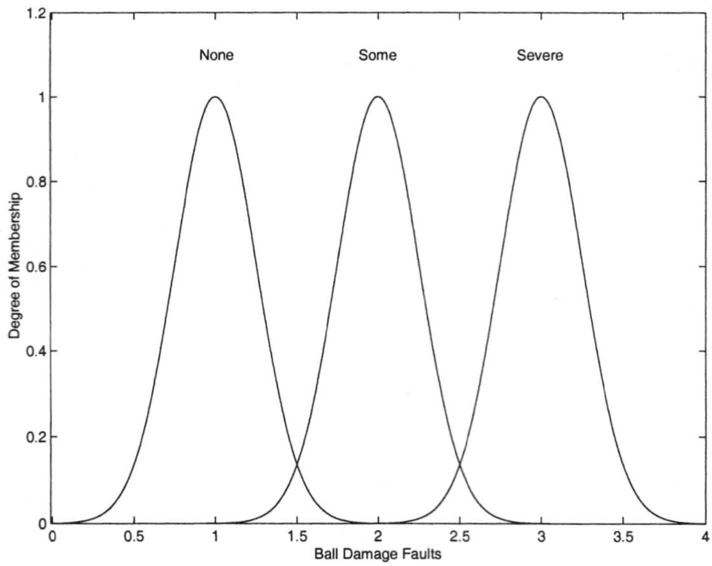

Fig. 18. Membership functions for the ball damage fault ($c_1 = 1, c_2 = 2, c_3 = 3, \sigma_1 = \sigma_2 = \sigma_3 = 0.25$).

Simulation data for our bearing fault diagnosis scheme is arranged in pairs, i.e., feature vector input *vs.* desired ball damage diagnosis output. The desired ball damage faults are determined by practicing diagnosis experts based on the corresponding feature vector. Moreover, simulation data is divided into two separate sets: training and validation data. Four examples of training data pairs are given in Table 1. Totally, there are 60 such data pairs for the training procedure. Another four validation data pairs are illustrated in Table 2, which will be used in the verification phase of our trained scheme.

Table 1. Examples of training data for the bearing fault diagnosis scheme.

x_{LOOSE}	x_{DAMAGE}	x_{TIME}	Ball Damage
8.48e-1	5.90e-2	2.97e-2	None
3.88e-1	2.53e-1	7.90e-2	Some
2.12e-1	3.30e-1	1.63e-1	Severe
1.74e-1	3.48e-1	2.16e-1	Severe

Table 2. Validation data for the bearing fault diagnosis scheme.

x_{LOOSE}	x_{DAMAGE}	x_{TIME}	Ball Damage
7.00e-1	1.72e-1	1.10e-1	None
5.89e-1	2.12e-1	1.40e-1	Some
4.20e-1	2.72e-1	2.00e-1	Severe
3.53e-1	2.90e-1	2.53e-1	Severe

Without significant loss of approximation accuracy, the discretion 'grades' number of the desired ball damage fault membership functions, m, is chosen to be 10. Since the output of a Gaussian type fuzzy membership function cannot be zero, we assign a very small value, 0.0001, to the initial 'grade'. With an even distribution between 0.0001 and 1, the common 'grades' of the fault membership functions are calculated as follows:

$$\{0.0001, 0.11, 0.22, 0.33, 0.44, 0.55, 0.66, 0.77, 0.88, 1.00\}.$$

Therefore, we can get the discrete representation of the three fault membership functions individually (left-half parts only). For instance, the '*None*' ball damage fault membership function is represented as:

$$\{-0.073, 0.48, 0.57, 0.63, 0.68, 0.73, 0.78, 0.82, 0.88, 1.00\}.$$

The learning rate of the modified ANFIS is selected to be $\eta = 0.001$. For simplicity, parameters of the membership functions in the premise part are *not* tuned in our simulations. We use the *batch* training mode for this off-line learning phase, as illustrated in Fig. 19. It can be found out that the modified ANFIS converges after about 500 iteration epochs. Next, the four validation data pairs in

Table 2 are employed to verify the effectiveness of the trained fault diagnosis scheme. The diagnosis results obtained thereafter are shown in Figs. 20, 21, 22, and 23, respectively, where dotted lines represent the 'standard' fault membership functions, and solid lines represent the actual diagnosis outputs.

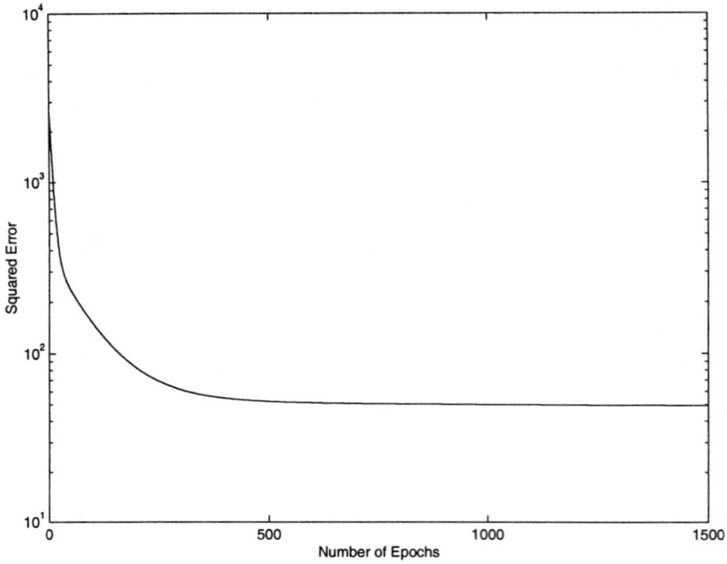

Fig. 19. Training process of the modified ANFIS.

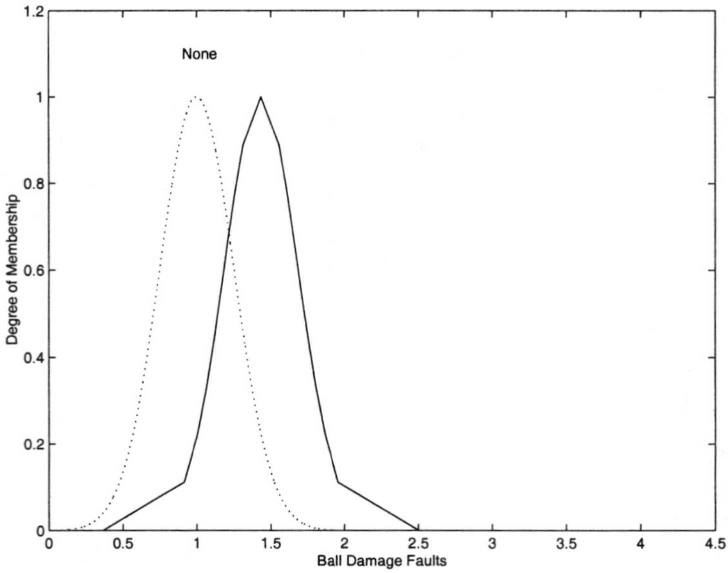

Fig. 20. Diagnosis result for '*None*' ball damage.

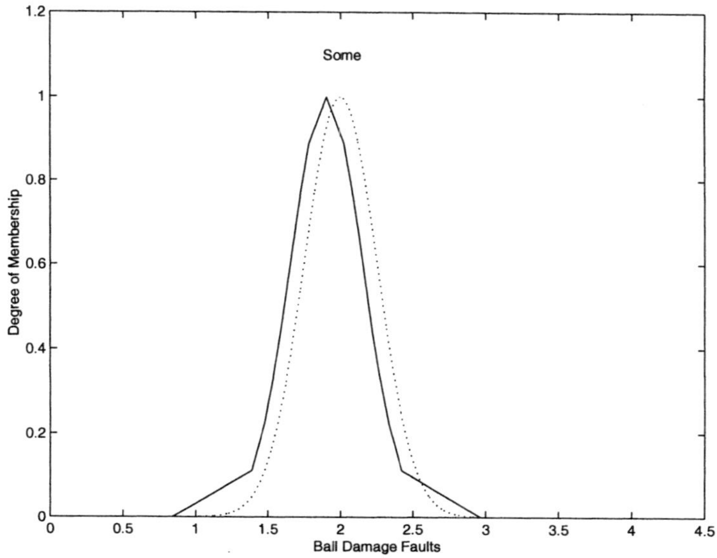

Fig. 21. Diagnosis result for '*Some*' ball damage.

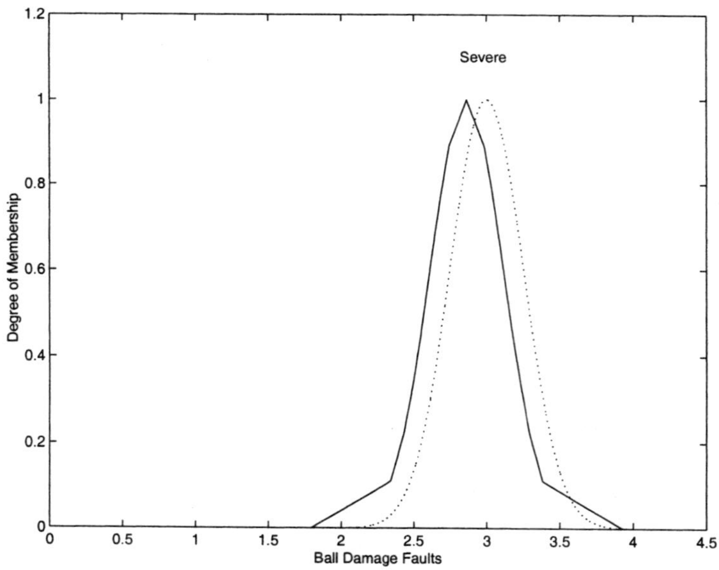

Fig. 22. Diagnosis result for '*Severe*' ball damage (Row 3 in Table 2).

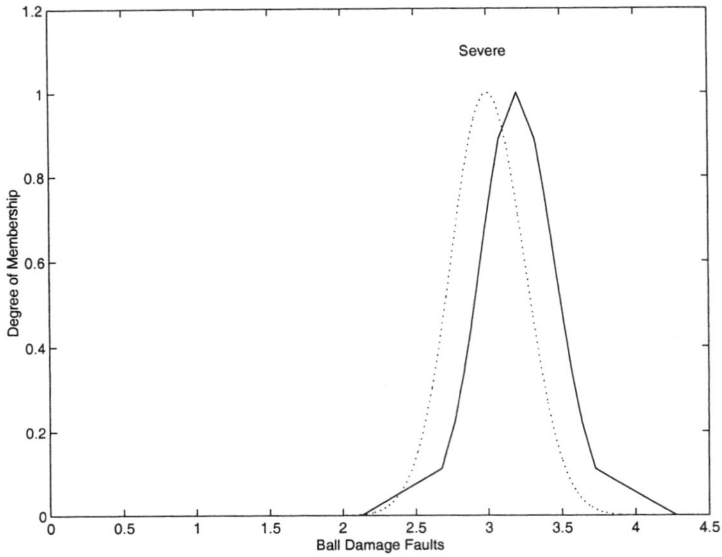

Fig. 23. Diagnosis result for '*Severe*' ball damage (Row 4 in Table 2).

Note that these linguistic diagnosis results in terms of fuzzy membership functions are reconstructed based on the output of the modified ANFIS and above preset 'grades' of ball damage fault membership functions in Fig. 18. In Figs. 20-23, comparing the actual diagnosis outputs with desired ball damage fault membership functions, we can find out that all the four verification diagnosis results are correct, although they are somewhat different from the 'prototype' fault descriptions given by the expert. That means our scheme has achieved a diagnosis accuracy of 100%. As we have discussed previously, the proposed diagnosis scheme can not only detect the ball damage fault but also evaluate its severity linguistically. This can be done by judging from the centers of those output diagnosis membership functions. For example, it is clearly visible that the ball damage fault demonstrated in Fig. 22 is less *severe* than that in Fig. 23, despite the fact that both of them are classified by the expert into the same '*Severe*' category in Table 2. The widths of the membership functions, on the other hand, indicate the *confidence* in the final fault diagnosis results. With simple calculation, we know that the width of the membership function in Fig. 21 is the smallest in these four cases. Therefore, it is concluded that the confidence of diagnosis revealed in Fig. 21 is the highest among them.

5.2.6 Conclusions

In this section, we first introduced a neural network-based scheme for fuzzy membership function representation as well as fuzzy rule approximation. A modified ANFIS with simple learning algorithm was then discussed. By applying the fusion of our neural networks-based fuzzy information processing method and the modified ANFIS model, we proposed a new motor fault diagnosis approach with *linguistic* output. The bearing fault diagnosis problem was used here as an illustrative example and testbed in our simulation experiments. As a matter of fact, the presented scheme can be easily generalized to other motor fault diagnosis cases, where prior knowledge from practicing operator plays an important role in the linguistic diagnosis decision.

6 Genetic Algorithms-Based Motor Fault Diagnosis

6.1 Introduction to Genetic Algorithms

In general, a genetic algorithm is a derivative-free and stochastic optimization method. Its orientation stems from the ideas of natural selection as well as evolutionary process [42]. For a specific optimization problem, the basic GA is summarized as follows [43] [44]:

Step 1: Choose an appropriate encoding method and evaluation function for possible solutions. The encoding method transforms the parameter space of solutions into bit-string representations. A bit-string is named *chromosome* in GA. The evaluation function is always case dependent, and can be applied to evaluate different solution candidates.

Step 2: Initialize a population of candidates (chromosomes) randomly.

Step 3: Evaluate these chromosomes and calculate their fitnesses based on the evaluation function given in *Step 1*.

Step 4: Create new chromosomes by applying *crossover* and *mutation* operators on the current population. Crossover is a unique GA operator that combines two parent chromosomes to produce offspring containing subparts from both parents. A probability term, P_c, determines its operation rate. Mutation is an operator that introduces variations into the chromosomes, and thus adds the property of random search to GA. The mutation operator acts in such a way that each bit of the bit-strings is replaced with a randomly generated bit on a probability bias, P_m to create new chromosomes. In fact, the aforementioned two probabilities for crossover and mutation operators, P_c and P_m, have a center effect on the convergence procedure of GA.

Step 5: Evaluate the new chromosomes produced in *Step 4*, and calculate their corresponding fitnesses as in *Step 3*.

Step 6: Combine the chromosomes from both *Step 2* and *Step 4* together, and merely select the ones with top fitnesses in order to keep the population scale fixed in each generation. In other words, only the mostly fitted chromosomes have the possibility of survival to the next generation.

Step 7: Repeat *Step 4* to *Step 7* until a preset criteria of the evaluation function is met.

As a universal solution to demanding optimization problems, GA has the remarkable features of *parallel search* and *near-global optimization* [42]. Note from the above brief description of GA, we can observe that the evaluation of chromosomes fitnesses and implementation of mutation and crossover operators among chromosomes can be done in parallel. In addition, GA needs less prior information of the problems to be solved than conventional optimization schemes, such as the steepest descent method, which often require the derivatives of objective functions. It is attractive to employ a genetic algorithm to optimize the parameters and structures of neural networks instead of using the back-propagation learning algorithm alone. As we know, the steepest descent-based method is easily trapped into local minima in a nonlinear search space of neural networks weights. Since GA is based on its *probabilistic* rather than *deterministic* search, genetic-neuro technique could overcome the severe shortcomings of applying pure back-propagation learning algorithm to train neural networks. More details of the working principle of GA can be found in [42].

6.2 Genetic Algorithms-Based Motor Fault Diagnosis

In principle, the training of all the motor fault diagnosis methods discussed above can be implemented using genetic algorithms. For instance, in [7], Vas introduced GA into the parameter estimation of an induction motor. In [45], Betta *et al.* discussed the use of GA to optimize a neural network-based induction motor fault diagnosis scheme, which is conceptually illustrated in Fig. 24. Actually, there are two GAs applied for the design and training of the neural network: *designer* GA and *trainer* GA. To put it to more details, the structure parameters of the neural network, i.e., number and dimension of hidden layers, as well as characteristics of the neuron transfer function, are first optimized by the designer GA. And, the connecting weights inside the neural network are optimized with the trainer GA. The application of these two GAs in the tuning procedure leads to the 'optimal' neural network structure and parameters for motor fault diagnosis. Particularly, this scheme is implemented on a TMS320C40 DSP chip to accelerate the GA calculation, which results in a diagnosis response time less than 300μs. The diagnosis performance is promising: the percentage of correct single-fault detection is higher than 98%. Moreover, it can also cope with double-fault, with correct diagnosis of both faults in about 66% of the considered cases and of at least one fault in about 100% of the cases. In Section 6.3, we will give a case study of the Elman neural network-based motor fault detection scheme, in which GA is applied to enhance its detection performance.

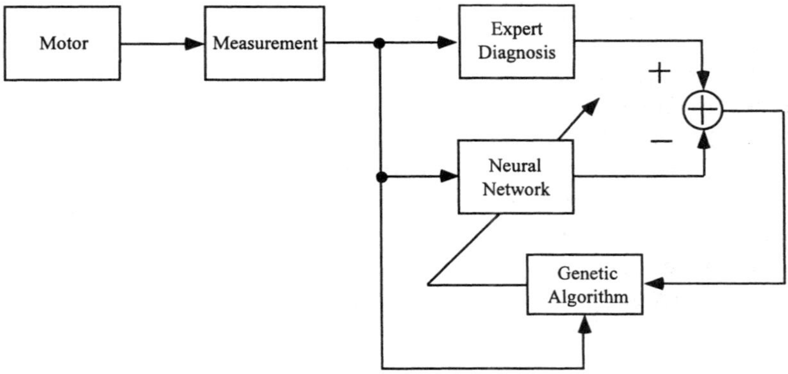

Fig. 24. Application of GA in neural network-based motor fault diagnosis.

6.3 Motor Fault Detection Using Elman Neural Network with Genetic Algorithm-Aided Training: A Case Study

6.3.1 Introduction

In this section, we investigate the fusion of neural networks (Elman neural network) and genetic algorithms with application in the motor fault detection [46]. Elman neural network has the powerful time series prediction capability because of its memory nodes and local recurrent connections. Motor faults are detected from changes in the expectation of squared feature signal prediction error. A GA-aided training strategy for Elman neural network is further introduced to improve the approximation accuracy, and thus achieve better detection performance. Analysis of a practical automobile transmission gear with an artificial fault is carried out to verify the effectiveness of our method. Encouraging fault detection results have been obtained without any prior information of the gear model.

This section is organized as follows: we first introduce some background knowledge of Elman neural network in Section 6.3.2. Our Elman neural network-based motor fault detection scheme is discussed in the following section. Taking advantage of the GA-aided optimization of initial outputs of context nodes, improved fault detection performance can be obtained. With the gear tooth fault detection as an illustrative example, we verify the validity of the proposed method in Section 6.3.4. Some conclusions and remarks are finally given in Section 6.3.5.

6.3.2 Elman Neural Network

Elman neural network is one kind of globally feedforward locally recurrent network model proposed by Elman [47]. It occupies a set of context nodes to store the *internal* states, and therefore has advantageous dynamic characteristics over static neural networks, such as the BP neural network and radial-basis function networks. The structure of an Elman neural network is illustrated in Fig. 25, where z^{-1} is a unit delay. It consists of four layers: input layer, hidden layer, context layer, and output layer. There are adjustable weights connecting every two adjacent layers. Generally, Elman neural network can be considered as a special type of feedforward neural network with additional memory neurons and local feedback. The distinct 'local connections' of context nodes inside Elman neural network make its output sensitive to not only the current but also history of input data, which is essentially useful in time series prediction.

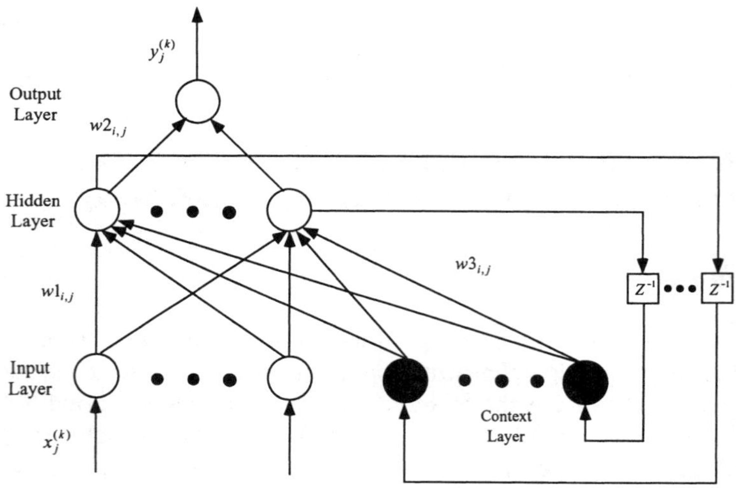

Fig. 25. Structure of an Elman neural network model.

The training algorithm for Elman neural network is similar to BP learning algorithm, both based on the gradient descent principle. However, the roles that the context weights as well as initial context nodes outputs play in the error back-propagation procedure must be taken into consideration in the derivation of this learning algorithm. From Fig. 25, at iteration *k*, we have the following relationship:

$$o\_c_i^{(k)} = net\_c_i^{(k)}, \tag{8}$$

where o_c_i and net_c_i are the output and input of context node *i*, respectively. Note the node activation function of the context nodes is *linear*. We denote $w1_{i,j}$

as the weight that connects node i in the input layer to node j in the hidden layer, $w2_{i,j}$ the weight that connects node i in the hidden layer to node j in the output layer, and $w3_{i,j}$ the weight that connects context node i to node j in the hidden layer. Let m, n, and l be the numbers of nodes in the input, output, and hidden layers, respectively. To calculate the input and output of hidden node j, we have

$$net\_h_j^{(k)} = \sum_{i=1}^{m} w1_{i,j} x_i^{(k)} + \sum_{i=1}^{l} w3_{i,j} o\_c_i^{(k)} , \tag{9}$$

and

$$o\_h_j^{(k)} = \mathrm{a}(net\_h_j^{(k)}) . \tag{10}$$

$x_i^{(k)}$ is the input of Elman neural network ($i = 1,2,\cdots,m$). net_h_i and o_h_i are the input and output of node i in the hidden layer ($i = 1,2,\cdots,l$). The nonlinear activation function in the hidden layer is represented by $\mathrm{a}(\cdot)$ here. Therefore, output of node j in the output layer, $y_j^{(k)}$, is given by:

$$y_j^{(k)} = \sum_{i=1}^{l} w2_{i,j} o\_h_i^{(k)} , \tag{11}$$

where $j = 1,2,\cdots,n$. We emphasize that the output of context node i at the kth training iteration is, in fact, *one-step-delayed* of the corresponding output of hidden node i:

$$net\_c_i^{(k)} = o\_h_i^{(k-1)} . \tag{12}$$

Actually, it is the existence of context nodes together with their initial outputs and variable weights connecting them with hidden nodes that distinguishes Elman neural network from conventional feedforward neural networks.

Suppose $Y_i^{(k)}$ is the desired output of Elman neural network at training step k, and we update $w1_{i,j}$, $w2_{i,j}$, and $w3_{i,j}$ to minimize the following approximation error E in the output layer:

$$w^{new}\zeta = w^{old}\zeta + \eta \Delta w \zeta, \ \zeta = 1,2,3 , \tag{13}$$

where η is the learning rate. E is defined as:

$$\mathrm{E}(w) = \frac{1}{2} \sum_{k=1}^{T} \sum_{i=1}^{n} (Y_i^{(k)} - y_i^{(k)})^2 . \tag{14}$$

Here, T is the length of the training sequence. $w1_{i,j}$ and $w2_{i,j}$ can be adjusted using the basic BP learning algorithm [35], because this portion of Elman neural network is just a regular feedforward network structure. Especially, when deducing the adaptation algorithm for weight $w3_{i,j}$ based on the derivative chain rules, we have:

$$\Delta w3_{i,j} = -\frac{\partial \mathrm{E}}{\partial o\_h_j}\frac{\partial o\_h_j}{\partial w3_{i,j}}. \tag{15}$$

At the kth iteration, we get:

$$\frac{\partial \mathrm{E}}{\partial o\_h_j^{(k)}} = \sum_{i=1}^{n}\frac{\partial \mathrm{E}}{\partial y_i^{(k)}}\frac{\partial y_i^{(k)}}{\partial o\_h_j^{(k)}} = -\sum_{i=1}^{n}\left[(Y_i^{(k)} - y_i^{(k)})w2_{j,i}^{(k)}\right] \tag{16}$$

and

$$\frac{\partial o\_h_j^{(k)}}{\partial w3_{i,j}^{(k)}} = \mathrm{a}'(net\_h_j^{(k)})\frac{\partial net\_h_j^{(k)}}{\partial w3_{i,j}^{(k)}} = \mathrm{a}'(net\_h_j^{(k)})o\_h_i^{(k-1)}. \tag{17}$$

In (17), $\mathrm{a}'(\cdot)$ is the first order derivative of $\mathrm{a}(\cdot)$. Now, $\Delta w3_{i,j}$ can be rewritten as:

$$\Delta w3_{i,j} = \mathrm{a}'(net\_h_j^{(k)})o\_h_i^{(k-1)}\sum_{k=1}^{T}\sum_{i=1}^{n}\left[(Y_i^{(k)} - y_i^{(k)})w2_{j,i}^{(k)}\right], \tag{18}$$

where $o\_h_i^{(0)}(i = 1,2,\cdots,l)$ are the 'initial states' of Elman neural network. Normally, they are just assumed to be:

$$o\_h_i^{(0)} = 0, \ (i = 1,2,\cdots,l) \tag{19}$$

for simplicity. This equivalently leads to $o\_c_i^{(1)} = 0,(i = 1,2,\cdots,l)$. From (9) to (11), it is clear that the initial outputs of context nodes, $o\_c_i^{(1)}$, have an important contribution to the output of Elman neural network. Further, based on (18), they play an indispensable role in the learning algorithm. Consequently, the overall performance of Elman neural network could be enhanced by *separately* optimizing $o\_c_i^{(1)}$ in some way.

Due to its powerful dynamical properties, Elman neural network has found numerous applications in such areas as time series prediction, system identification, and adaptive control. One of the distinguished advantages of applying Elman neural network in time series prediction is that the order of the time series is not needed for designing the Elman neural network-based predictors [48]. In addition, it is well suited for dealing with the intrinsic nonlinearity in the time series data. However, regular BP learning algorithm is often trapped into local minima, which may result in unavoidably large approximation errors deteriorating the prediction accuracy. Genetic algorithms introduced in Section 6.1 offer alternative solutions to this difficult problem. We will discuss employing them to optimize an Elman neural network-based motor fault detection scheme in the next section.

6.3.3 Motor Fault Detection Using Elman Neural Network with GA-Aided Training

A. Optimization of Elman Neural Network with GA

There are many examples of applying GA to optimize the parameters and structures of neural networks [49]-[51]. However, direct GA-based optimization of the weights in Elman neural network is time consuming, because the network size grows drastically with the number of hidden nodes. On the other hand, it is well known that initial values, e.g., initial weights, of neural networks also strongly affect the final training results [35]. Additionally, the initial outputs of context nodes in Elman neural network, $o\_c_i^{(1)}$, occupy a considerable role in the approximation accuracy. Normally, we let $o\_c_i^{(1)}=0$ as a general initialization, which could, unfortunately, cause the training to be trapped into poor local minima. Therefore, initial outputs of context nodes have to be taken into account for Elman neural network optimization. Since the initial outputs cannot be directly adjusted by any pure gradient descent-based learning algorithm, GA is suitable in optimizing these values. In brief, based on the combination of BP training and GA optimization, we propose a *hybrid* adaptation method for Elman neural network, as described below:

Step 1: Given a set of training data, the Elman neural network is first trained with *zero* initial context nodes outputs ($o\_c_i^{(1)}=0$) using BP learning. When the BP training procedure is detected to be stuck into a possible local minimum, i.e., no apparent approximation error reduction can be observed within a long enough iteration interval, it is terminated manually, and all the weights obtained so far are fixed.

Step 2: With the trained Elman neural network, GA is then applied to optimize the initial outputs of context nodes. To put it to more details, the approximation error of Elman neural network can be considered as the performance index for GA, refer to (8) - (15) in Section 6.3.2. Notice the fitness of each candidate of the initial context nodes outputs is calculated by subtracting its corresponding approximation error from a large constant:

$$Fitness = Constant - Approximation\_Error \,. \tag{20}$$

In principle, this step is a simple GA optimization phase that can be easily implemented using some available software packages [52].

Step 3: The ultimate Elman neural network is finally constructed based on the weights and initial context nodes outputs resulting from *Steps 1* and *2*, respectively.

Reasonably, with the GA-aided optimization of initial context nodes outputs, we can get a better Elman neural network than using the BP algorithm alone. However, it must be emphasized that in our method the initial outputs of context nodes optimized by GA are 'optimal' only under the constraint of *frozen* BP trained weights. Indeed, an optimal Elman neural network has to be acquired by applying GA to optimize both these two sets of parameters, weights and initial context nodes outputs, simultaneously.

B. Motor Fault Detection Using Elman Neural Network

It has been shown that the working conditions of a motor are implicitly reflected in its feature signal [12]. Therefore, the idea of our Elman neural network-based motor fault detection scheme is to first predict the time series of this feature signal with Elman neural network, and then inspect any suspicious change in the prediction characteristics, which is assumed to be caused by incipient faults. The training structure of feature signal prediction using Elman neural network is given in Fig. 26.

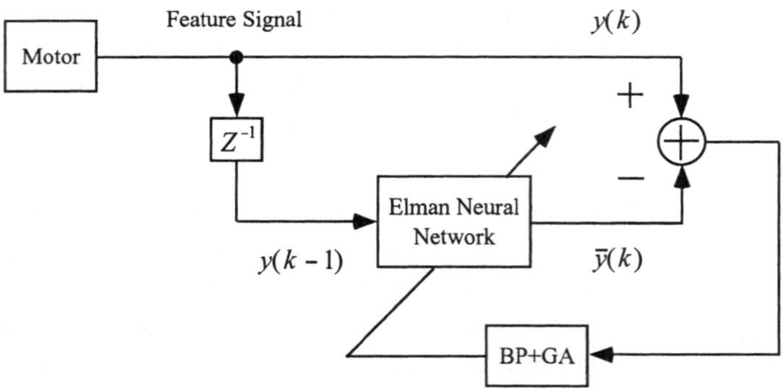

Fig. 26. Training structure of Elman neural network-based motor feature signal prediction.

In Fig. 26, $y(k)$ is the feature signal value at sample index k, and $\bar{y}(k)$ is the prediction output of Elman neural network. Note the training time series data here is collected from healthy motors only. Using the aforementioned hybrid learning algorithm (BP+GA), our Elman neural network is trained to give one step ahead prediction of this feature signal.

We denote $E_p(k)$ as the prediction error calculated by:

$$E_p(k) = y(k) - \overline{y}(k) . \tag{21}$$

The expectation of *squared* prediction error, $E_e(k)$, is defined in a recursive way:

$$E_e(k+1) = \frac{k}{k+1} E_e(k) + \frac{1}{k+1} E_p^2(k+1) . \tag{22}$$

The signs of the prediction errors are eliminated by their squared forms in (22). As a matter of fact, motor faults can be detected based on the change of $E_e(k)$, as illustrated in Fig. 27. More precisely, when employed on-line for healthy motors, $E_e(k)$ converges to a small constant, since the trained Elman neural network is capable of giving a satisfactory prediction of the feature signal. On the other hand, when the motor is working in *abnormal* conditions, $E_e(k)$ will significantly deviate from the nominal value. If this deviation grows above a preset threshold, we conclude that a fault has been detected. Computer simulations of an automobile transmission gear fault detection are made in the next section to verify the effectiveness of the proposed scheme.

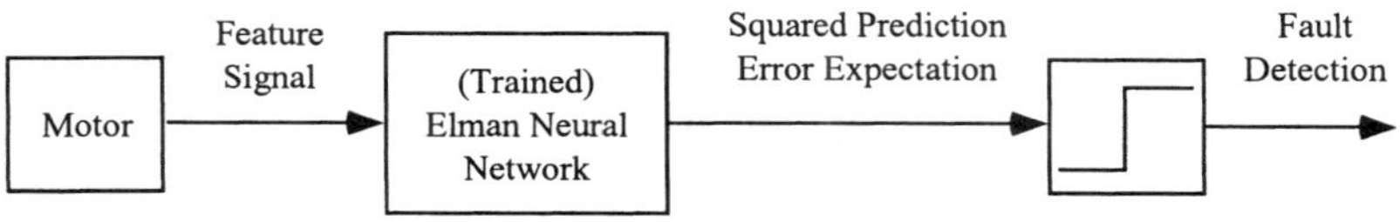

Fig. 27. Elman neural network-based motor fault detection.

6.3.4 Simulations

In this section, our method is examined in a practical automobile transmission gear fault detection application. The acoustic time series data, which represents the sound level of the working gear, was provided by a Japanese automobile manufacturer. Below is a brief summary of the experimental measurement conditions and used instrumentation [53]:

- 1000 rpm rotational velocity of the automobile engine;
- 5.0 Nm load torque;
- 20 μm cut in one of the gear teeth (an *artificial fault* machined by a technician);
- Integrated sound level meter (Ono Sokki LA-5110); and
- 5 ms sampling period.

Totally, we have two kinds of time series acquired from the sound level meter: one is based on the gear under normal conditions, and the other is measured with the minor artificial tooth cut fault. The beginning 1000 samples of these two time series are illustrated in Figs. 28 and 29, respectively. Moreover, the normal time series is divided into two separate sequences, each with 1000 samples. The first sequence is used for the BP training as well as GA optimization of Elman neural network, and the second (fresh data) is employed for verification of its prediction capability.

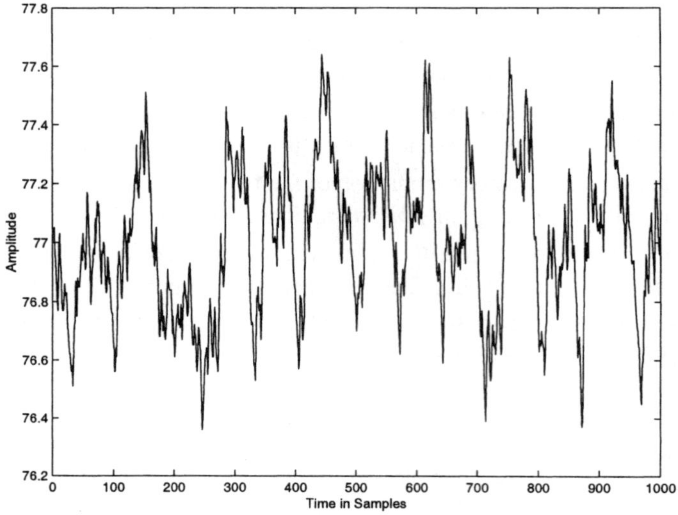

Fig. 28. Time series data from healthy gear.

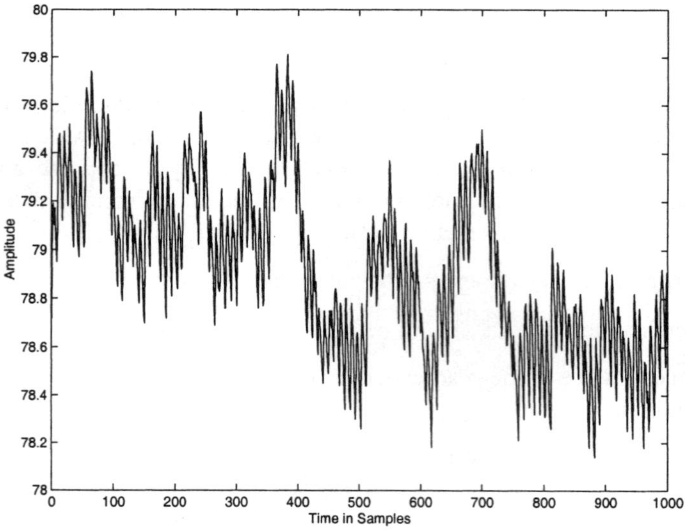

Fig. 29. Time series data from faulty gear.

As discussed in the previous section, Elman neural network is trained to provide a one step ahead prediction of the acoustic time series. Our Elman neural network-based predictor has one input node and ten hidden nodes. Basically, the number of hidden nodes has to be determined by *trial and error*. The training procedure of the Elman neural network with *zero* initial context nodes outputs using only back-propagation algorithm is shown in Fig. 30.

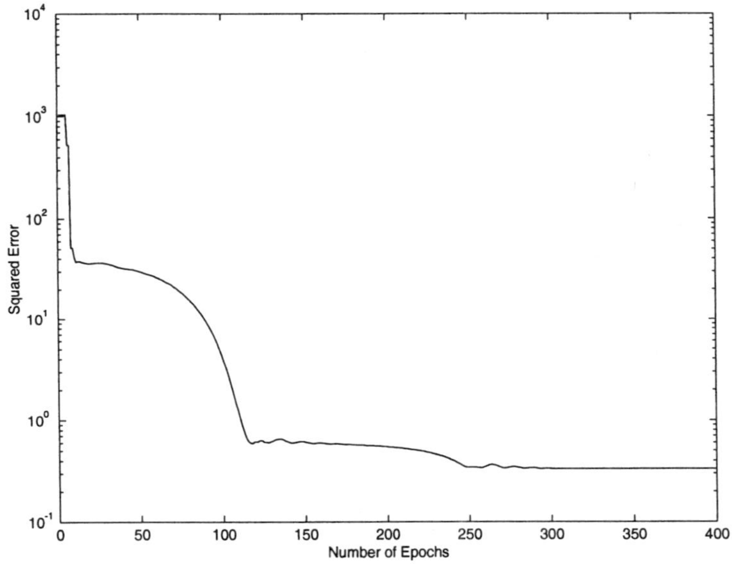

Fig. 30. Training process of Elman neural network using BP algorithm alone.

From Fig. 30, it is easy to find out that the training seems to be stuck into some local minimum after about 300 iteration steps. In this case, there is no use further extending the weight adaptation time. GA will be applied instead to increase the prediction accuracy by optimizing the initial outputs of context nodes. Since the training data (input and desired output) has been normalized in the *preprocessing* phase, we limit the range of possible solutions to the 'optimal' initial context nodes outputs within $[-1,1]$ in order to avoid causing harmful saturation on the hidden node functions. The constant in the calculation of fitness for each chromosome is set to be 10, refer to (20). We have

$$Fitness = 10 - Prediction\_Error \, . \tag{23}$$

In Fig. 30, the minimal prediction error resulting from standalone back-propagation learning algorithm is depicted as 0.3351, i.e., $Fitness = 9.6649$ for the initial chromosomes. A MATLAB-based GA toolbox package is utilized in our simulations [52]. The GA-evolved optimization process is illustrated in Fig. 31. We can conclude that after 25 generations, the fitness of the best chromosome is $Fitness = 9.8448$, which corresponds to a reduced prediction error of 0.1552.

That is to say, prediction accuracy of the Elman neural network is increased by more than 50% with the GA-optimized initial outputs of context nodes. The final values for the initial context nodes outputs, as given in Fig. 31, are:

$$\{0.023, 0.59, 0.35, 0.96, 0.73, 0.40, 0.24, -0.075, -0.66, -0.63\}.$$

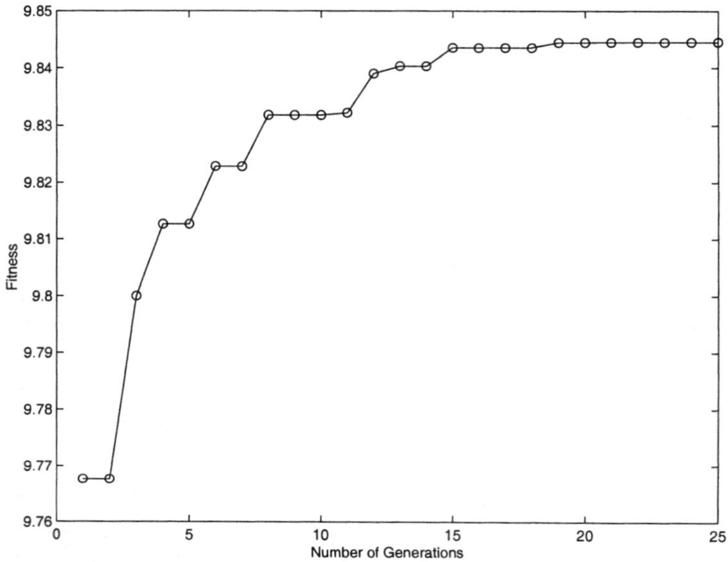

Fig. 31. GA-evolved optimization process of initial context nodes outputs.

The above finalized Elman neural network is then used on-line to produce one step ahead prediction of the gear time series. Figures 32 and 33 illustrate the prediction results in term of squared prediction error for the *testing* time series data from healthy and faulty gears, respectively. It is clearly visible that the squared prediction error of the healthy gear time series is much smaller than that of the faulty gear time series.

To demonstrate the effectiveness of our Elman neural network-based fault detection scheme, we cascade two 1000-sample sequences of time series from both healthy and faulty gears as the *validation* sequence. The detection result of the tooth cut fault is illustrated in Fig. 34 ($k = 2000$ in (22)). We can easily see that the expectation of squared prediction error of the first 1000 samples (normal data) quickly approaches to a constant value after the transient period. On the other hand, as for the next 1000 samples from the faulty gear (abnormal data), the expectation deviates tremendously from this constant with time, which indicates that the existing fault has been successfully detected. We point out that the gear time series in our simulations is considered to be *weakly* nonlinear. Nevertheless, the proposed Elman neural network-based fault detection scheme is flexible in coping with both linear and nonlinear time series data. Comparing with the fault detection method introduced in [53], our scheme has a simpler structure as well as algorithm. In addition, it does not need any on-line adaptation of neural network parameters, and is well suited for real-time motor fault detection applications.

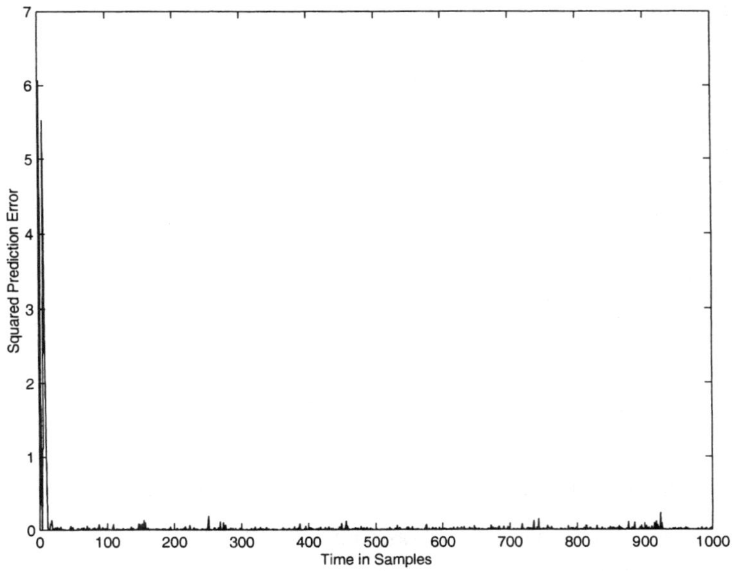

Fig. 32. Squared prediction error of normal time series.

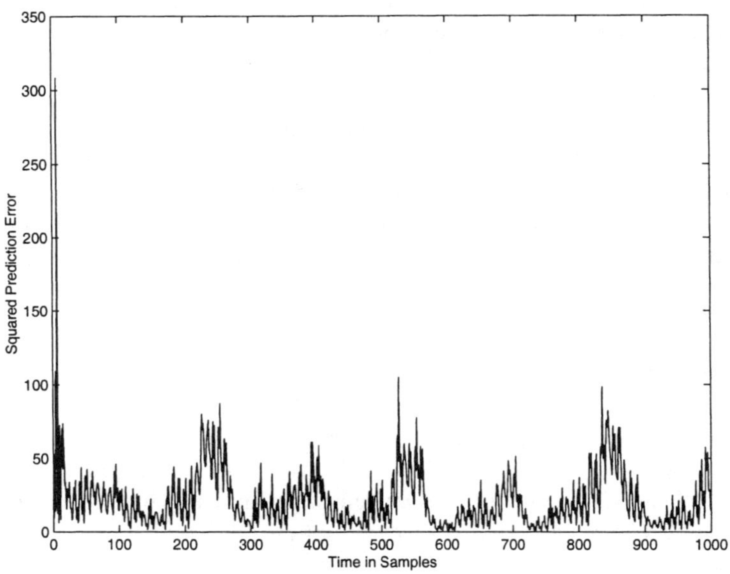

Fig. 33. Squared prediction error of abnormal time series.

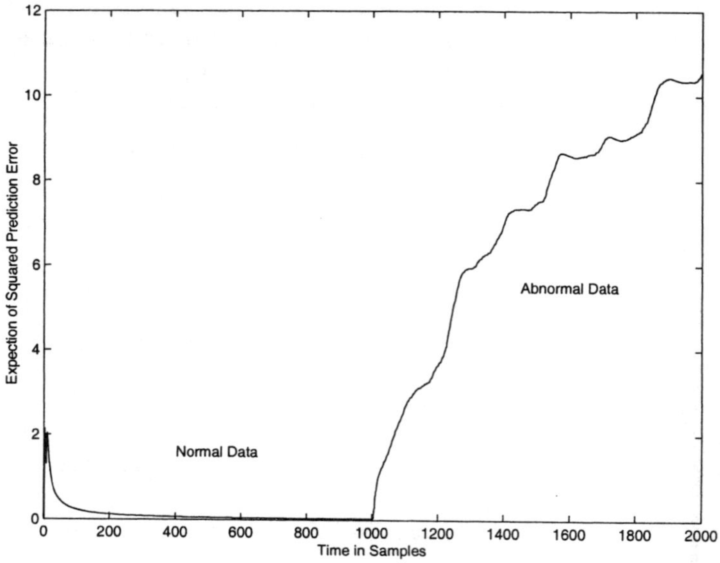

Fig. 34. Fault detection based on the expectation of squared prediction error.

As a case study, a motor fault detection scheme using Elman neural network with the GA-aided learning algorithm is presented in this section. By utilizing GA to optimize the initial context nodes outputs while retaining the back-propagation strategy for adaptation of connecting weights only, a *hybrid* Elman neural network training algorithm is proposed. The hybrid learning algorithm has the advantage of condensed search space for network optimization. Hence, the possibility of being trapped into poor local minima during training procedure is reduced. In our Elman neural network-based motor fault detection method, the expectation of squared prediction error of motor feature signal is used as the principal detection criteria. This scheme occupies intuitive principle and straightforward structure. Moreover, it is an appropriate approach to general fault detection problems. Simulation experiments with real-world data have been carried out to validate its efficiency in an automobile transmission gear system.

6.4 Conclusions

In conclusion, GA is only an auxiliary optimization method, and it cannot be applied independently in practice. The combination of GA with other motor fault diagnosis schemes has demonstrated enhanced performance with regard to global and near-global minima search. However, optimization with GA often involves heavy and time-consuming computation. Therefore, targeted at real-time use, fast GAs with parallel implementation to improve their convergence speed have to be developed.

7 Conclusions

In this chapter, we gave an overview on the recent progresses of soft computing methods-based motor fault diagnosis systems. Several typical motor fault diagnosis techniques using neural networks, fuzzy logic, fuzzy-neural, and genetic algorithms were discussed. Their advantages and drawbacks were summarized as well. Furthermore, two case study examples were shown in more details. Based on our observations, we conclude that emerging soft computing methods can provide us with alternative solutions over classical schemes to challenging motor fault diagnosis problems. Nevertheless, they are not supposed to *compete* with conventional methods. Instead, more accurate and robust diagnosis approaches should be developed based on the *fusion* of these two categories of methodologies, soft computing and hard computing [40].

References

1. Leonhard W. (1996), *Control of Electrical Drives*, Springer-Verlag, Berlin, Germany.
2. Chow M.Y. (1997), Motor fault detection and diagnosis, *IEEE Industrial Electronics Newsletter*, **44**, no. 4, pp. 4-7.
3. Romberg T.M., Black J.L., and Ledwidge T.J. (1996), *Signal Processing for Industrial Diagnostics*, John Wiley & Sons, West Sussex, UK.
4. Zadeh L.A. (1994), Fuzzy logic, neural networks, and soft computing, *Communications of the ACM*, **37**, no. 3, pp. 77-84.
5. Zadeh L.A. (1998), New frontiers in information/intelligent systems, in *IEEE International Conference on Systems, Man, and Cybernetics*, Panel Discussion, San Diego, CA, U.S.A.
6. Chow M.Y. (1997), *Methodologies of Using Neural Network and Fuzzy Logic Technologies for Motor Incipient Fault Detection*, World Scientific Publishing Co. Pte. Ltd, Singapore.
7. Vas P. (1999), *Artificial Intelligence-Based Electrical Machine and Drives: Applications of Fuzzy, Neural, Fuzzy-Neural, and Genetic-Algorithm-based Techniques*, Oxford University Press, New York, NY, U.S.A.
8. Filippetti F., Franceschini G., Tassoni C., and Vas P. (2000), Recent developments of induction motor drives fault diagnosis using AI techniques, *IEEE Transactions on Industrial Electronics*, **47**, no. 5, pp. 994-1004.
9. *Special Section on Motor Fault Detection and Diagnosis* (2000), *IEEE Transactions on Industrial Electronics*, **47**, no. 5, pp. 982-1107.
10. Sood A.K., Fahs A.A., and Henein N.A. (1985), Engine fault analysis, Parts I & II, *IEEE Transactions on Industrial Electronics*, **32**, pp. 294-307.
11. Frank P.M. (1990), Fault diagnosis in dynamic systems using analytical and knowledge-based redundancy—A survey and some new results, *Automatica*, **26**, no. 3, pp. 459-474.

12. Nandi S. and Toliyat H.A. (1999), Condition monitoring and fault diagnosis of electrical machines—A review, *Record of the IEEE Industry Applications Conference*, Phoenix, AZ, U.S.A., pp. 197-204.
13. Keyhani A. and Miri S.M. (1986), Observers for tracking of synchronous machine parameters and detection of incipient faults, *IEEE Transactions on Energy Conversion*, **1**, no. 2, pp. 184-190.
14. Lin C.-T. and Lee C.S.G. (1996), *Neural Fuzzy Systems: A Neuro-Fuzzy Synergism to Intelligent Systems*, Prentice-Hall, Upper Saddle River, NJ, U.S.A.
15. Jang J.-S.R., Sun C.-T., and Mizutani E. (1997), *Neuro-Fuzzy and Soft Computing: A Computational Approach to Learning and Machine Intelligence*, Prentice-Hall, Upper Saddle River, NJ, U.S.A.
16. Chow M.Y. and Yee S.O. (1991), Methodology for on-line incipient fault detection in single phase squirrel cage induction motors using artificial neural networks, *IEEE Transactions on Energy Conversion*, **6**, no. 3, pp. 536-545.
17. Chow M.Y. and Yee S.O. (1991), Application of neural networks to incipient fault detection in induction motors, *Journal of Neural Network Computing*, **2**, no. 3, pp. 26-32.
18. Chow M.Y., Mangum P.M., and Yee S.O. (1991), A neural network approach to real-time condition monitoring of induction motors, *IEEE Transactions on Industrial Electronics*, **38**, no. 6, pp. 448-453.
19. Chow M.Y., Sharpe R.N., and Hung J.C. (1993), On the application of artificial neural networks for motor fault detection—Parts I and II, *IEEE Transactions on Industrial Electronics*, **40**, no. 2, pp. 181-196.
20. Filippetti F., Franceschini G., and Tassoni C. (1995), Neural networks aided on-line diagnostics of induction motor rotor faults, *IEEE Transactions on Industry Applications*, **31**, no. 4, pp. 892-899.
21. Penman J. and Yin C.M. (1994), Feasibility of using unsupervised learning artificial neural networks for the condition monitoring of electrical machines, *IEE Proceedings, Part B*, **141**, pp. 317-322.
22. Tanaka M., Sakawa M., Shiromaru I., and Matsumoto T. (1995), Application of Kohonen's self-organizing network to the diagnosis system for rotating machinery, *Proceedings of the IEEE International Conference on Systems, Man, and Cybernetics*, Vancouver, Canada, pp. 4039-4044.
23. Nejjari H. and Benbouzid M.E.H. (1999), Application of fuzzy logic to induction motors condition monitoring, *IEEE Power Engineering Review*, **19**, no. 6, pp. 52-54.
24. Filippetti F., Franceschini G., Tassoni C., and Vas P. (1995), A fuzzy logic approach to on-line induction motor diagnostics based on stator current monitoring, *Proceedings of the Stockholm Powertek*, Stockholm, Sweden, pp. 150-161.
25. Goddu G., Li B., and Chow M.Y. (1998), Motor bearing fault diagnosis by a fundamental frequency amplitude based fuzzy decision system, *Proceedings of the 24th Annual Conference of the IEEE Industrial Electronics Society*, Aachen, Germany, pp. 1961-1965.
26. Goode P.V. and Chow M.Y. (1995), Using a neural/fuzzy system to extract heuristic knowledge of incipient faults in induction motors: Parts I and II, *IEEE Transactions on Industrial Electronics*, **42**, no. 2, pp. 131-146.
27. Fussel D. and Balle P. (1997), Combining neural-fuzzy and machine learning for fault diagnosis of a D.C. motor, *Proceedings of the American Control Conference*, Albuquerque, NM, U.S.A., pp. 37-41.

28. Filippetti F., Franceschini G., Tassoni C., and Vas P. (1997), Integrated condition monitoring and diagnosis of electrical machines using minimum configuration artificial intelligence, *Proceedings of the European Conference on Power Electronics and Applications*, Trondheim, Norway, pp. 2983-2988.
29. Altug S., Chow M.Y., and Trussell H.J., (1999), Fuzzy inference systems implemented on neural architectures for motor fault detection and diagnosis, *IEEE Transactions on Industrial Electronics*, **46**, no. 6, pp. 1069-1079.
30. Jang J.-S.R. (1993), ANFIS: Adaptive-network-based fuzzy inference system, *IEEE Transactions on Systems, Man, and Cybernetics*, **23**, no. 3, pp. 665-685.
31. Gao X.Z. and Ovaska S.J. (2001), Intelligent motor fault diagnosis using a modified ANFIS, *International Journal of Computers and Their Applications*, **8**, no. 1, pp. 42-51.
32. Belohlavek R. (1999), Feedforward networks with fuzzy signals, *Soft Computing*, **3**, no. 1, pp. 37-43.
33. Lin C.-T. and Lu Y.-C. (1995), A neural fuzzy system with linguistic teaching signals, *IEEE Transactions on Fuzzy Systems*, **3**, no. 2, pp. 169-188.
34. Gao X.Z. and Ovaska S.J. (2000), Fuzzy information processing with neural network, *Proceedings of the IEEE International Conference on Systems, Man, and Cybernetics*, Nashville, TN, U.S.A., pp. 3653-3659.
35. Widrow B. and Lehr M.A. (1990), 30 years of adaptive neural networks: Perceptron, madaline, and backpropagation, *Proceedings of IEEE*, **78**, no. 9, pp. 1415-1442.
36. Harris T.A. (1991), *Rolling Bearing Analysis*, John Wiley & Son, New York, NY, U.S.A.
37. Ohta H. and Sugimoto N. (1996), Vibration characteristics of tapered roller bearings, *Journal of Sound and Vibration*, **190**, no. 2, pp. 137-147.
38. Li B., Goddu G., and Chow M.Y. (1998), Detection of common motor bearing faults using frequency-domain vibration signals and a neural network based approach, *Proceedings of the American Control Conference*, Philadelphia, PA, U.S.A., pp. 2032-2036.
39. Schoen R.R., Habetler T.G., Kamran F., and Bertheld R.G. (1995), Motor bearing damage detection using stator current monitoring, *IEEE Transactions on Industry Applications*, **31**, no. 6, pp. 1274-1279.
40. Ovaska S.J., Dote Y., Furuhashi T., Kamiya A., and VanLandingham H.F. (1999), Fusion of soft computing and hard computing techniques: A review of applications, *Proceedings of the IEEE International Conference on Systems, Man, and Cybernetics*, Tokyo, Japan, pp. 370-375.
40. Lin C.T. (1994), *Neural Fuzzy Control Systems with Structure and Parameter Learning*, World Scientific, Singapore.
41. Goldberg D.E. (1989), *Genetic Algorithms in Search, Optimization, and Machine Learning*, Addison-Wesley, Reading, MA, U.S.A.
42. Man K.F., Tang K.S., and Kwong S. (1996), Genetic algorithms: Concepts and applications, *IEEE Transactions on Industrial Electronics*, **43**, no. 5, pp. 519-534.
43. Tang K.S., Man K.F., Kwong S., and He Q. (1996), Genetic algorithms and their applications, *IEEE Signal Processing*, **13**, no. 6, pp. 22-37.
44. Betta G., Liguori C., and Pietrosanto A. (1996), The use of genetic algorithms for advanced instrument fault detection and isolation schemes, *Proceedings of the IEEE Instrumentation and Measurement Technology Conference*, Brussels, Belgium, pp. 1129-1134.

45. Gao X.Z., Ovaska S.J., and Dote Y. (2000), Motor fault detection using Elman neural network with genetic algorithm-aided training, *Proceedings of the IEEE International Conference on Systems, Man, and Cybernetics*, Nashville, TN, U.S.A., pp. 2386-2392.
46. Elman J. (1990), Finding structure in time, *Cognitive Science*, **14**, pp. 179-211.
47. Gao X.Z. (1998), A Temporal Difference method-based prediction scheme applied to fading power signals, *Proceedings of the IEEE International Joint Conference on Neural Networks*, Anchorage, AK, U.S.A., pp. 1954-1959.
48. Angeline P.J. and Saunders G. (1994), An evolutionary algorithm that constructs recurrent neural networks, *IEEE Transactions on Neural Networks*, **5**, no. 1, pp. 54-65.
49. Brill F. and Brown D. (1992), Fast genetic selection of features for NN classifiers, *IEEE Transactions on Neural Networks*, **3**, no. 2, pp. 324-328.
50. Ichikawa Y. and Sawa T. (1992), Neural network application for direct feedback controllers, *IEEE Transactions on Neural Networks*, **3**, no. 2, pp. 224-231.
51. Houck C.R., Joines J.A., and Kay M.G. (1995), A genetic algorithm for function optimization: A Matlab implementation, Technical Report 95-09, Department of Industrial Engineering, North Carolina State University, Raleigh, NC, U.S.A.
52. Akhmetov D.F., Dote Y., and Ovaska S.J. (2001), Fuzzy neural network with general parameter adaptation for modeling of nonlinear time-series, *IEEE Transactions on Neural Networks*, **12**, no. 1, pp. 148-152. (Errata published in (2001) *IEEE Transactions on Neural Networks*, **12**, no. 2, p. 443.)

Neural Network Estimation and Neuro-Fuzzy Control in Converter-Fed Induction Motor Drives

Marian P. Kazmierkowski* and Teresa Orłowska-Kowalska†

* Warsaw University of Technology

† Wroclaw University of Technology

Keywords: *induction motor control, voltage source inverters, pulse width modulation, neural network estimation, neuro-fuzzy control, field-oriented control, direct torque control, neural network flux vector estimation, neural network speed estimation, neuro-fuzzy direct torque control, adaptive neuro-fuzzy inference system.*

1 Introduction

Advances in power electronics technology and the availability of low cost high-speed digital signal processors (DSPs) stimulated recent efforts in the design of complex control algorithms for converter-fed induction motor (IM) drives. Especially, high performance control methods as: field oriented control (FOC) [4], [10], direct torque control (DTC) [7], [31], adaptive, non-linear or sliding-mode control [15], [27], [35] require full information about state and output variables. However, sensors used in feedback loops increase costs and decrease reliability of the drive system, therefore, they should be avoided. Since the late 1980s, many efforts have been made to reconstruct such state variables of the induction motor as: rotor or stator flux vector, rotor speed, and load torque [11], [27], [35].

Artificial neural networks (NN) are well suited for application in IM control and estimation, because of their known advantages as: ability to approximate arbitrary nonlinear functions to any desired degree of accuracy, learning and generalization, fast parallel computation, robustness to input harmonic ripples and fault tolerance. These aspects are important in the case of nonlinear systems, like converter-fed drives, where linear control theory cannot be directly applied. Additionally, high efficiency power electronic converters used for IM supply operate in switch mode, which results in very noisy signals. For these reasons the NN are attractive for signal processing and control of IM drives [5], [30], [36].

This Chapter presents selected examples of NN-based speed and flux vector estimators as well as their evaluation from the point of view of sensitivity to motor parameter changes. Also, new high performance speed sensorless IM control schemes, based on FOC and DTC, are described. The simulation and experimental oscillograms that illustrate the performance of the discussed NN based control techniques are shown.

2 Neural Network Based Speed Estimation

2.1 General Problems of Speed Estimation in the Induction Motor Drives

Mechanical variables such as the rotor speed or position are usually measured by tachogenerators or incremental encoders and resolvers. As it was said before, these sensors mounted in the drive system cause higher installation cost, increase size of the drive motor, need the sensor cable and decrease reliability of the drive system [11], [27], [35]. So in the recent years there is a wide interest in the development of new methods for the induction motor speed estimation for the sensorless drives. Various methods of the rotor speed estimation were used, based on the mathematical models of the induction motor (IM) [11], [19], [20], [22] or nonlinear phenomena caused by rotor eccentricity or other motor saliences [18], as it was illustrated in Fig. 2.1.

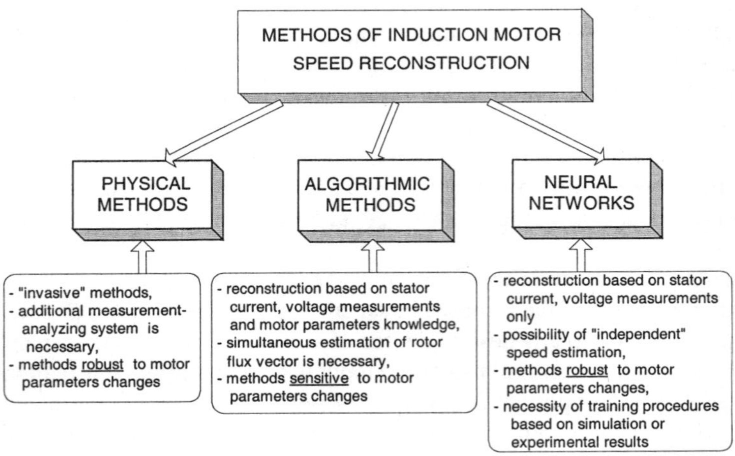

Fig. 2.1. Methods of speed reconstruction for the induction motor.

The methods based on nonlinear phenomena of IM require special measuring winding set up in the motor (invasive methods) and relatively advanced signal processing system [11], [27]. However their main advantage is robustness to motor parameter changes. In the contrary all approaches based on the motor mathematical models are sensitive to parameter identification errors [25]. So parameters of the model have to be exactly known or have to be adapted.

Neural networks (NN) offer an alternative way to handling this problem [5], [23], [36]. Because of their ability to mapping different kinds of nonlinearities and generalization properties, NN can be applied in the situation when system is not exactly identified. So its parameters could be not exactly known, but should be learned.

From the implementation point of view, the complexity of algorithmic and neural methods is comparable, because both require fast digital signal processor for reconstruction of the rotor speed and other necessary motor state variables based on easily measured motor signals, like stator currents and voltages. Physical methods need additionally measurement sensors mounted in the motor, so they should be treated as invasive methods.

Few authors have reported the application of NN for state variable reconstruction of the induction motor. Two kinds of neural network approaches can be used for speed estimation of induction motor:

- method based on neural modeling – on-line estimation [2], [3], [26], [28], [29];
- method based on neural identification – off-line estimation [5], [21], [22], [24], [36].

In the system, which works on-line, NN is used as a Model Reference Adaptive System (MRAS) [2] and mechanical speed is proportional to one of NN weights. The one layer linear NN is used as the Adaptive Model and the estimated speed is one of weight factors of this network. The block scheme is shown in Fig. 2.2a. In Fig. 2.2, the following symbols are used: $\underline{i}_s$ - stator current spatial vector, $\underline{\Psi}_r$ - rotor flux spatial vector, $\underline{u}_s$ - stator voltage vector, ω_m - rotor angular speed.

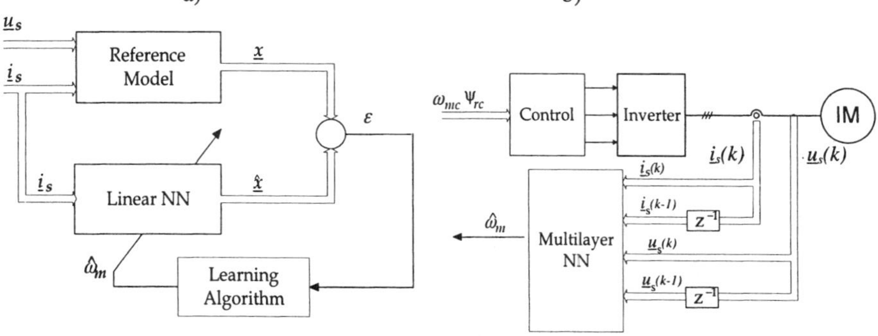

Fig. 2.2. Block scheme of NN-based speed estimators:
a) on-line estimator; b) off-line estimator.

In the off-line approach, a multilayer neural network is used and the motor speed is obtained as its output. The input vector to the NN consists of the line current and voltage vectors of the induction motor and/or their pre-processed values:

$$\underline{x}_{in} = \left[\underline{u}_s(k), \underline{u}_s(k-1), \underline{i}_s(k), \underline{i}_s(k-1), \ \ldots\right]^{\mathrm{T}} \tag{2.1}$$

The block diagram of this approach is presented in Fig. 2.2b. This approach needs an off-line training procedure of NN. In the further part of this chapter both approaches are described in detail, evaluated and tested in real system.

2.2 Speed Estimation Based on Neural Modeling

2.2.1 Some Remarks on the Mathematical Model of the Induction Motor

Mathematical model of the induction motor used in control problems is formulated with usual assumptions: the electric and magnetic symmetry of stator and rotor windings, the uniform air-gap, the sinusoidal field distribution and neglected effects of magnetic saturation, eddy-currents and hysteresis. In the control as well as state variables estimation problem the vector representation of this model is used, defined in a co-ordinate system rotating with angular speed ω_k. In per unit system and real-time representation the following vector equations describe the behavior of the squirrel-cage induction motor [15]:

- for stator and rotor electromagnetic equations:

$$\underline{u}_{sk} = r_s \underline{i}_{sk} + T_N \frac{d}{dt}\underline{\Psi}_{sk} + j\omega_k \underline{\Psi}_{sk} \tag{2.2}$$

$$0 = r_r \underline{i}_{rk} + T_N \frac{d}{dt}\underline{\Psi}_{rk} + j(\omega_k - \omega_m)\underline{\Psi}_{rk} \tag{2.3}$$

$$\underline{\Psi}_{sk} = x_s \underline{i}_{sk} + x_M \underline{i}_{rk} \tag{2.4}$$

$$\underline{\Psi}_{rk} = x_r \underline{i}_{rk} + x_M \underline{i}_{sk} \tag{2.5}$$

- for electromagnetic torque of the motor:

$$m_e = \frac{x_M}{x_r}\left|\underline{\Psi}_{rk} \times \underline{i}_{sk}\right| \operatorname{sign}\left(\underline{\Psi}_{rk} \times \underline{i}_{sk}\right) = \frac{x_M}{x_r}\operatorname{Im}\left\{\underline{\Psi}_{rk}\underline{i}_{sk}^{*}\right\} \tag{2.6}$$

- for the motion equation:

$$\frac{d\omega_m}{dt} = \frac{1}{T_M}\left(m_e - m_L\right) \tag{2.7}$$

where: $\underline{i}_{sk}$ - stator current spatial vector, $\underline{\Psi}_{rk}$ - the rotor flux spatial vector,
$\underline{u}_{sk}$ - stator voltage vector, (all vectors in rotating reference frames),
ω_m - rotor angular speed, r_s, r_r - stator and rotor resistance,
x_s, x_r - stator and rotor reactance, x_M - magnetizing reactance,
m_e , m_L - electromagnetic and load torque,
$T_N = 1/\omega_b$, $\omega_b = 2\pi f_{sN}$, , T_M - mechanical time constant.

This model will be used in next paragraphs for algorithms of speed estimation and field-oriented as well as direct torque control problems.

2.2.2 Description of the Neural Modeling Method

The neural modeling method used for IM speed estimation is based on an analogy between mathematical description of a simple linear perceptron and the differential equation for rotor flux simulator of the induction motor, written in a discrete form. This kind of speed estimation, based on parallel neural model, was described first by Ben-Brahim and Kurosawa [2] and is demonstrated in Fig. 2.3a.

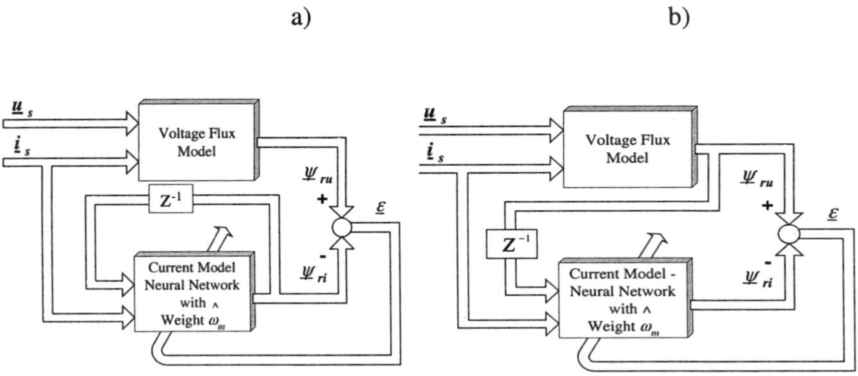

Fig. 2.3. Neural speed estimator based on neural modeling method.

In Fig. 2.3b a modification of this approach is presented, based on a series-parallel neural model. Both MRAS methods are similar from the point of view of estimator sensitivity to motor parameters changes, but the second solution works a little bit faster in transients.

The concept of above neural speed estimator is based on the comparison of two well known rotor flux models: 1 - traditional rotor flux simulator based on the stator differential equation and algebraic dependence between stator and rotor flux vectors; 2 - model based on the rotor differential equation which involved one of this equation parameter - rotor speed - as state variable. This second model may be regarded as a neural model with adjustable weights, which are simply related to the rotor speed.

Rotor flux simulator based on the voltage model, obtained from Eq. (2.2), (2.4) and (2.5), is the following:

$$\frac{d}{dt}\underline{\Psi}_{ru} = \frac{x_r}{x_M}\left(\underline{u}_s - r_s\underline{i}_s - x_s\sigma T_N \frac{d\underline{i}_s}{dt}\right)\frac{1}{T_N}. \tag{2.8}$$

Respectively the second model of rotor flux simulator based on the current model, obtained from Eq. (2.3), (2.4) and (2.5) is:

$$\frac{d}{dt}\underline{\Psi}_{ri} = \left[\frac{T_N}{T_r}\left(x_M \underline{i} - \underline{\Psi}_{ri}\right) + j\omega_m \underline{\Psi}_{ri}\right]\frac{1}{T_N}, \tag{2.9}$$

where: $\sigma = 1 - x_M^2 / x_s x_r$, $T_r = x_r T_N / r_r$ – rotor time constant.

The current model can be written in the following way for neural modeling purposes:

$$\hat{\underline{\Psi}}_{ri}(k) = \left(w_1 \mathbf{I} + w_2 \mathbf{J}\right) \hat{\underline{\Psi}}_{ri}(k-1) + w_3 \underline{i}_s(k-1), \tag{2.10}$$

where: $\mathbf{I} = \begin{bmatrix} 1 & 0 \\ 0 & 1 \end{bmatrix}$, $\mathbf{J} = \begin{bmatrix} 0 & -1 \\ 1 & 0 \end{bmatrix}$.

This model, transformed to the stationary reference (α, β) co-ordinate system [15], [35] can be treated as a simple connection of two neurons with linear activation functions, as it was illustrated in Fig. 2.4.

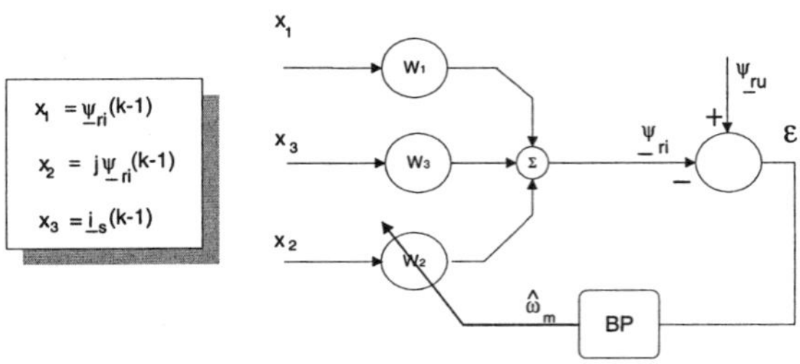

Fig. 2.4. Structure of linear NN used in on-line speed estimation method.

One of the weight coefficients of these neurons – w_2, is in fact the rotor speed. It can be derived using the backpropagation method (BP) [13], [36] of the error ε and thus the speed estimate can be obtained. After using of two (α, β) components of rotor flux vectors obtained from voltage and current models respectively, the BP algorithm for ω_m modification is the following:

$$\hat{\omega}_m[k] = \hat{\omega}_m[k-1] - \eta\left(-e_\alpha \Psi_{ri\beta} + e_\beta \Psi_{ri\alpha}\right)[k-1] + \mu\,\Delta\,\omega_m[k-1], \tag{2.11}$$

where: $e_\alpha = \Psi_{ru\alpha} - \Psi_{ri\alpha}$,
$e_\beta = \Psi_{ru\beta} - \Psi_{ri\beta}$

η - learning coefficient, μ - momentum factor.

As it was proved in [2] this neural identifier works very well for nominal parameters of the induction motor used in both models. In the case when rotor resistance is varying, the estimation errors occur, especially in the low speed region. But mathematical models Eq. (2.8) and Eq. (2.9) used in the estimation scheme depend on the other motor parameters also. In the case when activation function in each neuron is linear, as it was proposed in [2], [3], it is hardly to consider this estimator as real neural network with all its advantages, especially with robustness to noise and parameter disturbances. So the proposed estimator together with BP algorithm was checked to the changes of other motor parameters used in the voltage and current flux models.

2.2.3 Sensitivity of Speed Estimator to Motor Parameter Changes and Parameters of BP Algorithm

The neural speed estimator (Fig. 2.3a) was trained on-line in the wide range of motor speed and load torque changes based on simulation results obtained for converter-fed IM simulated in MATLAB-SIMULINK. Then the trained network was tested for speed and load torque reference signal different than used in the training procedures. These testing signals are presented in Fig. 2.5.

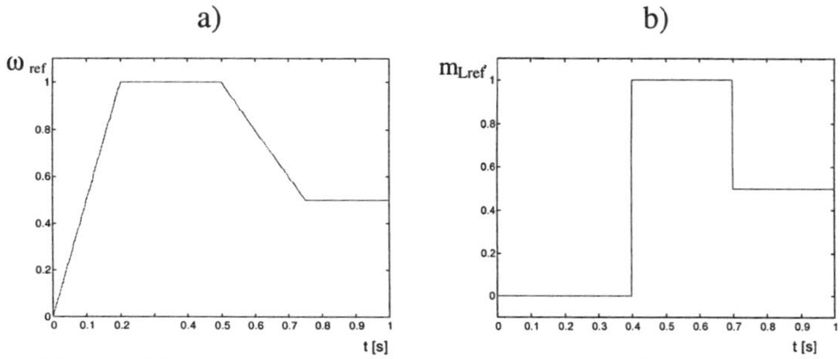

Fig. 2.5. Programmable changes of the speed (a) and load torque (b) for neural speed estimator testing.

Simulations were carried out using the Euler method with the integration step 100 μs, to have in simulation tests the similar conditions as in the real experiments. The transient and average estimation errors were investigated and presented in the case of motor parameter changes. It was shown that for the suitable choice of coefficients η and μ in BP algorithm (from the point of view of a minimal speed estimation error), it is possible to obtain a relatively low sensitivity of the estimator to motor parameter changes. In Fig. 2.6 selected examples of simulation tests for neural speed estimator with suitable transient errors are presented.

Similar tests were performed also for other motor parameter changes and average speed estimation errors during whole transient process were determined, to evaluate the estimator accuracy. These errors were calculated as follows:

$$Err\,[\%] = \sum_{k=1}^{n} \frac{e_{\omega_k}[\%]}{n}, \tag{2.12a}$$

where:

$$e_{\omega_k} = \frac{|\omega_{mk} - \hat{\omega}_{mk}|}{\omega_{mk}} \cdot 100\%, \tag{2.12b}$$

ω_m - rotor speed, $\hat{\omega}_m$ - estimated rotor speed, n - number of speed samples during transient process.

Figures 2.7-2.8 show the average speed estimation errors calculated for incorrect motor parameters. In all cases the speed and load torque of the motor were changed according to the testing transients from Fig. 2.5.

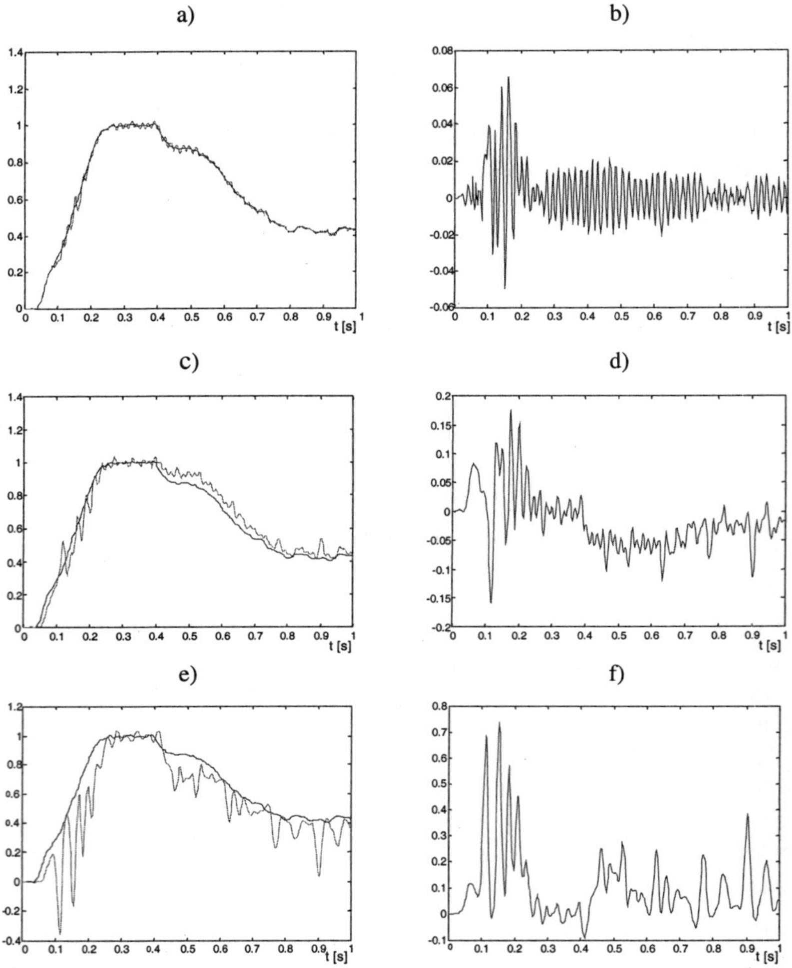

Fig. 2.6. Rotor speed (solid line), its estimate (dotted line) (a,c,e) and speed estimation error (b,d,f) for $\eta_e = 0.08$, $\mu_e = -0.00008$: a) $r_r = r_{rN}$, b) $r_r = 1.5r_{rN}$, c) $r_r = 0.5r_{rN}$.

It was shown that speed estimation error increases significantly when incorrect motor parameters occur: too low rotor resistance value, too high stator resistance and stator leakage reactance value, too low magnetizing reactance value. Identification errors of stator leakage reactance and stator resistance strongly influence transients of speed estimation errors. To minimize the speed ripples caused by wrong parameter values used in the speed estimator model, low-pass filters were used (see Fig. 2.7a and 2.8a,c). The optimal speed reconstruction was obtained for the 2nd-order Butterworth filter with cut-off angular frequency of 300 [$^{rad}/_{s}$]; in this case the average estimation errors calculated for the whole transient process were the smallest.

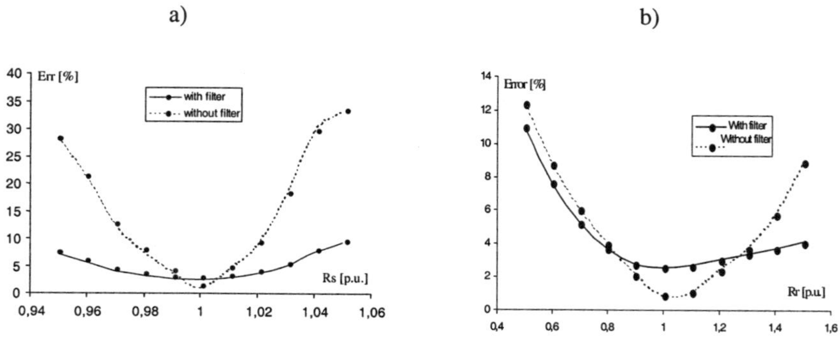

Fig. 2.7. Average value of speed estimation error versus stator (a) and rotor (b) resistance.

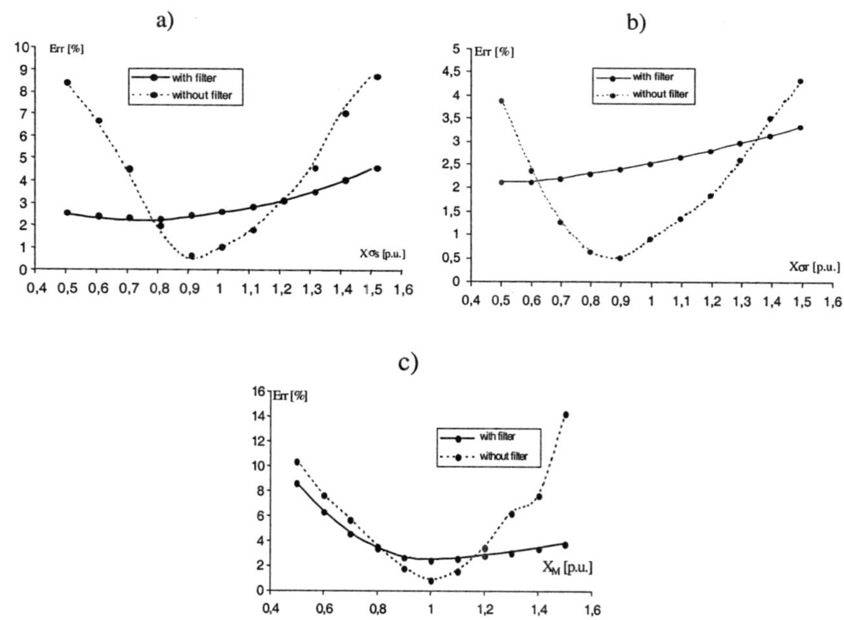

Fig. 2.8. Average value of speed estimation error versus stator (a), rotor (b) leakage reactance and magnetizing reactance (c).

It was observed during the simulation tests that the speed estimation errors depend hardly on the learning coefficient used in the backpropagation algorithm [26]. When this algorithm (Eq. (2.11)) is written in the discrete form, with sampling step T_s, the equivalent learning coefficient η_e is calculated in as follows: $\eta_e = \eta T_s$ (similarly $\mu_e = \mu T_s$). So it is clear that the value of learning coefficient η should be adapted to the sampling step used in simulation or practical implementation. When the numerical step is very small (for example: T_s = 20 μs), transient estimation errors obtained during the estimation process are also very small, less than 1.5% (for nominal motor parameters used in the voltage and current models of the estimator). For such small numerical step it is possible to apply a relatively high learning coefficient η in BP algorithm (Eq. (2.11)), which assures a good reconstruction of the motor speed (see Fig. 2.9a).

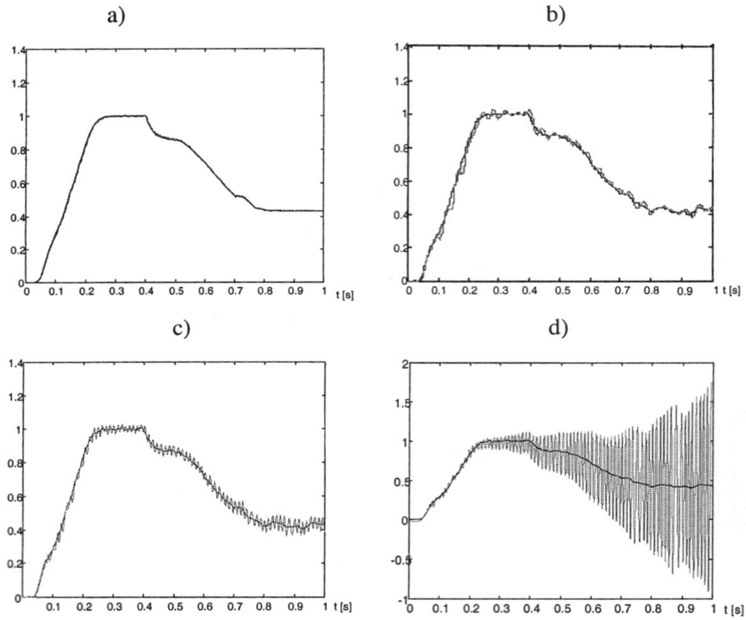

Fig. 2.9. Rotor speed and its estimate for different sampling steps and learning coefficients: a) T_s = 20 μs; η_e = 0.04 (η = 2000); b) T_s = 100 μs, η_e = 0.01 (η = 100), d) T_s = 100 μs, η_e = 0.08 (η = 800), c) T_s = 100 μs, η_e = 0.1 (η = 1000).

Generally, lower sampling step T_s gives better speed reconstruction for lower value of equivalent learning coefficient η_e. However, in DSP realization of such neural estimator, a limitation of the sampling step should be expected due to the computation power available. Taking into account this limitation, the optimization procedure was performed for searching the best value of equivalent learning coefficient η_e for practical sampling time T_s=100 μs (see Fig. 2.9b-d). This optimal value was 0.08 (η = 800) for the motor used in all experiments. Obtained speed reconstruction errors were minimal in the whole transient process for different motor parameter values (changed in the sensitivity analysis problem), as was shown in Fig. 2.6.

The influence of initial values of parameter $w_2(0)$ used in the BP algorithm is illustrated in Fig. 2.10. It shows rotor speed estimation quality for different initial values of neural network coefficient w_2 as well as for different values of learning coefficient η_e used in the algorithm.

a) b)

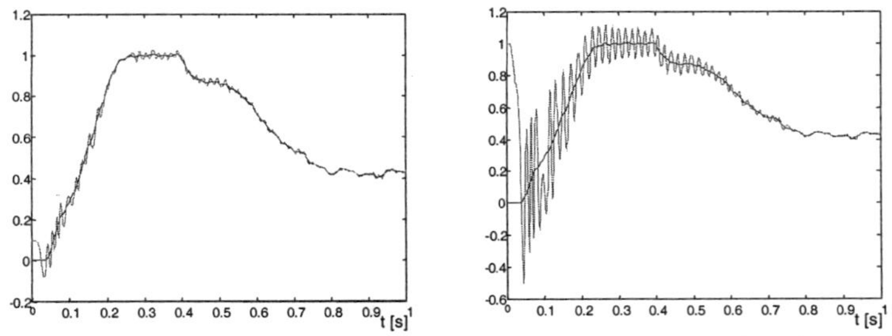

Fig. 2.10. Influence of initial value of $w_2(0) = \hat{\omega}_m(0)$ to speed estimation quality with $\eta_e = 0.08$; $\hat{\omega}_m(0) = 0.1$, b) $\hat{\omega}_m(0) = 1$.

a) b)

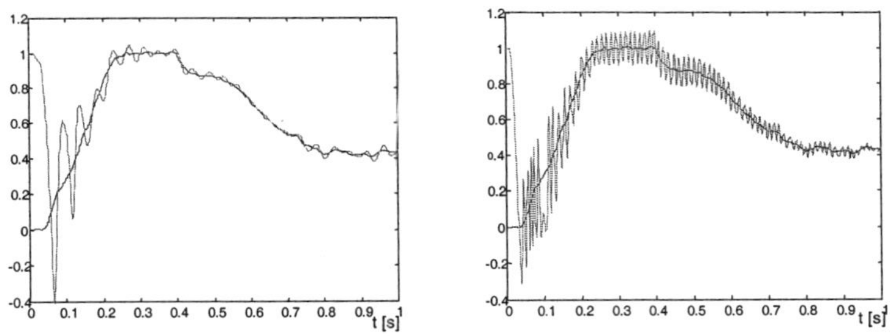

Fig. 2.11. Influence of learning factor value η on speed estimation quality for $\hat{\omega}_m(0) = 1$; a) $\eta_e = 0.015$, b) $\eta_e = 0.15$.

Simulated results show that the speed estimation scheme is stable independent on the initial value of $w_2(0) = \hat{\omega}_m(0)$ used in the BP algorithm (Eq. (2.11)). But suitable choice of the initial condition, close to the real speed, assures the smooth transients of the estimated rotor speed with small oscillations only (see Fig. 2.10a). From the other side, the proper choice of the learning coefficient enables minimization of estimated speed oscillations (Fig. 2.11).

2.2.4 Experimental Results

The proposed scheme was tested in the laboratory with induction motor supplied from PWM inverter and by DSP card with 32-bit signal processor TMS320C31. The optimized coefficients η and μ were chosen for the experimental drive. Good results were obtained for the wide speed range but only for nominal motor parameters used in the voltage and current flux model (Eq. (2.8) and (2.10)) of the neural speed estimator.

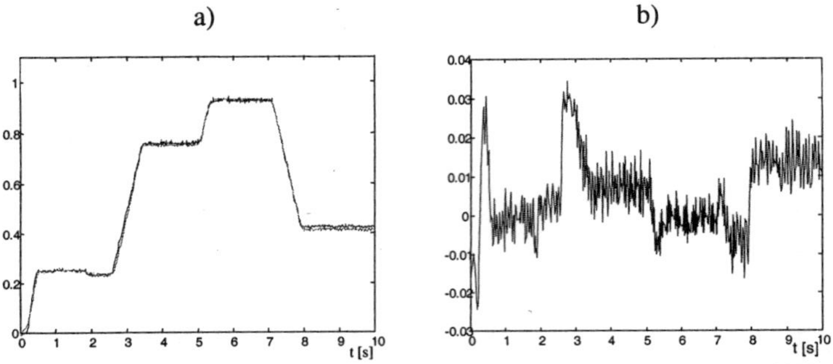

Fig. 2.12. Experimental results obtained for nominal motor parameters used in the speed estimator: a) real motor speed (solid line) and its estimate (dotted line); b) estimation error.

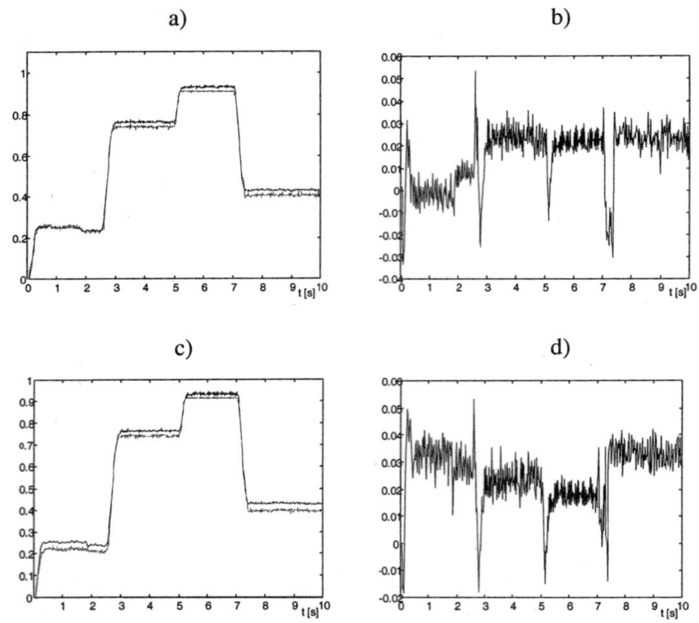

Fig. 2.13. Experimental results obtained for incorrect motor parameters used in the speed estimator : a,c,e) real motor speed (solid line) and its estimate (dotted line; b,d,f) estimation error; a,b) $r_r = 2\, r_{rN}$, c,d) $r_s = 2\, r_{sN}$.

Results of speed estimation obtained from the trained NN were compared to the real motor speed. The measured and estimated speed transients as well as estimation errors are presented in Fig. 2.12 (for nominal motor parameters). However, for wrong motor parameters used in flux models, speed estimation errors occurred, similarly as in the case of simulation (see Fig. 2.6-2.8). Selected examples of such experimental results are presented in Fig. 2.13.

2.2.5 Modification of Neural Modeling Method Based on Stator Current Error

After suitable transformation of IM mathematical model [29], [35] described by Eq. (2.2)-(2.5), one can obtain the following equation for stator current:

$$\sigma x_s \frac{d\underline{i}_s}{dt} = \frac{x_M}{x_r}\left(\frac{T_N}{T_r} - j\omega_m\right)\underline{\Psi}_r - \left(\frac{x_M^2 T_N}{x_r T_r} + r_s\right)\underline{i}_s + \underline{u}_s \tag{2.13}$$

The rotor flux vector can be easily calculated based on Eq. (2.8). The stator current value obtained from Eq. (2.13) can be compared with its measured value and thus motor speed can be estimated with help of the BP algorithm, similarly as in the scheme of Fig. 2.3b. The block diagram of the proposed modified speed estimation [29] is presented in Fig. 2.14.

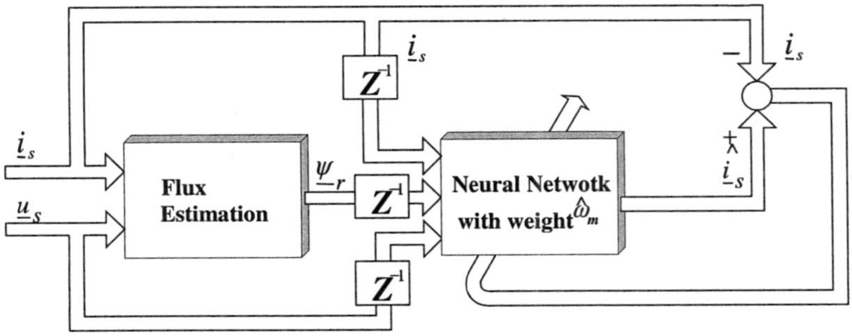

Fig. 2.14. Block diagram of modified neural modeling based speed estimation.

Therefore, the discrete model of stator current is as follows:

$$\hat{\underline{i}}_s(k) = \frac{x_M}{x_r \sigma x_s}\left(\frac{T_s}{T_r} - j\omega_m \frac{T_s}{T_N}\right)\underline{\Psi}_r(k-1) + \left[1 - \frac{1}{\sigma x_s}\left(\frac{x_M^2 T_s}{x_r T_r} + r_s \frac{T_s}{T_N}\right)\right]\hat{\underline{i}}_s(k-1) + \frac{T_s}{T_N \sigma x_s}\underline{u}_s(k-1) \tag{2.14}$$

This equation can be rewritten in NN form, with one weight w_2 proportional to mechanical speed ω_m, in the following way:

$$\hat{\underline{i}}_s(k) = w_1 \underline{\psi}_r(k-1) + w_2 j\underline{\psi}_r(k-1) + w_3 \underline{i}_s(k-1) + w_4 \underline{u}_s(k-1) \tag{2.15}$$

So, the linear NN can be designed, similarly as in the case of Eq. (2.9). It is presented in Fig. 2.15.

In this case not flux, but current estimation error is used for NN tuning:

$$\underline{e}(k) = \underline{i}_s(k) - \hat{\underline{i}}_s(k)\,. \tag{2.16}$$

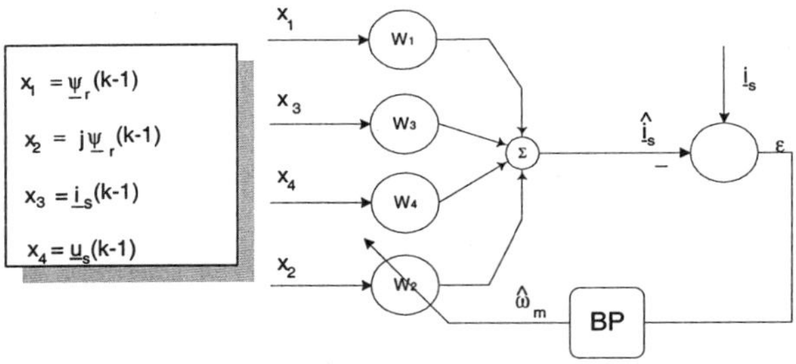

Fig. 2.15. Structure of linear NN used in speed estimation method from Eq. (2.14).

Therefore, the estimated speed can be calculated as follows:

$$w_2[k] = w_2[k-1] - \eta\left(-e_\alpha \Psi_{ri\beta} + e_\beta \Psi_{ri\alpha}\right) + \mu\,\Delta w_2[k-1], \tag{2.17}$$
$$\hat{\omega}_m = -\frac{T_N}{T_s}\frac{x_r}{x_M}\sigma x_s w_2 .$$

The structure of this new method is similar to the method presented in [2], [3]. The differences are as follows:
- different error signals for NN tuning,
- more NN weights,
- NN depends on a larger number of motor parameters.

The presented method was tested experimentally and by simulation. The influence of learning coefficient on the NN estimator behavior in the transient process is shown in Fig. 2.16 and in the steady state operation - in Fig. 2.17, respectively.

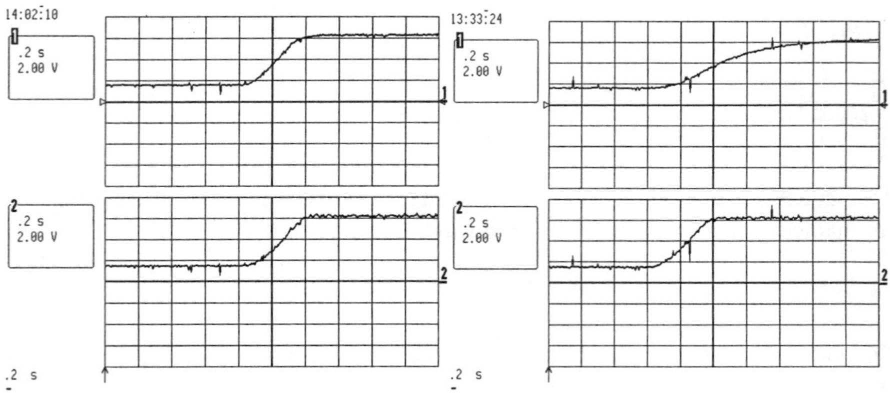

Fig. 2.16. Influence of learning rate on the NN estimator behavior in dynamic state of speed step response: lower fig. - speed from encoder, upper fig. - speed from estimator (left figure - results for $\eta = 0.01$, right figure - results for $\eta = 0.001$).

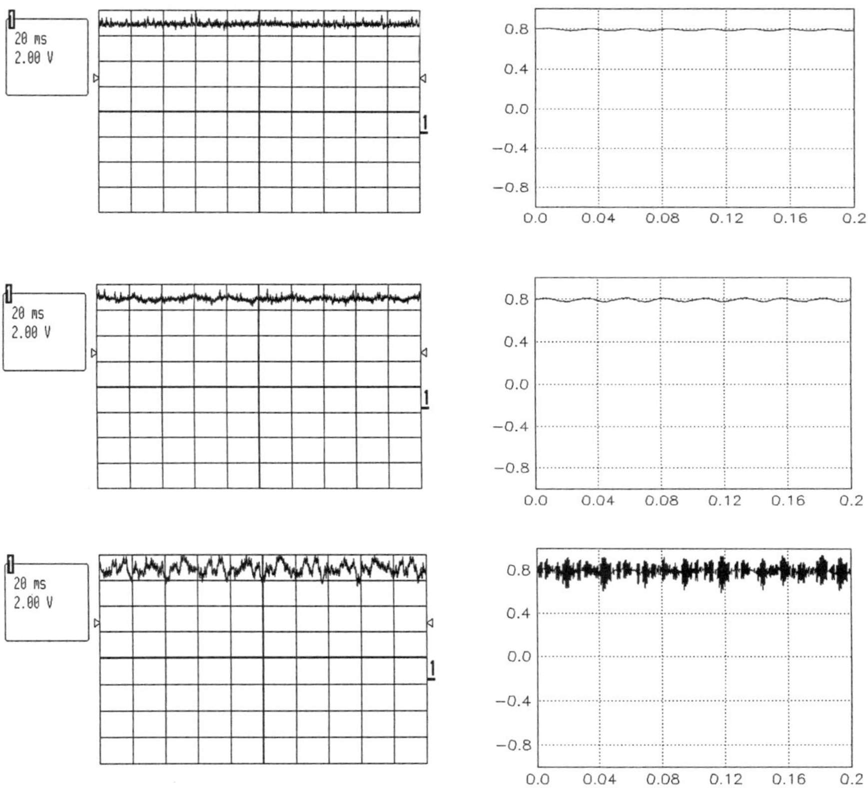

Fig. 2.17. Influence of learning rate on the NN estimator behavior in the steady state: η =0.02, η =0.2, η =2.0 from upper to lower figure, respectively (on the left: experimental results; on the right – simulation results).

It can be observed that the higher learning rate causes higher ripples in estimated speed. However, the dynamic behavior of the estimator with smaller learning rate could be not correct and the estimated speed could not follow the real mechanical speed of the induction motor. This situation is presented in Fig. 2.17 (right), where dynamic results of the NN speed estimator are shown. So the learning rate should be a compromise between steady-state ripples and dynamical tracking behavior, to achieve fast response and to avoid high ripples in the steady-state operation of the estimator. In the system, presented in this section, the value of the learning rate was assumed $\eta = 0.1$ and is similar to the value used in the Section 2.2.3.

2.3 Speed Estimation Based on Neural Identification

2.3.1 State of the Art in the Field of Multilayer Networks

Neural networks can be used for IM state variables reconstruction problem in the different way also. It could be assumed that the NN of specially designed structure should approximate the nonlinear dynamical properties of the IM and reconstruct its chosen state variables based on easily measured input signals of the motor and training procedure performed off-line. So the NN will realize the identification problem. The identification procedure can be solved in two ways:

1 - with help of a feedforward multilayer NN, which uses the measured inputs and outputs of the plant in few previous steps as inputs of the neural network - similarly as in the problem of neural modeling of the system;

2 - with help of a recurrent network (with feedback loop) which uses the estimated outputs of the network as its actual inputs (Fig. 2.18).

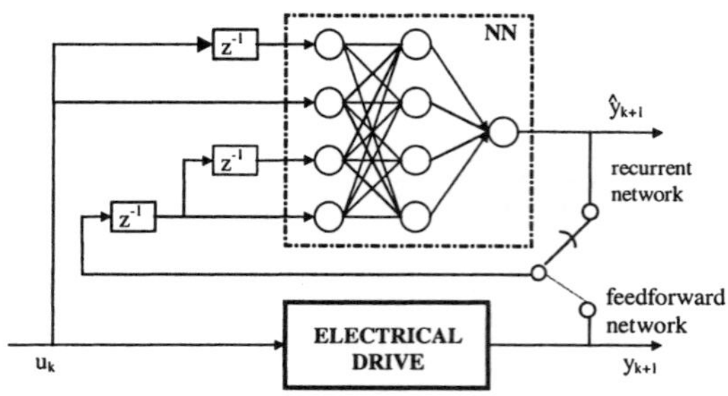

Fig. 2.18. Schematic diagram of IM state variables identification by multilayer NN.

This concept can be used directly in the task of IM speed or rotor flux estimation and NN can be trained based on measurement of motor current and voltages. But the main problem is the choice of proper structure of neural network. There are no general rules which can be used for this purpose.

Very recently, in theoretical attempts of the neural network field, there are a number of researchers who have experimented with an architecture of neural network that is somewhere in between a feedforward only multilayer perceptron-type architecture, and a full recurrent network architecture. For lack of a more generic name, this class of architectures was called a *locally-recurrent-globally-feedforward (LRGF)* architecture [34] and is presented in Fig. 2.19. The general *LRGF* architecture may be considered as a generalization of the classic McCulloch–Pitts' neuron [34] from an input – output point of view. Instead of constant weight synapses, one has synapses that are modeled by linear transfer functions. In addition, the feedback from the output is also modeled by a linear transfer function.

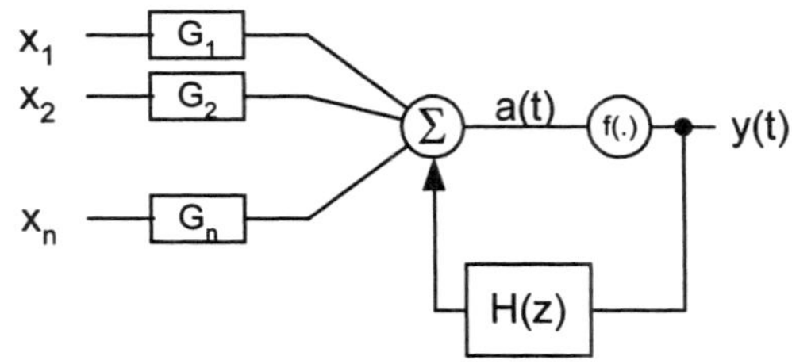

Fig. 2.19. Generalized *LRGF* network architecture.

This generalized neuron can then be used in a multilayer perceptron globally feedforward-type architecture. From the above it appears that there are three major *locally-recurrent-globally-feedforward (LRGF)* network architectures (see Fig. 2.19): with local synapse feedback, local activation feedback and local output feedback [34]. Based on theoretical works in the field of NN it is difficult to say which architecture – with global or local feedback – is better and more efficient in modeling a given nonlinear dynamic system. At present it is not known which *LRGF* architecture should be used in a particular situation. These two last statements were the main reasons why the authors have decided to perform some experiments with different structures of *LRGF* networks in their specific application, e.g., in angular speed and rotor flux vector reconstruction problem for the induction motor drive.

2.3.2 Practical Solutions and Applications of Compound Network Structures

Based on the work of Tsoi and Back in the field of *LRGF* networks [34] and their conclusion concerning the efficiency of NN in modeling a given nonlinear dynamical system, it was assumed that the application of NN architectures with different feedback configuration can results in better IM speed reconstruction. In this paragraph chosen testing results of different network architectures with compound feedback connections are presented. Based on the general concept presented in Fig. 2.19, NN with different internal structures were designed: *Multilayer Feedforward Network, Cascade Network, Local Output Feedback Network, Inter-layer Feedback Network, Global Feedback Network* [25]. If one connects each structure described above to another, so called Compound Network will become, which is presented in Fig. 2.20.

It is possible to obtain 256 different network structures. The main goal was the analysis of the network structure influence on the estimation quality of the IM state variables. The assumption was, that the induction motor speed should be reconstructed based on measurement of stator line currents, as network inputs, which could be pre-processed in a different way.

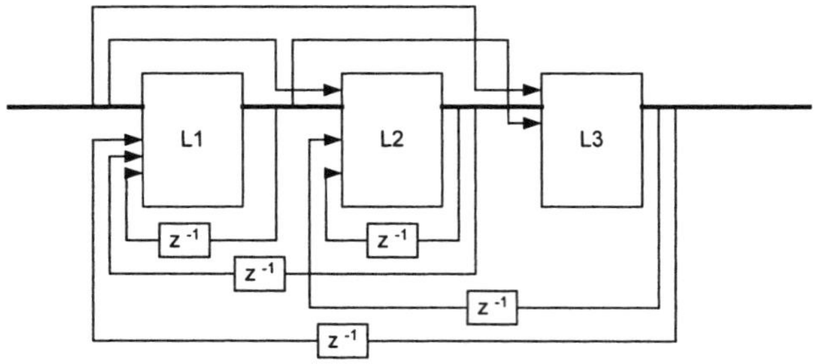

Fig. 2.20. Compound Multilayer Network with all possible connections (L1, L2 – 1st and 2nd hidden layers, L3 – output layer).

Over 100 different NN structures, with internal connections described above were tested. Also number and form of input variables was changed: the same network structure was trained for different number of delayed input currents, also nonlinearly pre-processed (i.e., into the α-β co-ordinates of stator current vector, their squared values or stator current magnitude, etc.).

2.3.3 Simulation Results for Speed Reconstruction Based on Multilayer NN

Neural networks of different internal structure, with nonlinear (sigmoidal) activation functions of neurons used in the hidden layers and linear output neuron were applied. These NNs were trained based on training patterns, prepared as input/output data table obtained by simulations of the induction motor transients during line-start of the motor and load-torque m_L step changes, using MATLAB-SIMULINK and Neural Network Toolbox. The backpropagation algorithm with Levenberg-Marquardt's modification was used for the training procedures. NNs were trained for constant and varying rotor resistance r_r values: $(0.8, 1.0, 1.2)r_{rN}$ and various load torque values: $(0, 0.4, 0.8, 1.2)m_L$. Then the trained networks were tested for different regimes from those trained, to evaluate the speed estimation performance of different neural network architectures.

In Fig. 2.21 (left side) transients of the rotor speed estimation by the feedforward NN (FNN) with relatively simple structure – with two hidden layers, consisting of 5 and 3 neurons respectively, are demonstrated. The estimation is completely bad for this simple architecture, but for the bigger number of neurons used in each hidden layer (Fig. 2.21 – right side) or for feedforward architecture replaced by the cascade network (see Fig. 2.22 – left side), the estimation quality is much better.

The same simple 2-5-3-1 network with inter-layer feedback connections from input to the 1st or 2nd hidden layer was also tested, but the obtained results were acceptable only in the steady-state; during starting process of the motor as well as load-torque change, the network presented errors in the range of 6% (in relation to actual motor speed). This result is presented in Fig. 2.22 (right side). The similar effect was obtained for the network with global feedback (recurrent network). In contrary, the steady-state value of rotor speed was estimated with very small error (about ± 1%).

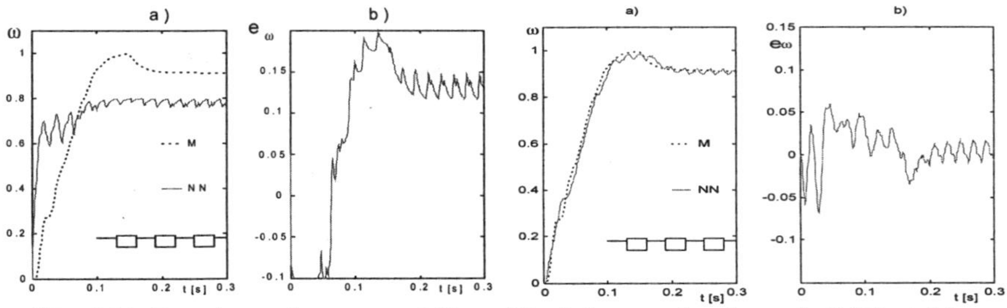

Fig. 2.21. Transients of rotor speed (dotted line), its neural estimate (solid line) (a) and estimation error ε_ω (b) for 2-5-3-1 FNN with 2 pairs of current delayed inputs (left side) and for 2-10-5-1 FNN with 1 pair of current delayed inputs (right side).

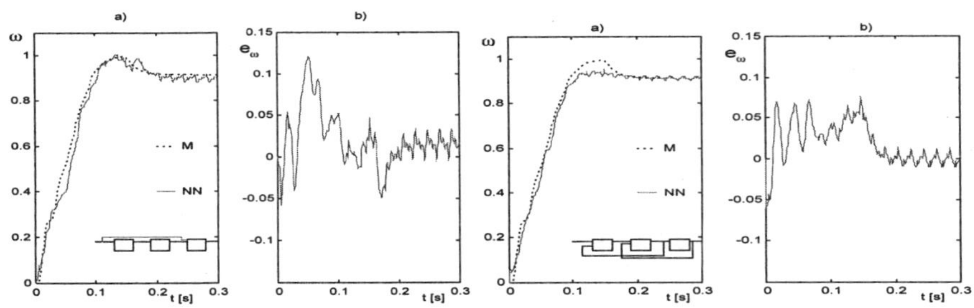

Fig. 2.22. Transients of rotor speed (dotted line), its neural estimate (solid line) (a) and estimation error ε_ω (b) for 2-5-3-1 cascade (left side) and for 2-5-3-1 NN with inter-layer connections (right side) and 2 pairs of current delayed inputs.

The significant improvement was achieved when additional, pre-processed input signal was introduced to the network; it was the stator current magnitude calculated based on measured two line currents. This additional signal has improved the results of speed estimation not only in the case of Elman network (Fig. 2.23 – left side) but even for the feedforward network (FNN) with small number of neurons in hidden layers (Fig. 2.23 – right side).

Similar results were obtained for different network structures with increased number of delayed current inputs and with increased number of neurons in the hidden layers. But it should be noted, that when number of neurons in the 1st layer was greater than 15 and in the 2nd layer – greater than 7 (for this specific induction motor drive), the quality of the speed estimation intended to decrease because the NN lost a generalization ability due to too many degrees of freedom.

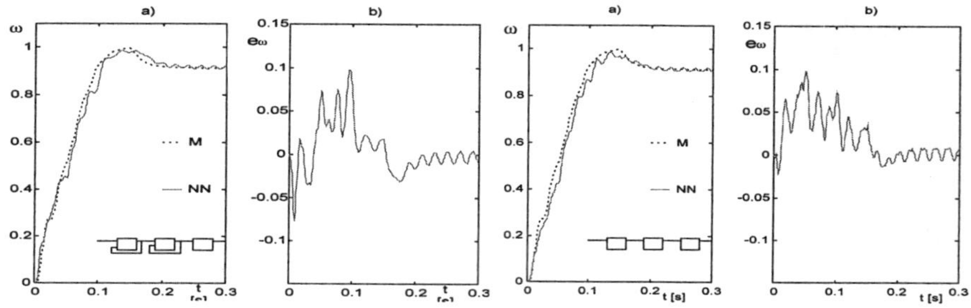

Fig. 2.23. Transients of the rotor speed (dotted line), its neural estimate (solid line) (a) and estimation error ε_ω (b)for the 3-5-3-1 Elman NN (left side) and for the 3-5-3-1 FNN (right side) with current magnitude as additional input and 2 pairs of current delayed inputs.

Networks with additional pre-processed input information, including the squared values of $i_{s\alpha}$, $i_{s\beta}$ components of stator current vector were also tested. But the obtained results were comparable to those with stator current magnitude added to the network input.

From testing of over 100 different NN structures and comparison of the quality of induction motor speed reconstruction based on stator current measurement only, the following conclusions can be formulated:

- the best speed reconstruction is obtained for the network with additional input information including stator current magnitude and/or squared values of current components;
- the best results with above additional input were obtained for 2-hidden layer feedforward and Elman networks; comparable results were obtained also for cascade networks;
- cascade networks have good performances in the speed estimation problem, when connection between input and 2nd hidden or output layer is designed;
- the larger number of delayed input current values improves the speed estimation quality.

2.3.4 Experimental Results of the Induction Motor Speed Estimation by Multilyer NN

Various structures of NN, with various numbers of neurons in hidden layers were tested in experiments and results of rotor speed estimation were compared. In the case of sinusoidal supply (like in the simulation tests), the information about stator currents was sufficient for proper speed estimation. But in the case of the stator frequency changes (converter supply), the additional input signal of NN was necessary for proper speed reconstruction. Two concepts were tested:

1 - the additional I_s/f_s (stator current magnitude/stator frequency) transducer, realized numerically;

2 - the pre-processed current and/or voltage values (their magnitudes or squared values).

Chosen examples of rotor speed estimation realized by different structures of feedforward and Elman-type NNs are presented in following figures. Experiments were performed in the laboratory rig equipped with DSP for IM supplied from PWM inverter.

In Fig. 2.24 the speed estimation for the frequency-start of IM to $0.5\omega_N$ and then for change of the reference frequency value to nominal speed ω_N is presented, for feedforward and Elman networks. The information about actual stator current frequency was presented to NN as an additional input.

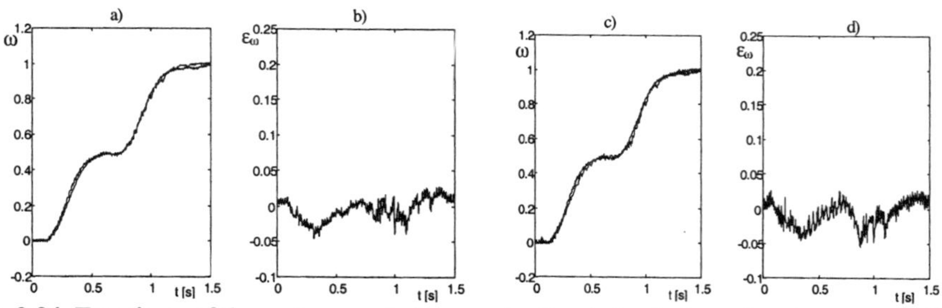

Fig. 2.24. Transients of the motor speed, its neural estimate (a,c), the estimation error (b,d) during frequency changes: a,b - for 7-7-15-1; c,d - for 7-20-1 FNN (including 2 delayed pairs of input currents and additional frequency input).

From these experimental results, that in the case of feedforward NN with only one hidden layer, the obtained transient estimation errors are not greater than ±4% what is almost the same result as in the case of two-hidden layer network.

For comparison, other types of neural estimators based on recurrent networks of Elman type were tested in experiments for the same input information. Results of speed reconstruction are presented in Fig. 2.25.

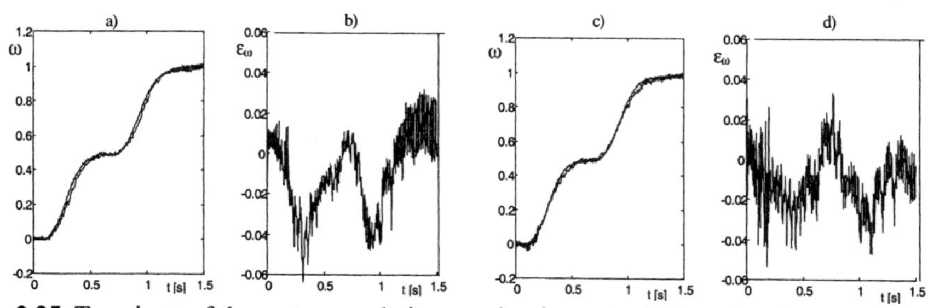

Fig. 2.25. Transients of the motor speed, its neural estimate (a,c), the estimation error (b,d) during frequency-changes: a,b - for 7-15-7-1; c,d - for 7-15-1 Elman network (including 2 delayed pairs of input currents and additional frequency input).

The obtained results are similar in the case of these two types of NN structures. It seems however that in the case of Elman networks it is possible to obtain the proper speed reconstruction with help of simplest NN structures (less neurons in the hidden layer), what is important in practical implementations.

To obtain better speed estimation, the different concept was tested, which was mentioned before. As the additional NN input information, magnitudes of stator current and voltage were introduced. It significantly improved the speed reconstruction and obtained average errors were in the range of 1-2%. In Fig. 2.26 the most characteristic results are presented for feedforward networks with one and two hidden layers. In all NN presented here number "4" in input layer means that the inputs signals are the following: stator currents $i_{s\alpha}$, $i_{s\beta}$ in stationary $\alpha-\beta$ co-ordinates, stator current magnitude and stator voltage magnitude.

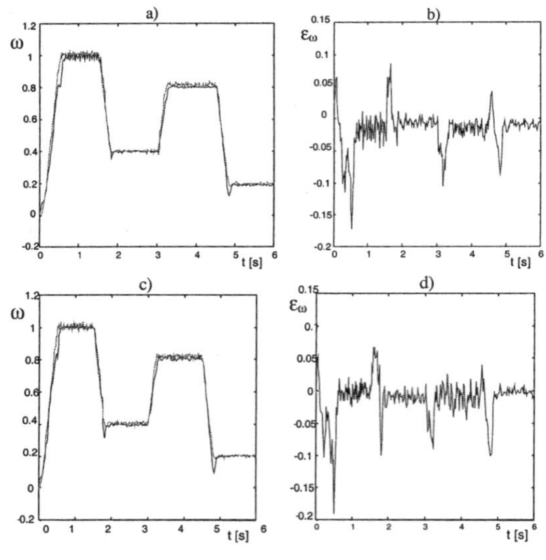

Fig. 2.26. Transients of the motor speed, its neural estimate (a,c) and the estimation error (b,d) during frequency changes for 4-5-1 (a,b) and for 4-5-5-1 (c,d) feedforward NN.

The best results were obtained for neural network with two hidden layers (Fig. 2.26c), but even in the case of very simple one-hidden layer network (Fig. 2.26a), speed estimation results are acceptable. It means that proposed concept with additional pre-processed input signals of neural networks is good solution for converter-fed IM speed reconstruction in the wide range of motor speed changes.

The experiments performed with two concepts of neural speed estimation, based respectively on neural modeling method (on-line estimator) and neural identification method (off-line-estimator), indicate that:

- In the case of exactly known motor parameters, the application of the on-line neural speed estimator, based on MRAS concept and BP algorithm is more convenient due to the simplicity of implementation. It was proved by simulation and experiments that in the case of not exactly known motor parameters it is quite difficult to adapt the suitable values of learning and momentum factors in BP algorithm, to obtain the accurate speed estimation. Moreover, taking into account that in the real motor its parameters are changing during the drive operation (due to temperature and saturation effects), the parameters used in the mathematical model of such neural estimator should be on-line adapted to the real motor parameters. Only in such condition the good accuracy of speed estimation will be obtained;
- The neural speed estimator based on neural identification method seems to be more convenient when motor parameters are not exactly known and are not on-line identified during the drive operation. The multilayer neural network can be trained off-line for a wide range of motor parameter changes in simulation or based on measured motor signals. Thus the obtained neural speed estimator can be robust to motor parameter changes. However, the practical realization is more complicated for this approach, because an implementation in DSP requires optimization of machine codes to perform fast processing of network input variables.

3 NN-Based Flux Estimation

3.1 General Remarks on Flux Estimation of the Induction Motor

In order to control an induction motor requiring high dynamic performance, an accurate knowledge of the magnitude and position of the rotor flux is necessary, irrespective of the operating point [4], [15], [35]. Various methods of the rotor flux estimation are used like: simulators, flux observers, Kalman filters [11], [20]. All these approaches use models of the motor which are sensitive to parameter errors [19], [25], similarly as in the case of the IM speed estimation. Neural networks offer a good alternative solution for this purpose.

Recently, the application of neural networks for induction motor state variable reconstruction was reported by few authors [5], [21], [33]. In [5] the feedforward NN with two hidden layers was used for the estimation of the rotor flux magnitude and position as well as the motor torque based on measured stator current and stator flux calculated from the difference between stator voltage and voltage drop on the stator resistance. These variables were estimated for the vector controlled induction motor in the condition when rotor flux magnitude was kept constant and reference speed of the motor was changed according to the presumed curve. In [33] authors tried to estimate the rotor flux magnitude and phase during transients such as line-start operation or step-load torque changes of the motor on the base of stator current measurements. However, simulation studies were performed for constant motor parameters only. The different approach was taken into account in [21]: feedforward and recurrent NNs were trained for constant as well as for changed motor parameters and obtained results were tested, similarly as in the case of IM speed estimation described in the Section 2.3.

3.2 Flux Estimation in the Induction Motor Drives

3.2.1 Case of Constant Motor Parameters

Various structures of NN, with various activation functions of neurons in hidden layers and different input signals were analyzed and the results of the estimation of rotor flux amplitude and phase were compared. The same approach, as demonstrated in previous section in Fig. 2.18, was taken for design of optimal structure of NN in different operating condition of the induction motor.

For the rotor flux vector estimation two different NNs were used: one - for the flux vector magnitude Ψ_r, the second - for the flux vector position $sin\gamma_\psi$. The feedforward networks of various internal structure were trained for different load torque transients: from no load during IM line-start operation to 1.2 m_{LN} with 0.1 or 0.4 step, for constant motor parameters. Then neural simulators were tested for various sequences of load torque changes presented in Fig. 3.1.

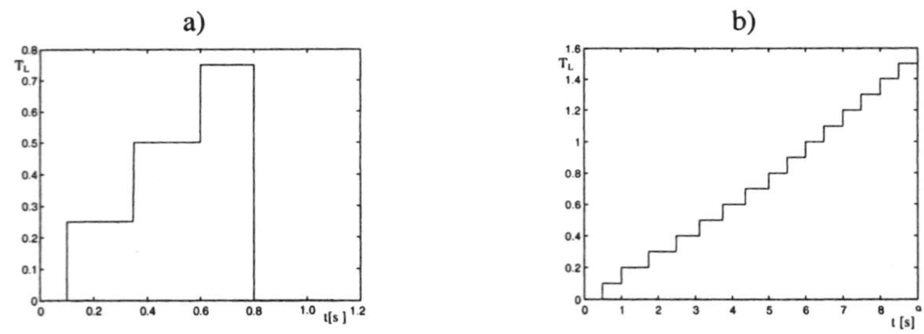

Fig. 3.1. The load torque change programs for neural networks testing.

The net 3-25-1 with sigmoidal activation functions in the hidden layer and with linear activation function in the output layer was tested for various sequences of load torque changes. The number of hidden layer neurons were determined by trial and error. Two components of the stator current vector in (α-β) reference frames and rotor speed were used as NN input signals. Results of the rotor flux magnitude estimation are presented in Fig. 3.2. Similar results were obtained for the network with two hidden layers 3-8-5-1. Flux magnitude in these figures is presented only for the time period after decrease of the rotor flux transients, because the neural net was not able to reconstruct correctly the transients process during line-start operation of the motor. Also the greater number of hidden layers did not improve the quality of the reconstruction process [21].

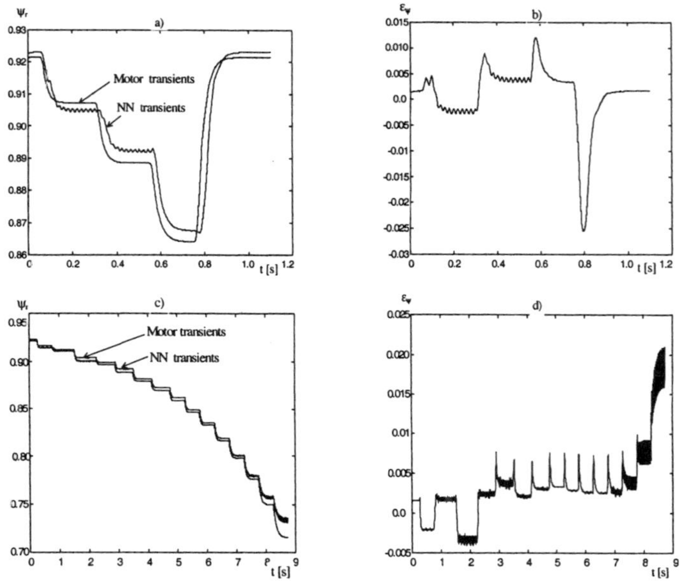

Fig. 3.2. Transients of the rotor flux magnitude Ψ_r, its neural estimate (a, c) and estimation error (b,d) during load change tests for the feedforwad 3-25-1 NN.

Proper results were obtained for the network with historical inputs (delayed current values). The estimation of the flux was very good in the beginning of the transient process as well as during load torque changes of the motor, even for the network with 2 inputs only – the stator current vector components. In Fig. 3.3 transients of the rotor flux magnitude and its neural estimate during the IM line-start operation are presented.

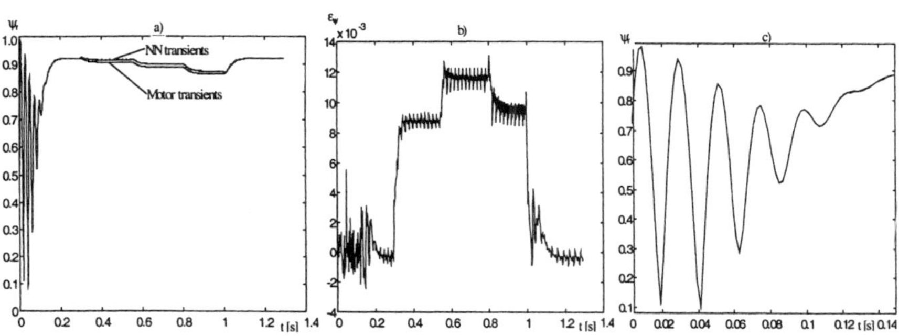

Fig. 3.3. Transients of the rotor flux magnitude Ψ_r, its neural estimate (a,c) and the estimation error (b) during load changes for 2-15-7-1 NN with 4 pairs of delayed current inputs (for constant r_r).

The two-hidden-layer 2-15-7-1 NN was trained with help of input and output vectors of 1000 data each and with 4 historical inputs in each current component vector. As it results from Fig. 3.3c, in which the detailed transient for the first few milliseconds of the rotor flux magnitude is demonstrated, this NN estimator works very well. Also the maximum estimation error did not exceed the value of 1.2% during load torque changes, when neural network was tested for the load torque in the range (0-1.2)m_{LN} - see Fig. 3.3a,c. This range of load torque changes was used in training procedure. In the next figure the results of the neural flux estimation are presented in the case when load torque exceeds the training range. Even in this case the estimation error did not exceed the value of 1.5% (Fig. 3.4b).

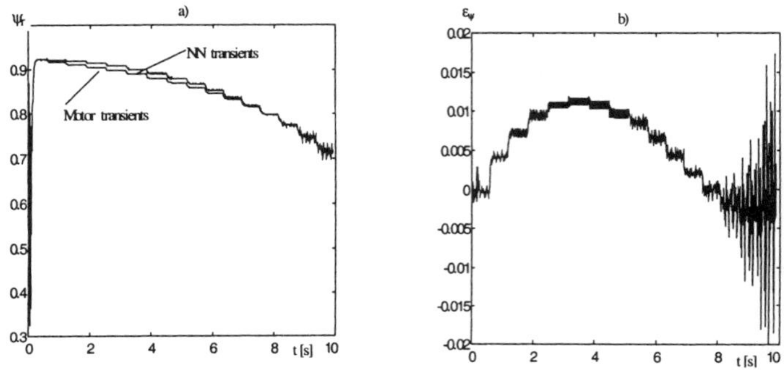

Fig. 3.4. Transients of the rotor flux magnitude Ψ_r, its neural estimate (a) and the estimation error (b) during load changes for feedforward 2-15-7-1 NN with delayed current inputs (for constant r_r).

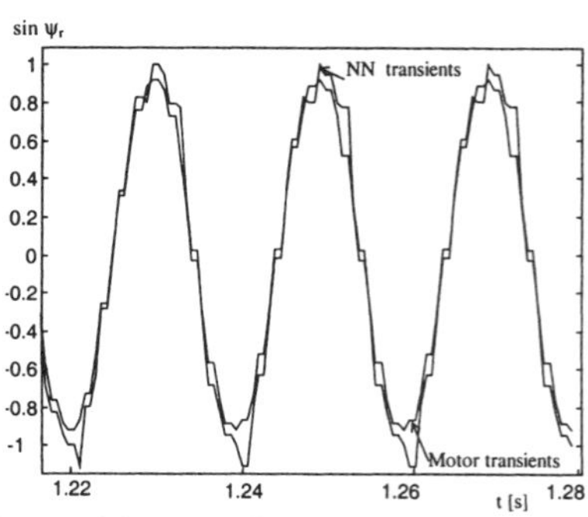

Fig. 3.5. Transients of the rotor flux phase $sin\gamma_{\psi}$ and its neural estimate for 2-15-7-1 feedforward NN.

Fig. 3.5 illustrates the ability of NN estimator to the reconstruction of the rotor flux vector position of the induction motor. The experiment was performed for

NN of the same structure 2-15-7-1, trained with help of 500 data only in each training vector. Also for this case the estimator worked quite well.

It is worth of mention that for the same structure of NN with historical inputs, magnitude and phase of the rotor flux vector were reconstructed based only on current measurements.

3.2.2 Case of Changed Motor Parameters

The same structure of feedforward NN was used for rotor flux estimation in the case of changing parameters of the induction motor. The obtained results were much worse than for constant rotor parameters, especially in the beginning of line-start transient process of the motor. So simulation tests, similar to those for speed estimation were performed (see Section 2.3.3) and neural networks of different internal structure were trained for different motor parameters and load torque changes during transient process of the motor.

The compound NN with inter-layer feedback and cascade connections did not work properly in the case of flux estimation during starting operation of the motor, even for relatively big number of neurons in hidden layers. The best results in this class of NN were obtained for two-layer cascade network with external feedback, but transient errors in the beginning of line-start operation as well as during load torque changes were in the range of a few percent.

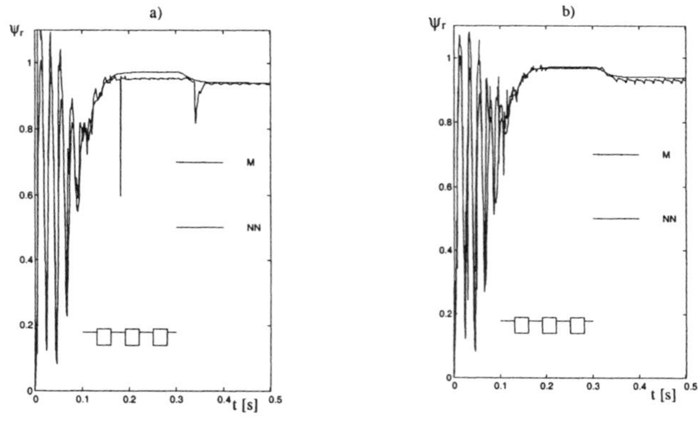

Fig. 3.6. Transients of rotor flux magnitude Ψ_r and its neural estimate for rotor resistance r_r and load torque changes for feedforward NN with 2 delayed current inputs: a) 2-8-3-1 NN; b) 2-15-7-1 NN with current magnitude as additional input.

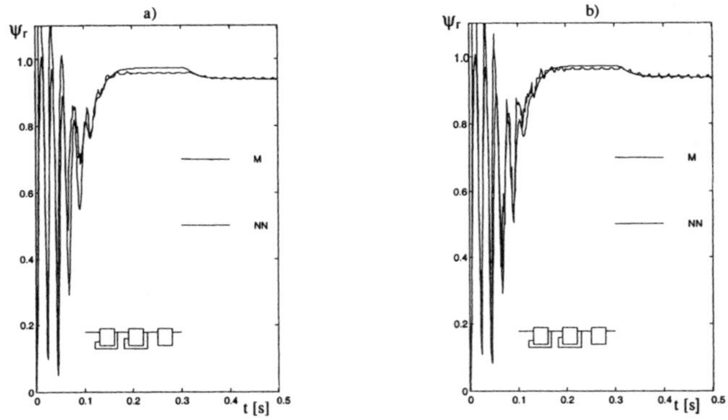

Fig. 3.7. Transients of the rotor flux magnitude Ψ_r and its neural estimate during line-start for rotor resistance r_r and load torque changes for Elman NN with 2 delayed current inputs: a) for 2-15-7-1 NN; b) for 2-8-3-1 NN with current magnitude as additional input.

Better results were obtained for feedforward and Elman type NNs, especially when additional pre-processed current input - stator current magnitude - was introduced. In Fig. 3.6 selected examples of the flux magnitude estimation are presented in the case of feedforward networks and in Fig. 3.7 – for Elman network respectively.

NN structure from Fig. 3.6a was not able to reconstruct the flux magnitude during load torque changes of the motor. But the larger number of neurons in the hidden layers as well as additional input information about stator current magnitude improved significantly the flux estimation in the whole transient process (Fig. 3.6b). Recurrent networks of Elman type, with sigmoidal neurons in its hidden (recurrent) layers and linear neuron in its output layer, were also taken into account. In Fig. 3.7a the results of flux reconstruction are presented in the case of 2-15-7-1 Elman-type NN with 2 delayed current inputs. The better flux estimation was obtained for smaller network 3-8-3-1, but only in the case, when additional input information about current magnitude was presented to the network.

It is also possible to train NN for proper estimation of rotor flux magnitude even for so specific transient process, as the first period of line-start operation of the induction motor. Such case is presented in Fig. 3.8 for the feedforward 2-15-7-1 NN trained for different rotor resistance r_r values from the range $(0.6\text{-}1.3)r_{rN}$.

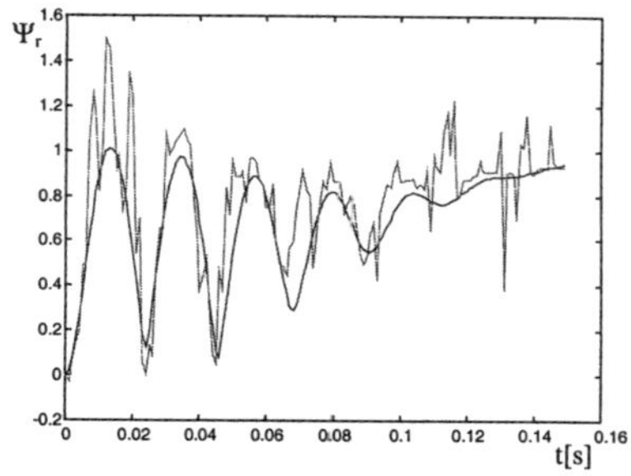

Fig. 3.8. Transients of the rotor flux magnitude Ψ_r and its neural estimate for rotor resistance r_r changes for 2-15-7-1 feedforward NN with 2 delayed current inputs.

As results from presented simulation tests for the rotor flux estimation, similarly as in the case of motor speed, the feedforward and Elman type neural networks are the best solution in the case of varying motor parameters. In comparison with speed estimation problem, in the case of flux estimation the more complicated network's structure (with two hidden layers and larger number of neurons in the hidden layers or more delayed input signals) is necessary. It is caused by the more complicated and fast dynamics of the rotor flux vector during the induction motor transients. But these networks, after suitable training procedure (for wide range of the drive operation modes and motor parameter changes), can estimate properly the rotor flux based on two line stator currents only. Additional information about pre-processed stator currents and/or stator voltage improves the estimation process and enable minimization of neurons number used in hidden layers.
All these NN estimators need a large amount of arithmetic operations and function evaluations. So their computational complexity is greater than for conventional, algorithmic methods. But due to their robustness to drive parameter and operation mode changes, they can be an interesting solution in the near future, when specialized neuroprocessor will be available.

4 Sensorless Field Oriented Control of Induction Motor with NN Speed Estimation

4.1 Basics of Field Oriented Control (FOC)

The most popular high performance induction motor control method known as *Field Oriented Control* (FOC) or *Vector Control* has been proposed by Hasse [10] and Blaschke [4] and has constantly been developed and improved by other researchers [15], [35]. In this method the motor equations (2.1)-(2.6) are (rewritten) transformed in a coordinate system x-y that rotates with the rotor flux vector ψ_r (Fig. 4.1). Thus, the angular speed of the coordinate system is $\omega_K = \omega_S$.

Under these assumptions, substituting the rotor current vector from the rotor voltage Eq. (2.3) by Eq. (2.5) we obtain a differential equation for the rotor flux vector:

$$\frac{d\psi_r}{dt} = -\frac{\psi_r}{T_r} - j(\omega_s - \omega_m)\frac{\psi_r}{T_N} + \frac{x_M}{T_r}\mathbf{i}_s \tag{4.1}$$

where: $T_r = \frac{x_r}{r_r} T_N$ - is the rotor time constant.

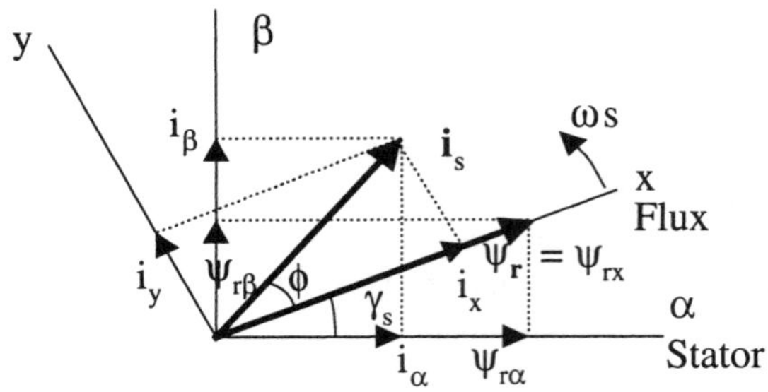

Fig. 4.1. Field oriented control: vector diagram.

For the rotor field oriented coordinates x-y (Fig. 4.1) we have

$$\psi_{rx} = \psi_r\,; \qquad \psi_{ry} = 0 \tag{4.2}$$

and Eq. (4.1) can be rewritten as follows

$$\frac{d\psi_r}{dt} = -\frac{\psi_r}{T_r} + \frac{x_M}{T_r} i_{sx} \tag{4.3}$$

$$0 = -(\omega_s - \omega_m)\frac{\psi_r}{T_N} + \frac{x_M}{T_r} i_{sy} \tag{4.4}$$

Eq. (4.3) describes the influence of the flux current components i_{sx} on the rotor flux magnitude ψ_r. The motor torque, according to Eq. (2.5), can be expressed as follows:

$$m = \frac{x_M}{x_r} i_{sy}\psi_r \tag{4.5}$$

Finally, the dynamics of current controlled induction motor can be described as follows:

$$T_N \frac{d\psi_r}{dt} = -\frac{r_r}{x_r}\psi_r + \frac{r_r}{x_r}x_M i_{s\alpha}$$

$$T_M \frac{d\omega_m}{dt} = \frac{x_M}{x_r}\psi_r i_{sy} - m_L \qquad (4.6)$$

$$\omega_r = \omega_{s\psi} - \omega_m = \frac{r_r x_M}{x_r}\frac{i_{sy}}{\psi_r}$$

Where: ω_r is slip frequency, and m_L is a load torque.
The block diagram, which represents equations Eq. (4.6), is shown in Fig. 4.2.

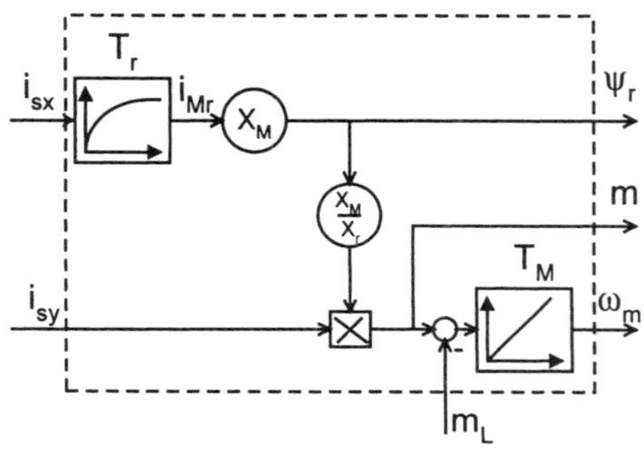

Fig. 4.2. Field oriented control: block diagram of IM in x-y field coordinates.

When compare this scheme with the separately excited dc motor, we can note the following analogies:
- i_{SX} component of the stator current vector in induction motor corresponds to the exciting current in the separately excited dc motor,
- i_{Sy} component of the rotor current vector in induction motor corresponds to the armature current in the separately excited dc motor
The basic control properties of induction motor can be summarized as:

- The relationship between i_{sx} component and rotor flux magnitude ψ_r is not static. The dynamic is represented by the first order system with the rotor time constant T_r. Therefore, i_{sx} component is equal to the magnetizing current

$$i_{Mr} = \frac{\psi_r}{x_M} \qquad (4.7)$$

  only in steady state;
- The torque m and flux magnitude ψ_r are dependent on the stator current vector $\mathbf{i}_s$ and cannot be controlled independently;
- Decoupled control must guarantee, that the change of i_{sx} component does not influence on i_{sy} component and vice versa. In this case we can control rotor flux using i_{sx} , and torque using i_{sy} . If we additionally keep the value of i_{sx} as constant, the situation is analogous to the separately excited dc motor.

4.2 NN Speed Sensorless FOC Scheme

The basics of the field-oriented control have been presented in the previous section. The main feature of this approach is the coordinate transformation, which allows to recalculate the decoupled field oriented coordinates i_{sx}, i_{sy} of the vector stator current into the fixed stator frame $i_{s\alpha}i_{s\beta}$:

$$i_{s\alpha} = i_{sx}\cos\gamma_s - i_{sy}\sin\gamma_s \tag{4.8a}$$

$$i_{s\beta} = i_{sx}\sin\gamma_s - i_{sy}\cos\gamma_s \tag{4.8b}$$

In indirect FOC system (Fig. 4.3.) the rotor flux angle γ_s is calculated using the reference values of flux and torque currents (the slip frequency model) and the value of mechanical speed ω_m.

$$\gamma_s = \int(\omega_m + \omega_r)dt\,; \tag{4.9}$$

and

$$\omega_r = T_N \frac{r_r}{x_r}\frac{i_{syc}}{i_{sxc}} \tag{4.10}$$

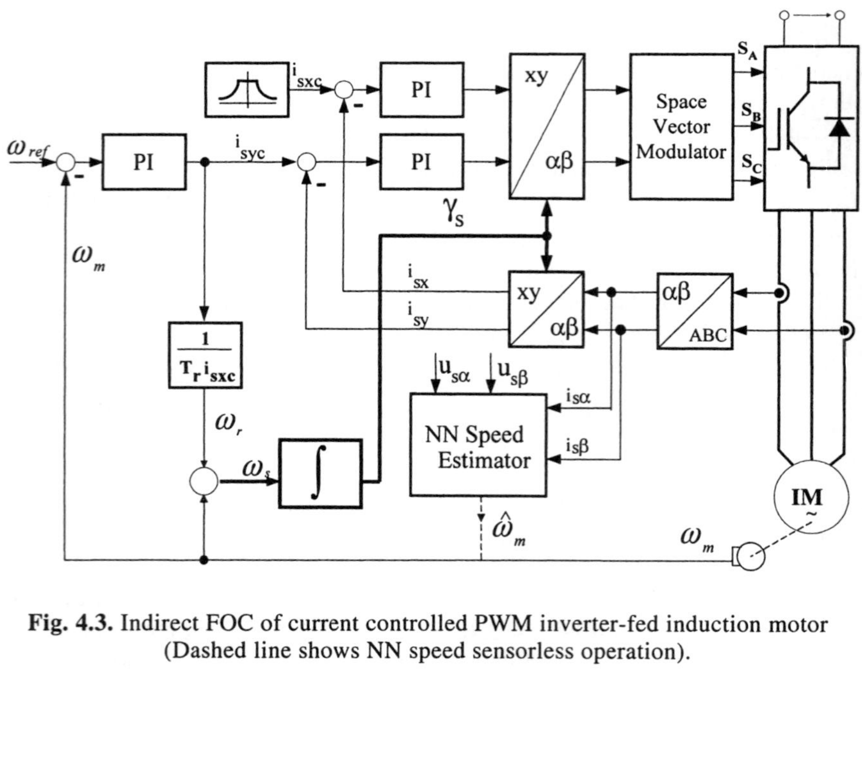

Fig. 4.3. Indirect FOC of current controlled PWM inverter-fed induction motor (Dashed line shows NN speed sensorless operation).

There are two generic types of FOC schemes called *direct* and *indirect* FOC. Most of the literature adopts the following definition [10], [15], [27] [35]: direct FOC refers to the case where flux vector position γ_s is measured or estimated, while indirect FOC refers to an implementation where the flux vector position γ_s is calculated from the reference values (feed-forward control) and mechanical speed or position measurement (Fig. 4.3).

The structure of the NN based speed estimator has been implemented according to the scheme presented in the Section 2.1 (see Fig. 2.14)

4.3 Results

To verify the NN speed sensorless FOC concept, a simulation program and the laboratory setup with four poles 3kW – induction motor drive with dSPACE DS1102 laboratory control board [37] was constructed (Fig. 4.4).

The system is based on Texas Instrument DSPs TMS320C31 and TMS320P14. The first (main) processor implements the FOC and speed estimation algorithm, whereas the second provides the vector modulation. The board is equipped with four analog to digital converters (two 16-bit and two 12-bit), four digital to analog converters, and the input for an encoder. The software is written in C language.

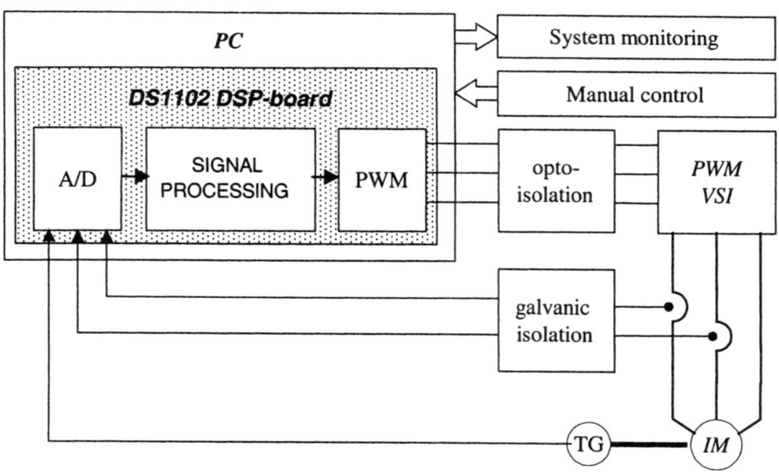

Fig. 4.4. Laboratory control unit for experimental tests.

The steady state operations of the tuned system are presented in Fig. 4.5. The sampling time in the experimental system has been set to 170 µs. The result has been obtained for half of the nominal speed (the induction motor data are given in Appendix).

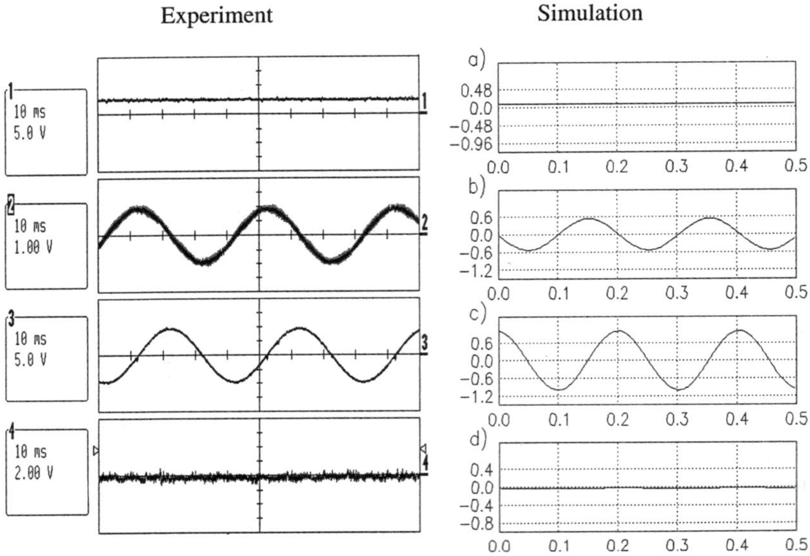

Fig. 4.5. Steady state operation of the induction motor controlled via FOC with NN based speed estimator from Fig. 2.14 ($\omega_m = 0.5$; $m_L = 0$).

(1- estimated speed, 2 - stator phase current, 3 - rotor flux, 4 - electromagnetic torque)

The speed transients to the step changes are presented in Fig. 4.6. Figure 4.6 a presents the startup of the motor in low speed region (small signal behavior), whereas Fig. 4.6 b illustrates half rated speed reversal. It can be seen, that with constant flux current component i_{sx} the flux amplitude is kept constant during torque transients. So, the flux and torque are dynamically decoupled. The torque response time is about 5 ms and is mainly determinate by current control loop design [15]. The dynamic behavior of the drive for a step change of the load torque is shown in Fig. 4.7. Also here the flux magnitude is kept constant, and speed and torque transients are well damped. The speed controller is designed according to the symmetry criterion [15].

Note that there is a good agreement between experimental and simulated oscillograms presented in Fig. 4.5 – 4.7.

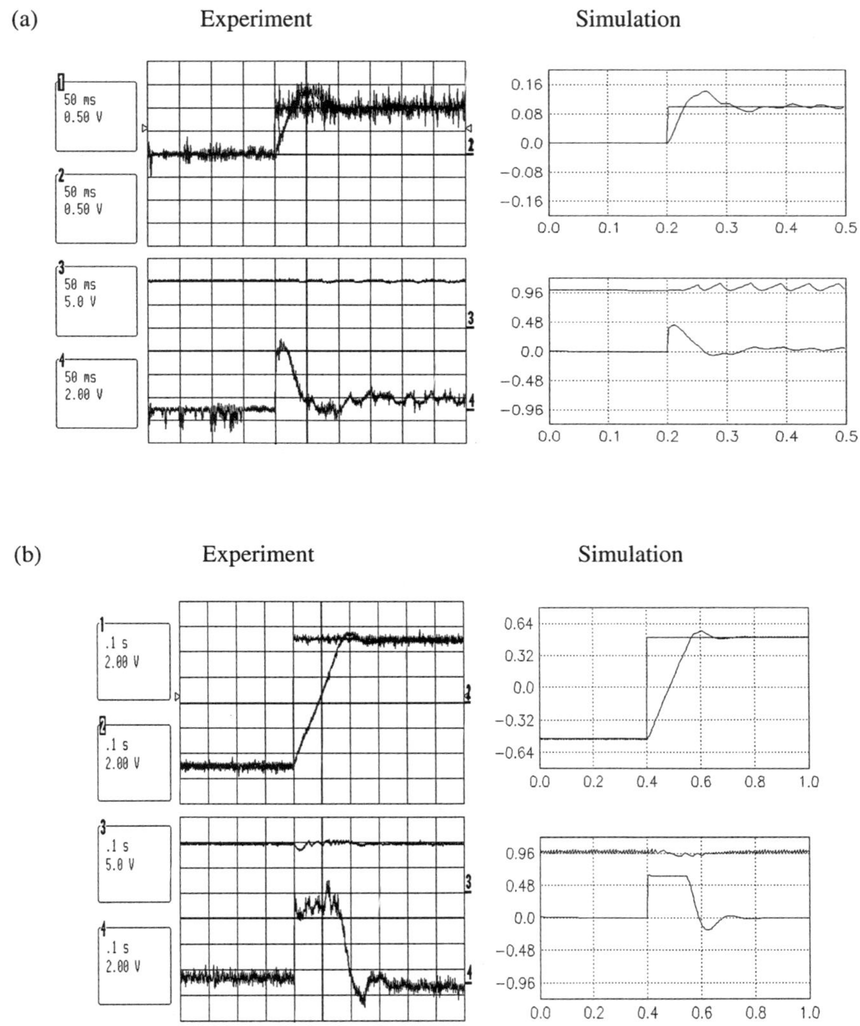

Fig. 4.6. Speed step response for the induction motor controlled via FOC with NN speed estimator from Fig. 2.14 a startup $\omega_m = 0.0 \rightarrow 0.1$; (b) speed reversal $\omega_m = -0.5 \rightarrow 0.5$ (1 - reference speed, 2 - mechanical speed, 3 - amplitude of rotor flux, 4 - electromagnetic torque).

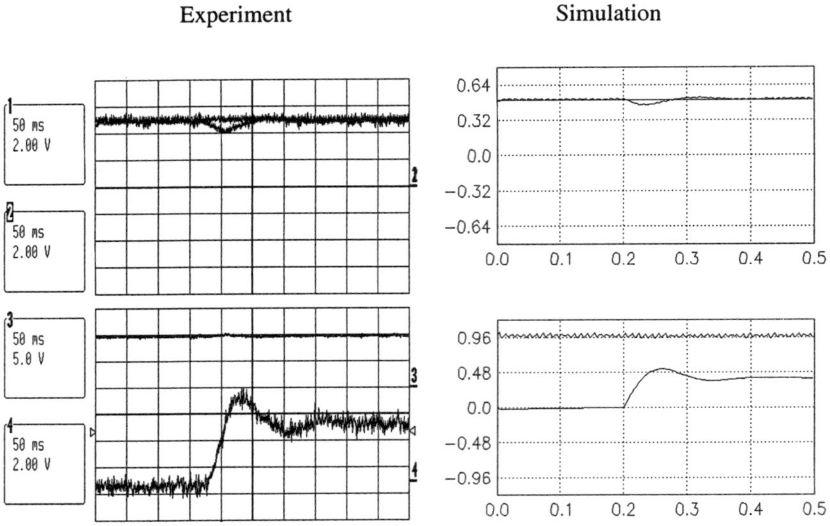

Fig. 4.7. Response to load step change for the induction motor controlled via FOC with NN speed estimator from Fig. 2.14 ($m_L = 0.0 \rightarrow 0.4$)

(1 - reference speed, 2 - mechanical speed, 3 - amplitude of rotor flux, 4 - electromagnetic torque).

4.4 Conclusions

The research results presented in this Section based on the simulation and experimental investigations can be summarized as follows:

- Application of NN speed estimators for PWM inverter-fed induction motors creates many new schemes because architecture and type of NN can be choosen freely (see Section 3.2).
- One of the simplest *on line* operated NN estimator schemes is based on combination of the voltage flux model and current estimatior (Fig. 2.14).
- The speed sensorless FOC scheme operated with a new NN speed estimator, instead of mechanical speed sensor, shows very good performance in steady state and in dynamical states.
- The drive keeps all well-known advantages of FOC scheme: allows the independent control of rotor flux and motor torque, and has very fast torque response.
- There is no problem with start and stop as well as operation in the low speed region (1-2% of nominal speed).
- The unique advantage of the simple NN based speed sensorless scheme lies in use of backpropagation algorithm, which has an intrinsic filtering function. Anti-noise filters used in conventional (e.g. MRAS) estimators deteriorate the control performance, especially, in higher speed region.

5 Neuro-Fuzzy Direct Torque Control (NF-DTC) of Induction Motor

5.1 Introduction

When, in the mid 1980s, it appeared that high performance induction motor control systems would be standardized on the basis of the field oriented control (FOC) philosophy (see Section 4.1), there appeared the innovative studies of Depenbrock [7] and Takahashi and Nogouchi [31], which depart from idea of coordinate transformation and the analogy with DC motor control. These innovators propose to replace motor decoupling via nonlinear coordinate transformation with bang-bang self control, which goes together very well with on-off operation of inverter semiconductor power devices. In contrast to FOC systems, where the flux and torque is controlled by acting on current vector components, Direct Torque Control (DTC) has been developed to regulate flux and torque directly, while currents and voltages are regulated indirectly.

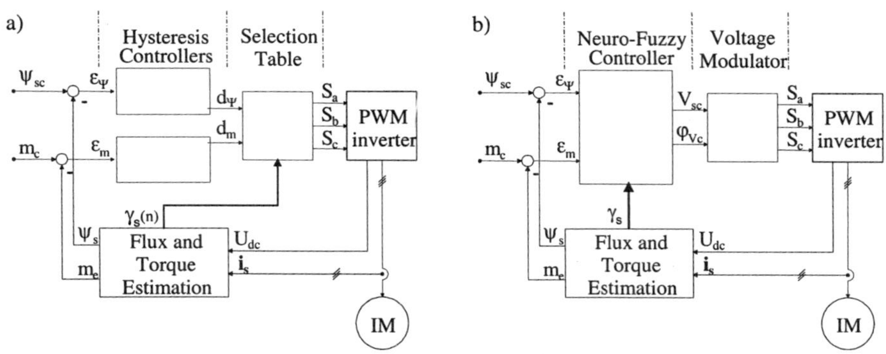

Fig. 5.1. a) Direct Torque Control (DTC) of PWM inverter-fed induction motor drive, b) Direct Torque Neuro-Fuzzy Control of PWM inverter-fed induction motor.

Fig. 5.1 shows two system configurations for the DTC controlled induction motor drive. Both systems use stator flux vector and torque estimators on a PWM inverter fed drive. The stator flux amplitude Ψ_{Sc} and the electromagnetic torque m_c are the command signals which are compared with the estimated ψ_S and m_e values respectively, giving instantaneous flux error ε_ψ and torque error ε_m, as shown in the figure. In the conventional scheme (Fig. 5.1(a)) [31] the ε_ψ and ε_m signals are delivered to two hysteresis comparators. The corresponding digitized output variables d_ψ, d_m and the stator flux position sector $\gamma_S(n)$ create a digital word, which selects the appropriate voltage vector from the switching table. Thus, the selection table generates pulses S_a, S_b, S_c for the power switches.

Among the well-known disadvantages of the DTC scheme are [6], [35]: variable switching frequency, violence of polarity consistency rules, current and torque distortion caused by sector changes, start and low speed operation problems, high sampling frequency needed for digital implementation of hysteresis comparators. All the above difficulties can be eliminated when, instead of the selection table, a voltage modulator is applied.

In this Chapter, a controller based on *Adaptive Neuro-Fuzzy Inference System* (ANFIS) [12-14] for voltage source PWM inverter-fed induction motor is presented (Fig. 5.1b). The ANFIS structure has been selected because of its flexibility and possibility to use for a wide rande of control tasks [14], [36]. This controller combines fuzzy logic and artificial neural networks for decoupled flux and torque control in connection with space vector modulator. The error signals ε_ψ and ε_m are delivered to the Neuro-Fuzzy (NF) controller, which also uses information on the position (γ_S) of the actual stator flux vector. The NF controller determinates the stator voltage command vector in polar coordinates $\mathbf{v}_c = [V_c, \varphi_{Vc}]$ for the voltage modulator, which finally generates the pulses S_a, S_b, S_c to control transistor switches of the PWM inverter.

5.2 Neuro-Fuzzy Controller Scheme

Internal structure of the NF controller of Fig. 5.1b is shown in Fig. 5.2. The system based on the ANFIS structure consist of five layers.

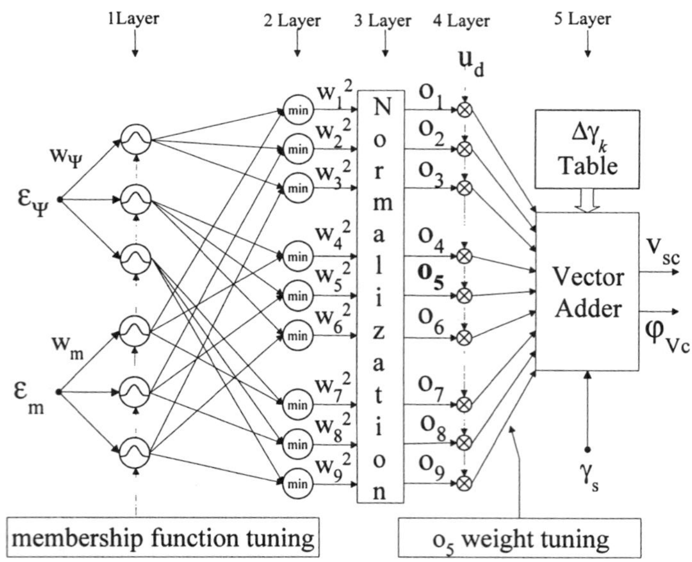

Fig. 5.2. Neuro-Fuzzy Controller.

Layer 1:

Sampled flux ε_Ψ and torque ε_m errors multiply by w_Ψ and w_m weights as:

$$\varepsilon_\Psi' = w_\psi \varepsilon_\Psi, \tag{5.1}$$

$$\varepsilon_m' = w_m \varepsilon_m, \tag{5.2}$$

are delivered to the membership functions in both inputs. To simplify the DSP calculations, the functions are triangular shaped as shown in Fig. 5.3.

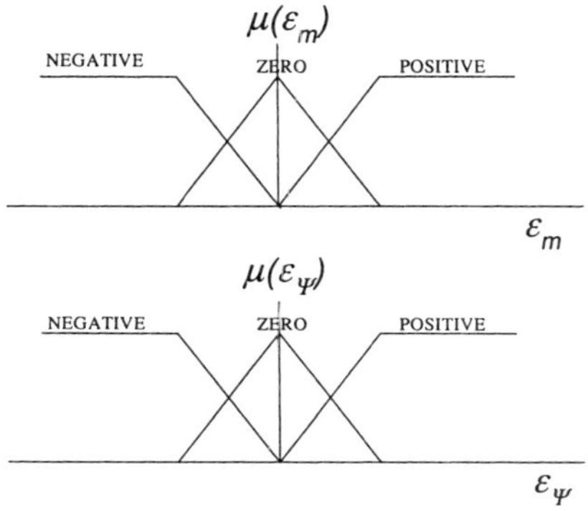

Fig. 5.3. Triangular membership function sets.

The first part outputs are calculated based on equation:

$$O_{mi}^{\ 1} = \mu_{A_{mi}}(w_m e_m), \tag{5.3}$$

$$O_{\Psi j}^{\ 1} = \mu_{A_{\Psi j}}(w_\Psi e_\Psi), \tag{5.4}$$

where $O_{mi}^{\ 1}$, $O_{\Psi j}^{\ 1}$ - first layer output signals,

i=1,2,3 - node number for the torque error,

j=1,2,3 - node number for the flux error,

$\mu_{A_{mi}}(w_m e_m)$ - triangular membership function for the torque error,

$\mu_{A_{\Psi j}}(w_\Psi e_\Psi)$ - triangular membership function for the flux error,

w_Ψ - stator flux error input weight,

w_m - torque error input weight,

The number of membership function is I, J for torque and flux error respectively.

Layer 2:

The second layer calculates the minimum what correspond to the classical fuzzy logic system. The calculation can be written as:

$$w_k^{\ 2} = \min\left[\mu_{A\,mi}(w_m e_m), \mu_{A\Psi j}(w_\Psi e_\Psi)\right], \tag{5.5}$$

where $w_k^{\ 2}$ - second layer output signals,

$k = IJ$ - node number for present layer,

There is not every node connected together. The connections are between outputs of membership functions with different input.

Layer 3:

In the third layer the output values are normalized, in that way that the following equation is fulfilled:

$$o_k^{\;3} = \frac{w_k^{\;2}}{\sum\limits_{K} w_k^{\;2}}\,, \qquad (5.6)$$

where $o_k^{\;3}$ - third layer output signals.

Layer 4:

The weight calculated in this layer as:

$$v_{sck} = o_k^{\;3} \cdot u_d\,, \qquad (5.7)$$

where: v_{sck} – amplitude of kth component of the reference voltage vector.
When the weight $o_k^{\;3}$ is active ($o_k^{\;3} > 0$), then the regulator chooses the increment angle $\Delta\gamma_i$ value from Table 1. The increment $\Delta\gamma_k$ is not needed when $o_k^{\;3} = 0$, because the multiplication $o_k^{\;3} * u_d$ is equal to zero.

Table 1. An example of reference voltage increment angle selection.

ε_ψ	P	P	P	Z	Z	Z	N	N	N
ε_m	P	Z	N	P	Z	N	P	Z	N
$\Delta\gamma_K$	$+\frac{\pi}{4}$	0	$-\frac{\pi}{4}$	$+\frac{\pi}{2}$	$+\frac{\pi}{2}$	$-\frac{\pi}{2}$	$+\frac{3\pi}{4}$	$+\pi$	$-\frac{3\pi}{4}$

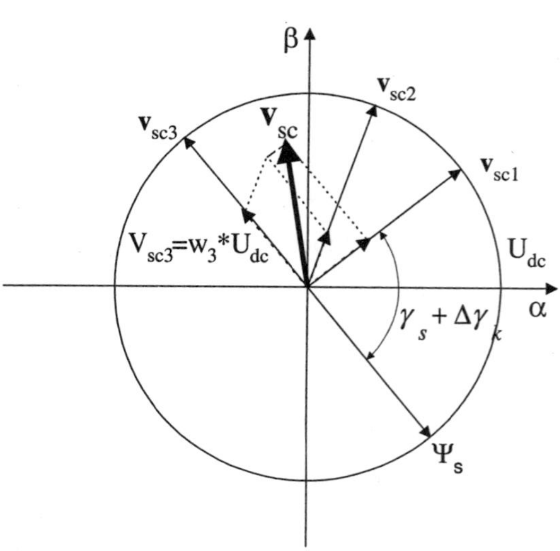

Fig. 5.4. Reference voltage calculation method.

Layer 5:
The reference voltage vector $\mathbf{v}_{sc}$ is a vector sum of the reference voltage vector components that:

$$\mathbf{v}_{sc} = \sum_{k}^{K} \mathbf{v}_{sck} \quad , \tag{5.8}$$

where $\mathbf{v}_{sck} = v_{sck}\, e^{j\varphi_k}$ - reference voltage component vector,

φ_k - reference voltage component vector phase.

The angle of the reference voltage vector is calculated from the following equation:

$$\varphi_{V_c} = \gamma_s + \Delta\gamma_k \tag{5.9}$$

where: γ_s - actual angle of the stator flux vector, $\Delta\gamma_k$ – increment angle (from the Table 1).

There are four non-zero output signals from the first layer (two for each input) during the steady state operations. It results with four generated voltage vectors components ($\mathbf{v}_{sck}$) in every sampling time. Vectors $\mathbf{v}_{sck}$ are added to each other and the result, $\mathbf{v}_{sc}$ voltage vector, is delivered to the space vector modulator. An example of the reference voltage $\mathbf{v}_{sc}$ calculation is presented in Fig. 5.4 (for simplicity, instead of four, there are only three $\mathbf{v}_{sck}$ non-zero vectors used for illustration). The space vector modulator calculates switching states *Sa*, *Sb* and *Sc* according to the well-known algorithm [15].

5.3 Construction of Increment Selection Table

Let consider the induction motor voltage equation in stator flux oriented system:

$$\mathbf{u_s} = r_s \mathbf{i_s} + T_N \frac{d\boldsymbol{\psi}_\mathbf{s}}{dt} + j\omega_s \boldsymbol{\psi}_\mathbf{s} \, . \tag{5.10}$$

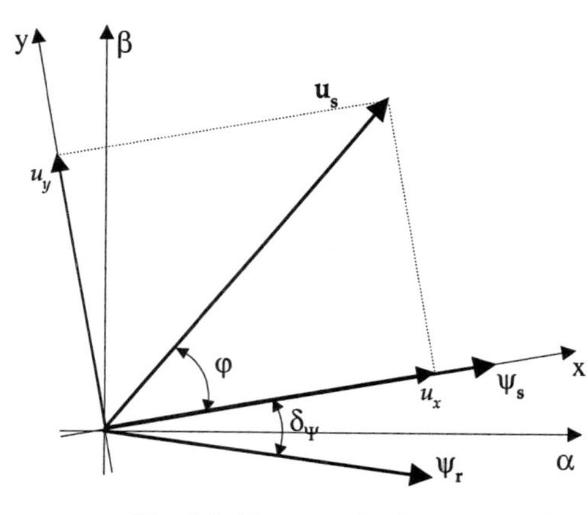

Fig. 5.5. Fluxes and voltage angles in stator flux oriented system.

The equation can be resolve into components x and y (as in Fig. 5.5):

$$u_{sx} = r_s i_{sx} + T_N \frac{d\psi_s}{dt}, \tag{5.11}$$

$$u_{sy} = r_s i_{sy} + \omega_s \psi_s, \tag{5.12}$$

Furthermore, let us take into consideration the torque equation:

$$m_e = \mathrm{Im}\left(\boldsymbol{\psi}_s^{*} \mathbf{i}_s\right), \tag{5.13}$$

what in stator flux coordinate system can be written as:

$$m_e = \psi_s i_{sy}. \tag{5.14}$$

By taking into account equations (5.11), (5.12) and (5.13) and assuming that:

$$u_{sx} >> r_s i_{sx}, \tag{5.15}$$

one obtains:

$$\psi_s = \frac{1}{T_N} \int_0^t u_{sx} dt + \psi_{s0}, \tag{5.16}$$

$$m_e = \psi_s \frac{u_{sy} - \omega_s \psi_s}{r_s}, \tag{5.17}$$

what further for:

$$u_{sx} = \mathrm{u_s} \cos\varphi, \tag{5.18}$$

$$u_{sy} = \mathrm{u}_s \sin\varphi, \tag{5.19}$$

can be write as:

$$\psi_s = \frac{1}{T_N} \int_0^t (\mathrm{u_s} \cos\varphi) dt + \psi_{s0}, \tag{5.20}$$

$$m_e = \psi_s \frac{\mathrm{u_s} \sin\varphi - \omega_s \psi_s}{r_s}. \tag{5.21}$$

It should be noticed that the increment angle $\Delta\gamma_k = \varphi$.

It can be seen from the equation (5.21), that the output torque is depended on the speed. Moreover, it can be seen that the x and y components of the stator voltage can control the stator flux and torque respectively. Unfortunately, the torque is also depended on the stator flux amplitude, what further means that the torque is not decoupled from the stator flux. However, the NF-DTC method controls the stator flux precisely and ensure almost full decoupling.

During the POSITIVE$_m$ or NEGATIVE$_m$ torque error and ZERO$_\Psi$ flux error there is needed only torque control. For the increment angle $\Delta\gamma_k = \pi/2$ and $\Delta\gamma_k = -\pi/2$ (see Table 1) the equations (5.20) and (5.21) will transform into the form:

$$\psi_s = \psi_{s0}, \tag{5.22}$$

$$m_e = \psi_s \frac{\pm \mathrm{u_s} - \omega_s \psi_s}{r_s}, \tag{5.23}$$

what results in torque control without changes of the stator flux amplitude. During the $ZERO_m$ torque error and $POSITIVE_\Psi$ or $NEGATIVE_\Psi$ flux the angle $\Delta\gamma_k$ is set to 0 or 2π. The flux is controled in accordance with the equation:

$$\psi_s = \frac{1}{T_N}\int_0^t u_s dt + \psi_{s0} \,. \tag{5.24}$$

However, it has also an influence on the output torque according to:

$$m_e = \frac{\omega_s \, \psi_s{}^2}{r_s} \,. \tag{5.25}$$

The increment angle for both non zero errors of the flux and torque (pairs: $POSITIVE_\Psi$ and $POSITIVE_m$, $POSITIVE_\Psi$ and $NEGATIVE_m$, $NEGATIVE_\Psi$ and $POSITIVE_m$, $NEGATIVE_\Psi$ and $NEGATIVE_m$) is chosen as a compromise between increments for separate flux and torque errors (last two points). For example, for the $POSITIVE_\Psi$ flux error and $ZERO_m$ torque error and for $POSITIVE_m$ torque error and $ZERO_\Psi$ flux error there has been chosen respectively $\Delta\gamma_k = 0$ and $\Delta\gamma_k = \pi/2$. It means that for the $POSITIVE_\Psi$ flux error and $POSITIVE_m$ torque error there is chosen middle value between 0 and $\pi/2$ increments and is equal to $\pi/4$. For the $ZERO_m$ torque error and $ZERO_\Psi$ flux error the increment angle is $\Delta\gamma_k = \pi/2$ and the torque is expressed by equation (5.25).

5.4 Self-Tuning Procedure

There are many methods of tuning of fuzzy systems and neural networks. The ANFIS structure [12-14] used for inverted pendulum stabilization is very similar to the presented NF-DTC controller. The controller has been tuned automatically by least square estimation algorithm (for output membership function) and back propagation algorithm (for output and input membership function). The NF-DTC can be tuned in the same way. However, we propose another simple and effective off line tuning method. The NF-DTC system contains three membership functions for each input. The tuning of the membership functions width corresponds to scaling of the flux and torque errors. The scaling factors are w_Ψ and w_m weights.

The NF-DTC is a non-linear high order system. Therefore, it is very helpful to use simulation for controller design. Fig. 5.6 presents computed flux and torque error as function of the input weights. The surfaces have also been verified experimentally. It can be seen that there are clearly defined minimums without any other local minimum points.

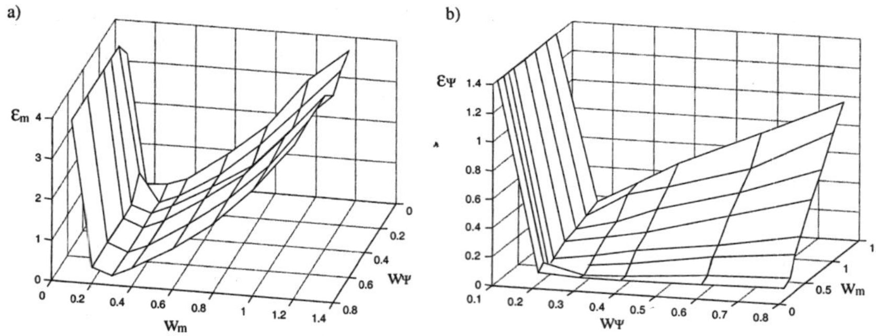

Fig. 5.6. The torque (a) and flux (b) error tuning surfaces.

This is because, for small weights the controller chooses high value of the reference voltage amplitude, which results in high flux and torque errors (ripples). From the other hand, if the weights are too big, the steady state errors increase. This tendency allows to use a simple gradient method to find the optimal working point, which guarantees minimal flux and torque errors. However, the torque and flux are not fully decoupled. It can be seen from the Eqs. (5.20) and (5.21) that, for non-zero synchronous angular speed, the changes of the flux magnitude Ψ_s influences the output torque, while the torque change does not influence the flux magnitude. That is why flux error minimum should be found first, before searching the torque error minimum. The *off-line* tuning process of the system is presented in Fig. 5.7. Note that the tuning surfaces in Fig. 5.6 have a general validity. The numerical results may little differ according to motor parameters and supply voltage. Also the final flux and torque errors depend on the chosen inverter switching frequency (Fig. 5.8).

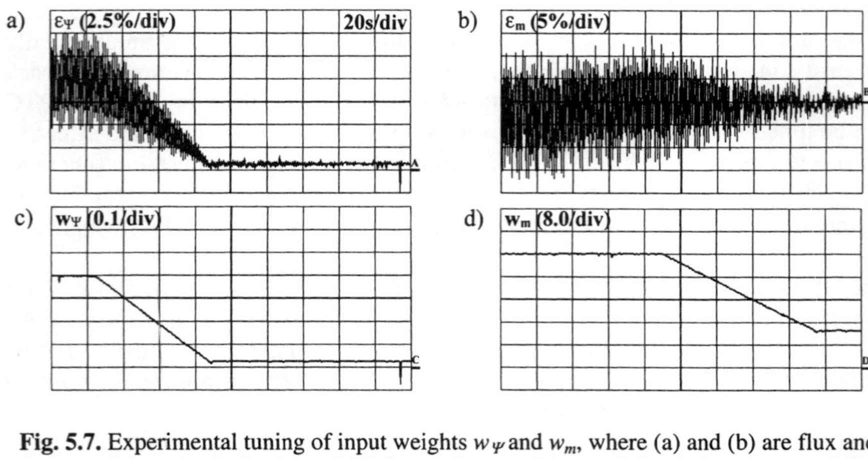

Fig. 5.7. Experimental tuning of input weights w_Ψ and w_m, where (a) and (b) are flux and torque errors respectively, (c) and (d) flux and torque input weights during system tuning.

The NF-DTC scheme guarantees very fast flux and torque responses. It is thanks to the lack of integration. Unfortunately, this property causes constant torque error in steady state operation. One of the solutions is adding an integration block. However, as consequence, the torque response will become slow. In the NF-DTC, instead of integration, the weight o_5 can be used to reach zero torque error at the steady state. This weight decides about the amplitude of the reference voltage vector. Therefore, instead of calculating by the controller, the value of the output weight o_5 is calculated from the equation:

$$o_5 = k_\omega \omega_m + k_m m_c + k_\Psi (1 - \Psi_{sc}) \tag{5.26}$$

where the k_ω, k_Ψ and k_m are experimentally chosen factors to compensate for the steady state torque error (see Appendix).

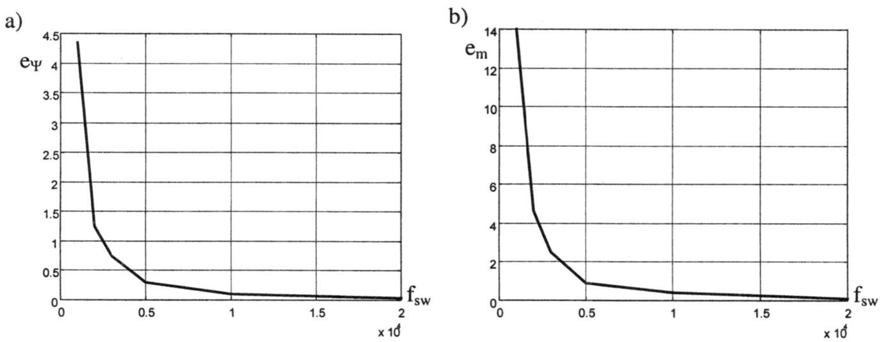

Fig. 5.8. The flux (a) and torque (b) error versus switching frequency in the tuned system.

5.5 Results

To verify the proposed NF-DTC concept, a simulation program as well as a laboratory setup with a four-pole 3kW induction motor drive and dSPACE DS1102 laboratory control board was constructed (see Fig. 4.4). The steady state operations of the tuned system are presented in Fig. 5.9 a-c. The sampling period was set to 500 μs, which gave flux and torque errors in the range of 1% and 3.5%, respectively. The result has been obtained for half of the nominal speed. The stator current is not distorted by the sector changes, as in the conventional DTC method [1], [6], [16], [31] and the stator flux trajectory is circular.

The torque transients to the step changes are presented in Fig. 5.10b. It can be seen, that for the constant stator flux amplitude, the flux and torque are fully decoupled and the flux amplitude is not distorted during torque steps. The stator current response is also presented in the figure. The response time is about 3 ms, what give similar dynamic as in the conventional DTC method [15], [31], [35].

The four quadrant speed transient for fast speed ramp reversal is presented in Fig. 5.10c. When a speed sensor is used the system is stable in whole speed range (including zero speed at full load). However, instabilities can only occur for sensorless operation in zero speed region. The lower speed control range depends on the used flux and speed estimators' quality. In the laboratory setup the speed sensorless stable operation has been observed over 1-2% of nominal speed.

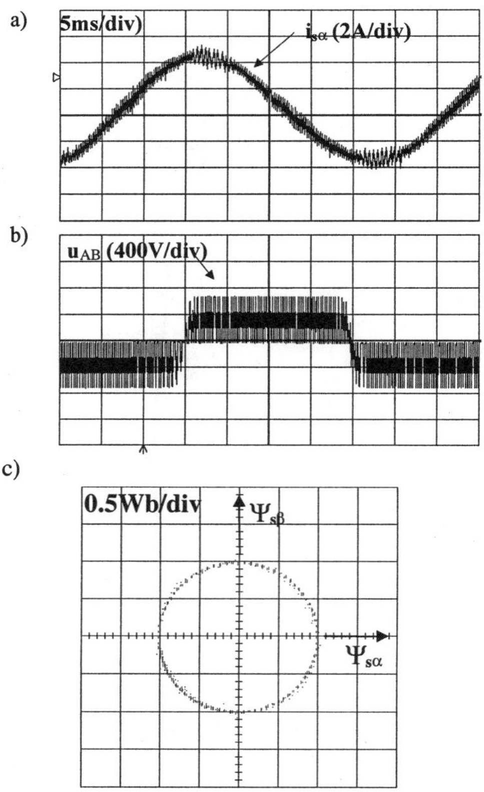

Fig. 5.9. Experimental results for the steady state operation for the tuned NF-DTC system, (a) stator current, (b) line-to-line voltage PWM, (c) stator flux vector path.

5.6 Conclusions

The application of neuro-fuzzy approach for Direct Torque Control of PWM inverter-fed induction motor has been investigated through DSP based experimental implementation. The design and tuning procedure are described. Compared to DTC the presented NF-DTC scheme has the following features and advantages: constant switching frequency and uni-polar motor voltage thanks to applied space vector modulator, lower sampling time is required, reliable start and low speed operations, torque and current harmonics mainly depended on sampling time, fast torque and flux responses, simple tuning procedure.

(a)

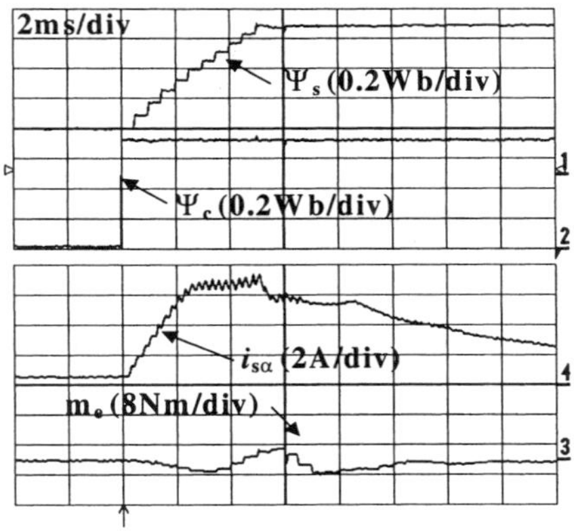

(b)

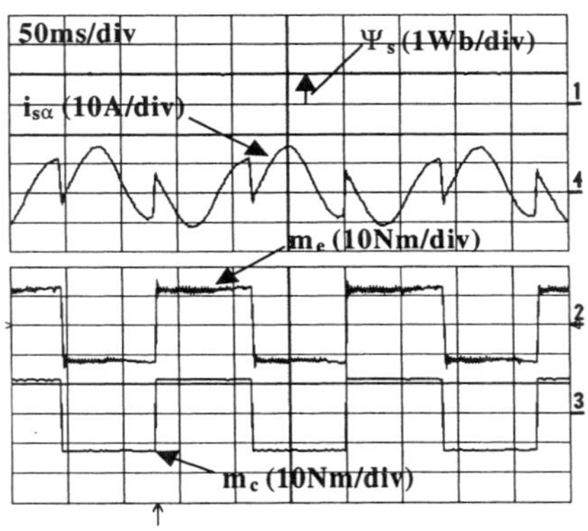

(c)

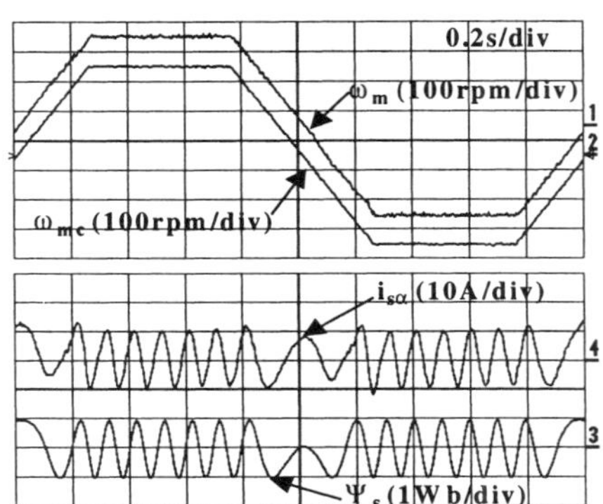

Fig. 5.10. Experimental oscillograms for: a) motor magnetization at zero speed b) the torque transients to the step changes, c) four-quadrant operation, where: Ψ_s- stator flux amplitude, Ψ_c- reference stator flux amplitude, $i_{s\alpha}$– stator current, m_e- output torque, m_c- reference torque, ω_{mc}- reference rotor speed, ω_m- rotor speed.

6 Summary and Conclusion

This Chapter has presented several applications of neural network and neuro-fuzzy systems for estimation and control in high performance PWM inverter-fed induction motor drives. Key conclusion include the following:

- For the IM speed estimation two kinds neural network approaches can be used: *on-line* trained neural modeling, *off-line* trained neural identification.
- Speed estimators based on neural modeling are simple for implementation, however, are sensitive to motor parameter changes.
- Contrary, speed estimators based on neural identification are less sensitive to motor parameter changes, but have usually more complicated architecture and require long training time with large data sets.
- Similar conclusions are valid for flux vector estimation.
- The indirect sensorless field oriented control scheme with on-line trained NN speed estimator (see Section 4) shows good steady state and dynamic performance. Thanks to applied back propagation algorithm, which has an inherent filtering function, NN estimator performs well with noisy signals. The drive operation is stable in both low and high speed regions.
- Combining both fuzzy logic and neural networks allows achieving all of the advantages of both systems: human expert knowledge can be used to build the initial structure of the controller, and *on-line* or *off-line* learning process can improve underdone parts of the structure. The ANFIS scheme applied for direct torque control (DTC) allows to decouple and linearize the dynamic structure of the IM even in the case when a fixed switching frequency space vector modulator is applied (see Section 5).

There are no doubts that neural network, fuzzy logic and neuro-fuzy systems will offer a new interesting perspective for future research. At present, however, they represent only some alternative solutions to existing estimations and control techniques, and their specific applications areas in power electronics cannot be clearly defined. It is belived that thanks to continuous developments in digital signal processing technology, artificial intelligence techniques will have strong impact on drive control, estimation and monitoring in the coming decades.

Appendix

<u>Motor parameters (per unit):</u>

r_s (stator resistance)	0.0384	U_{dc} (DC link voltage)	1.5
r_r (rotor resistance)	0.0577	k_ω (speed scaling factor)	1.33
x_s (stator reactance)	4.1621	k_m (torque scaling factor)	0.13
x_r (rotor reactance)	4.1621	k_ψ (flux scaling factor)	0.65
x_m (main reactance)	4.007		

References

1. Buja G. (1998), A new control strategy of the induction motor drives: the direct flux and torque control, *IEEE Industrial Electronics Society Newsletter*, **45**, no.4, pp. 14-16.
2. Ben-Brahim L. and Kurosawa R. (1993), Identification of induction motor speed using neural networks, *Proceedings of the Power Conversion Conf.*, Yokohama, Japan, pp. 689-694.
3. Ben-Brahim L., Tadakuma S., and Akdag A. (1999), Speed control of induction motor without rational transducers, *IEEE Trans. on Industry Applications*, **35**, no. 4, pp. 844-850.
4. Blaschke F. (1972), Das Verfahren der Feldorientirung zur Regeleung der Asynchron-maschine, *Siemens Forschungs und Entwicklungs-berichte*, **1** no. 1, pp. 184-193.
5. Bose B.K. (1994), Expert system, fuzzy logic and neural network applications in power electronics and motion control, *Proceedings of the IEEE*, **82**, no. 8, pp. 1303-1323.
6. Casadei D., Grandi G., Serra G., and Tani A. (1994), Effects of flux and torque hysteresis band amplitude in direct torque control of induction machines, *Proceedings of the IEEE Int. Conf. on Industrial Electronics Control and Instrumentation*, Bologna, Italy, pp. 299-305.
7. Depenbrock M. (1988), Direct self control of inverter-fed induction machines, *IEEE Trans. on Power Electronics*, **3**, no. 4, pp. 420-429.
8. Grabowski P.Z., Kazmierkowski M.P., Bose B.K., and Blaabjerg F. (2000), A simple direct-torque neuro-fuzzy control of PWM-inverter-fed induction motor drive, *IEEE Trans. on Industrial Electronics*, **47**, no. 4, pp. 863-870.
9. Grabowski P.Z. (1999), Direct flux and torque neuro-fuzzy control of inverter-fed induction motor drives, *Ph.D. Thesis, Institute of Control and Industrial Electronics, Warsaw University of Technology,* Warsaw, Poland.
10. Hasse K. (1972), Drehzahlgelverfahren fur schnelle Umkehrantriebe mit stromrichter-gespeisten Asynchron-Kurzchlusslaufermotoren, *Regelungstechnik*, **20**, pp. 60-66.
11. Holtz J. (1995), State of the art of controlled AC drives without speed sensor, *Proceedings of the Int. Conf. on Power Electronics and Drive Systems*, Singapore, pp. 1-6.
12. Jang J.-S.R. (1993), ANFIS: Adaptive-network-based fuzzy inference system, *IEEE Trans. on Systems, Man and Cybernetics*, **23**, no. 3, pp. 665-684.
13. Jang J-S.R. (1992), Self-learning fuzzy controllers based on temporal back propagation, *IEEE Trans. on Neural Networks*, **3**, no. 5, pp. 714-723.
14. Jang J.-S.R. and Sun C.T. (1995), Neuro-fuzzy modeling and control, *Proceedings of the IEEE*, **83**, no. 3, pp. 378-406.
15. Kazmierkowski M.P. and Tunia H. (1994), *Automatic Control of Converter-Fed Drives,* Elsevier, Amsterdam-London-New York-Tokyo.
16. Kazmierkowski M.P. and Kasprowicz A. (1995), Improved direct torque and flux vector control of PWM inverter-fed induction motor drives, *IEEE Trans. on Industrial Electronics*, **45**, no. 4, pp. 344-350.
17. Kazmierkowski M.P. and Malesani L. (1998), Current control techniques for three-phase voltage-source PWM converters: a survey, *IEEE Trans. on Industrial Electronics*, **45**, no. 5, pp. 691-703.
18. Lorenz R.D. (2000), Sensorlees drive control methods for stable high performance zero speed operation, *Proceedings of 9th Conference on Power Electronics and Motion Control,* Kosice, Slovak Republic, pp. 1.1-1.11.
19. Orlowska-Kowalska T. (1989), Application of extended observer for flux and rotor time – constant estimation in induction motor drives, *IEE Proceedings, Part D,* **136**, pp. 324-330.
20. Orlowska-Kowalska T. (1992), State and parameter observers in the induction motor control systems, *Proceedings of the International Conference on Electrical Machines*, Manchester, UK, pp. 582-586.

21. Orlowska-Kowalska T. and Kowalski C.T. (1996), Neural network based flux estimator for the induction motor, *Proceedings of the Int. Conf. on Power Electronics and Motion Control,* Budapest, Hungary, pp. 187-191.
22. Orlowska-Kowalska T. and Migas P. (1998), Neural speed estimator for the induction motor drive, *Proceedings of the 8th Int. Conf. on Power Electronics and Motion Control,* Prague, Czech Republic, pp. 7.89-7.94.
23. Orlowska-Kowalska T. (1998), Artificial neural networks in the electrical drives control – a survey, *Archives of Electrical Engineering*, **47**, no. 2, pp. 127-143.
24. Orlowska-Kowalska T. and Migas P. (1999), Influence of neural network structure on the induction motor speed estimation quality, *Proceedings of 13th Int. Conf. on Electrical Drives and Power Electronics,* Kosice - Stara Lesna, Slovak Republic, pp. 23-31.
25. Orlowska-Kowalska T. and Wojsznis P. (2000), Comparative study of different flux estimators for sensorless induction motor drive, *Archives of Electrical Engineering*, **49**, no. 1, pp. 49-63.
26. Orlowska-Kowalska T. and Pawlak M. (2000), Induction motor speed estimation based on neural modeling method, *Archives of Electrical Engineering*, **49**, no. 1, pp. 35-48.
27. Rajashekara K., Kawamura A., and Matsue K. (1996), *Sensorless Control of AC Motor Drives: Speed and Position Sensorless Operation;* IEEE Press, Piscataway, NJ, USA.
28. Sobczuk D.L. and Grabowski P.Z. (1998), DSP implementation of neural network speed estimator for inverter fed induction motor, *Proceedings of the IEEE Int. Conf. on Industrial Electronics Control and Instrumentation* Aachen, Germany, pp. 981-985.
29. Sobczuk D. L. (1999), Application of artificial neural networks for control of PWM inverter-fed induction motor drives, *Ph.D. Thesis, Institute of Control and Industrial Electronics, Warsaw University of Technology*, Warsaw, Poland.
30. Simoes K. and Bose B.K. (1995), Neural network based estimation of feedback signals for a vector controlled induction motor drive, *IEEE Trans. on Industry Applications*, **31**, no. 3, pp. 620-639.
31. Takahashi I. and Noguchi T. (1986), A new quick-response and high efficiency control strategy of an induction machine, *IEEE Trans. on Industry Applications*, **22**, no. 5, pp. 820-827.
32. Tamai S., Sugimoto H., and Yano M. (1987), Speed sensorless vector control of induction motor with model reference adaptive system, *Proceedings of the IEEE Industry Applications Society Annual Meeting,* pp. 189-195.
33. Toh A.K.P. and Nowicki E.P. (1994), A flux estimator for field oriented control of an induction motor using an artificial neural network, *Proceedings of the IEEE Industry Applications Society Annual Meeting,* pp. 585-592.
34. Tsoi A.C. and Back A. (1994), Locally recurrent globally feedforward networks: a critical review of architectures, *IEEE Trans. on Neural Networks*, **5**, no. 2, pp. 229-239.
35. Vas P. (1998), *Sensorless Vector and Direct Torque Control*, Oxford University Press, Oxford-New York-Tokyo.
36. Vas P. (1999), *Artificial-Intelligence-Based Electrical Machines and Drives*, Oxford University Press, Oxford, UK.
37. dSPACE Manuals (1998) Technologie Park 25, Paderborn, Germany.

High Robustness of Fuzzy Logic Systems in Switched Reluctance Motor Drives

Adrian David Cheok

National University of Singapore

Keywords: *robustness, fuzzy logic model, fuzzy predictive filter, estimation, motor drives.*

1 Introduction

The switched reluctance (SR) motor has advantages in practical variable speed drive systems due to an intrinsic simplicity and ruggedness that makes it well suited for many commercial applications. Generally the SR drive controller requires rotor position feedback, because excitation of the SR motor phases need to be synchronized with the rotor angle. Position sensors are commonly employed to obtain rotor position measurements, however in many systems advantages can be found in eliminating these sensors. These benefits include the elimination of electrical connections to the sensors, reduced size and cost, low maintenance, and insusceptibility to environmental factors.

Hence, a diverse range of indirect or sensorless position estimation methods has previously been proposed [1] - [3]. The previous estimation methods can be classified into two major groups. In one group, low amplitude test signals are inserted into the motor phase windings to derive rotor position information, whilst in the other group the actual motor excitation waveforms are monitored without the use of additional signals in the motor. Mathematical model based schemes are also classified under the latter group.

A major concern of position estimation using these schemes is that feedback signals measured from the motor are required to calculate the motor shaft position. Therefore the accuracy of the estimated position is dependent on the accuracy of the measured feedback signals. However, motor drives are electromagnetically noisy environments, and in addition practical measurement circuits of electronic signals are imperfect. Thus the feedback signals are generally corrupted by noise and errors.

Hence a rarely stated disadvantage of sensorless position estimation as opposed to sensor based position estimation is that the sensors have a guaranteed accuracy under normal conditions, and usually a high reliability. Furthermore the sensors are robust to electromagnetic noise.

Therefore, although it is commonly stated that position sensors tend to reduce the reliability of a drive system, the previously developed estimation schemes may not be useful in industrial applications unless their reliability and robustness against noise and error is proven, or at least thoroughly understood.

Thus in this Chapter, the high robustness and reliability of a new rotor position estimator for the SR motor is described. The position estimator uses fuzzy logic based modeling, estimation, and prediction in a novel manner to provide a high system reliability against feedback signal noise and error.

The first part of this Chapter will analyze the noise and error commonly found in practical motor drives, and it will be shown that the feedback noise affects the accuracy of rotor position estimation. Secondly, the new fuzzy logic based position estimation algorithm will be described, and a theoretical analysis of the robustness of the scheme will be given. Finally experimental results will be shown with the fuzzy logic based position estimator operating with erroneous and noisy signals.

2 Noise and Errors in SR Motor Position Estimation

As discussed in [2], essentially all position estimation schemes for the SR motor are based on the relationship of the motor phase magnetic flux linkage with rotor position. The measured flux linkage characteristics of an SR motor are shown in Figure 1. In the figure, zero degrees represents the aligned rotor position and thirty degrees represents the unaligned position. It can clearly be seen that the flux linkage $\psi(\theta, i)$ is a non-linear function of rotor position θ and current i. Therefore, if the flux linkage can be determined, together with current measurement, then the rotor position may be estimated. Furthermore, inductance $L(\theta)$ (which can be derived from the flux linkage if it is assumed that the motor magnetic circuit is linear) can also be used to determine position.

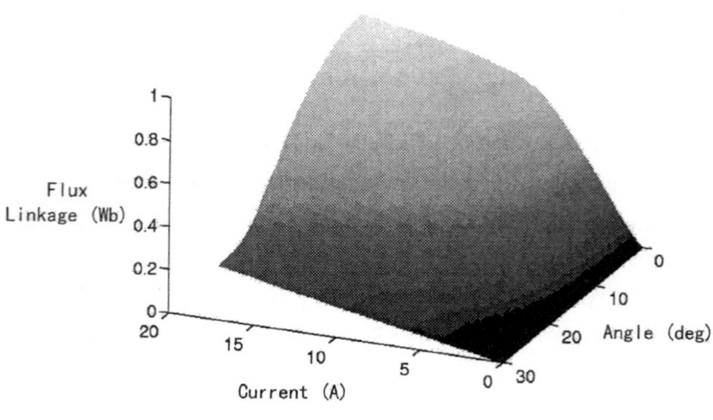

Fig. 1. Measured magnetization characteristics of SR motor (as flux linkage changes from 0 to 1, surface gray-scale changes from dark to light).

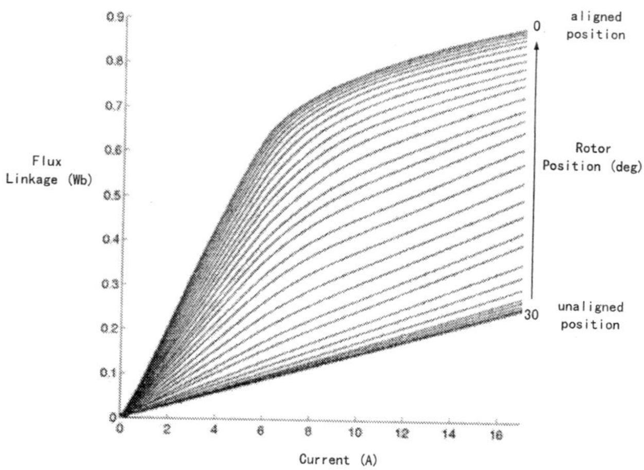

Fig. 2. Flux linkage versus current curves for SR motor.

Existing literature on the effects of errors in previously developed SR motor position estimation schemes has been relatively sparse. In [4] only the time and amplitude quantization errors due to digital implementation of a specific linear inductance based scheme was discussed. The issue was further discussed in [5], where specific types of SR motor physical designs were proposed to improve the robustness of position estimation to error. However for simplicity all error analysis were based on linear or piecewise-linear motor magnetization characteristics.

In our error analysis no assumptions will be made about the motor linearity, and thus the results may be applied to any sensorless position estimation scheme in general.

Hence, consider the estimation of position using the non-linear magnetization function $\psi(\theta, i)$, from measured values of flux linkage ψ and current i. Generally flux linkage is impractical to measure directly, and thus integration of the total voltage across an excited phase is a common method to estimate phase flux:

$$\psi = \int (v - iR)dt \tag{1}$$

where: ψ = phase flux linkage, R = phase resistance, v = phase voltage, i = phase current.

Analog integrators can be used to estimate the flux linkages of the motor but these often have the problem of drift in the output signal due to temperature sensitivity and the need for compensation [6]. Therefore the integration may be performed digitally as shown in equation (2). This requires digitized phase voltage and current signals that are measured using analog to digital (A/D) converters.

$$\psi(n+1) = \psi(n) + \Delta T\left(v(n) - Ri(n) + v(n-1) - Ri(n-1)\right)/2 \quad (2)$$

$$\psi(0) = 0 \quad (3)$$

Where: n = sample number, ΔT = sampling period.

Hence the feedback signals that are used for the estimation of rotor position in practice are the measured motor currents and voltage, and estimated or calculated flux linkage.

Generally the feedback signals will contain some error component which can be defined as the difference between the actual physical motor quantities of current, voltage, and flux linkage, and the corresponding feedback values used for position estimation. The error components can be expressed as:

$$\Delta v = v_m - v \quad (4)$$

$$\Delta i = i_m - i \quad (5)$$

$$\Delta\psi = \psi_e - \psi \quad (6)$$

Where: Δv, Δi, $\Delta\psi$ are the error components, v_m, i_m, and ψ_e are the measured and estimated values, and v, i, ψ are the physical values of voltage, current, and flux linkage respectively in the motor.

The source of the feedback error components in practical motor drives can be classified under various groups, as shown in Table 1.

Table 1. Sources of Feedback Error.

Measurement Errors	a) Sensor offset error.
	b) Sensor scaling error.
Parameter Variation	a) Variation of motor phase winding resistance value, primarily due to temperature variation.
Measurement Noise	a) Noise due to capacitive coupling of measuring circuits and associated cabling with switching power semiconductor circuit.
	b) Electromagnetic interference from motor and power circuit on the measuring circuit.
	c) Noise in measuring circuits (including (A/D) converter circuits) due to non-ideal circuit layout and grounding.
	d) Noise in A/D circuits due to noise generated from any connected microprocessor digital bus lines.
	e) Crosstalk between signals due to coupling in signal cabling.
Integration Errors	a) Drift and temperature variation (analog integrators).
	b) Integration offset errors (analog and digital integration).
	c) Time quantization errors due to finite sampling period (digital integration).
	d) Amplitude quantization errors due to A/D conversion.

Using these classifications, an expression for the total flux linkage error due to the various error producing sources can be found. The total flux linkage error at some sample k will be equal to:

$$\Delta\psi(k) = \Delta\psi_o(k) + \Delta\psi_a(k) + \Delta\psi_{tq}(k) + \Delta\psi_{aq}(k) + \Delta\psi_\mu(k) + \Delta\psi_r(k) \quad (7)$$

where: $\Delta\psi$ = Total flux linkage error, $\Delta\psi_o$ = Offset error, $\Delta\psi_a$ = Amplitude error, $\Delta\psi_{tq}$ = Time quantization error, $\Delta\psi_{aq}$ = Amplitude quantization error, $\Delta\psi_\mu$ = Mean signal noise error, $\Delta\psi_r$ = Error due to change in resistance.

In references [5] and [7], the analytical expressions of the various flux linkage error components in equation (7) were given. For reference, Table 2 summarizes some of the results.

Table 2. Analytical Error Expressions.

1. Due to dc offset in voltage and current measurements	$\Delta\psi_o(k) = k\Delta T\,(v_o - Ri_o)$
2. Due to amplitude or scaling error	$\Delta\psi_e(k) = \sum_{n=0}^{k-1} \Delta T\,(A_v\,(v(n) + v(n-1)) - RA_i\,(i(n) + i(n-1)))\,/2$
3. Due to time quantization of A/D converters	$\Delta\psi_{tq}(k) = \frac{k\Delta T P^2 M}{12}$
4. Due to amplitude quantization of A/D converters	$-k\Delta T\left(R\frac{2I_{max}}{2^n-1}\right) \le \Delta\psi_{aq}(k) \le k\Delta T\left(\frac{2V_{max}}{2^n-1}\right)$
5. Due to noise	$\Delta\psi_\mu(k) = k\Delta T\,(\mu_v - R\mu_i)$
6. Due to resistance variation	$\Delta\psi_r(k) = k\Delta T\Delta R\sum_{n=0}^{k-1}(i(n) + i(n-1))\,/2$
where: R = phase resistance, ΔR = change in phase resistance, v = phase voltage, i = phase current, v_o = voltage measurement dc offset error, i_o = current measurement dc offset error, A_v = voltage measurement amplitude error, A_i = current measurement amplitude error, P = A/D sample period, M = highest value of second derivative of measured signals, I_{max} = maximum amplitude of A/D converter current measurement, V_{max} = maximum amplitude of A/D converter voltage measurement, n = number of A/D converter bits, μ_v = mean level of voltage feedback noise, μ_i = mean level of current feedback noise.	

If there are flux linkage estimation errors due to one or more of the above explained effects, there will be a corresponding error in the rotor position estimation. Regardless of the various sources of the total flux linkage estimation error expressed in equation (7), the corresponding angle error at sample time k will be given by:

$$\Delta\theta(k)_\Psi = \Delta\psi(k)\left.\frac{\partial\theta}{\partial\psi}\right|_{i(k),\psi(k)} \quad (8)$$

where: $\Delta\theta(k)_\Psi$ = the angle error due to error in the flux linkage feedback, $i(k)$ = the actual motor phase current, $\psi(k)$ = the actual motor phase flux linkage.

Therefore, for a given flux error, the angle error will be proportional to the partial derivative of the rotor position function with respect to flux linkage at the point $(i(k), \psi(k))$.

Although quantitative analysis of this error will be detailed below, it is interesting to examine the angle error qualitatively. In Figure 2 the magnetization characteristics of Figure 1 are plotted on a two-dimensional plane with curves of constant rotor position. From the curves, it can be seen that when the angle is near the aligned position, the unaligned position, or the motor phase current is low, the angle curves are tightly bunched up. This entails that the slope of the function in the three-dimensional plane is high. Hence in these areas, the partial derivative of angle with respect to flux will be high. Under these conditions, according to equation (8), small errors in the flux linkage will result in large errors in the position estimate.

As was discussed above, the current feedback signal is also affected by measurement error and noise. Similarly to equation (8) the current feedback error at sample k can be written as:

$$\Delta i(k) = \Delta i_o(k) + \Delta i_a(k) + \Delta i_\mu(k) \tag{9}$$

where: $\Delta i(k)$ = Total current feedback error, $\Delta i_o(k)$ = Current measurement offset error, $\Delta i_a(k)$ = Current measurement amplitude error, $\Delta i_\mu(k)$ = Current measurement mean signal noise.

Hence, even if the flux linkage could somehow be measured without error, the current measurement error will lead to a corresponding rotor position estimation error. For a given current error at sample time k the angle error can be written as:

$$\Delta\theta(k)_I = \Delta i(k) \left.\frac{\partial\theta}{\partial i}\right|_{i(k),\psi(k)} \tag{10}$$

where: $\Delta\theta(k)_I$ = the angle error due to error in the current feedback.

By examining again the contour curves of angle with respect to current and flux as shown in Figure 2, it can be seen that similarly to the case of flux linkage error, the angle estimation error is most sensitive to current feedback error near the aligned and unaligned rotor positions and when the motor current is low. This results in the fact that in these areas the partial derivative of angle with respect to current will be high.

Generally there will be error in both the flux linkage and current feedback. To consider the effect of simultaneous flux linkage and current errors on the position estimation the following expression may be employed:

$$\Delta\theta(k) = \Delta\psi(k) \left.\frac{\partial\theta}{\partial\psi}\right|_{i(k),\psi(k)} + \Delta i(k) \left.\frac{\partial\theta}{\partial i}\right|_{i(k),\psi(k)} \tag{11}$$

2.1 Quantitative Error Analysis

It was seen in the above analysis (equations (7), (9), and Table 2) that the instantaneous feedback error of the flux linkage and current signals are complex and non-linear functions. Furthermore, it was seen that to numerically calculate the exact instantaneous error requires detailed knowledge of the hardware parameters (e.g., offset error, amplitude error, resistance variation) which may be time and temperature variant. Additionally, the signal parameters will also need to be known (e.g., mean noise, measured signal time-derivatives) which are generally time varying and system dependent.

Therefore, to perform a quantitative error analysis it is normally not useful or feasible to predict the instantaneous feedback errors $\Delta\psi(k)$ and $\Delta i(k)$. Instead it will be useful to quantitatively analyze the resultant position error $\Delta\theta(k)$ for given feedback signal error amplitudes, regardless of their source, using equations (8) and (10).

Effect of Flux Linkage Error on Rotor Position Error To examine the effect of flux linkage error on the rotor position estimation, the rotor position as a function of flux linkage at various values of constant current is plotted, as shown in Figure 3 (a). These curves are derived from the motor magnetization characteristics previously shown in Figure 1.

From the functions shown in Figure 3 (a), the function $\left.\frac{\partial\theta}{\partial\psi}\right|_{(i,\psi)}$ can be found. Hence, from the partial derivative functions, and using equation (8), the quantitative value of position error for a given error in the flux linkage estimate can be calculated.

As an example consider a flux linkage error that has an amplitude of +10% of the maximum operating flux linkage (0.7 Wb) of the SR motor used in the research. Figure 3 (b) plots the calculated numerical value of the resultant angle estimation error.

It can be confirmed from the curves that, as discussed above, at low current, or near the aligned and unaligned angles, the angle estimation is more sensitive to errors in the flux linkage feedback. Fundamentally, this is due to the fact that the partial derivative, or slope of the curves shown in Figure 3 (a), is higher in these regions. It can also be noted that as the function $\left.\frac{\partial\theta}{\partial\psi}\right|_{(i,\psi)}$ is negative the resultant angle error for positive flux error is negative, and vice-versa. When the rotor angle error is negative, this entails that the estimated angle lags the actual rotor angle of the motor.

The *minimum* resultant error amplitude for a +10% error in the flux linkage for each of the curves shown in Figure 3 (a) is shown in Table 3. The results are shown in both absolute degrees, and as a percentage of an electrical cycle of 30 degrees. It should be emphasized that, as seen in Figure 3 (b), these are the *minimum* error values corresponding to the points when the derivative or slopes of the curves in 3 (a) are at their minimum. However,

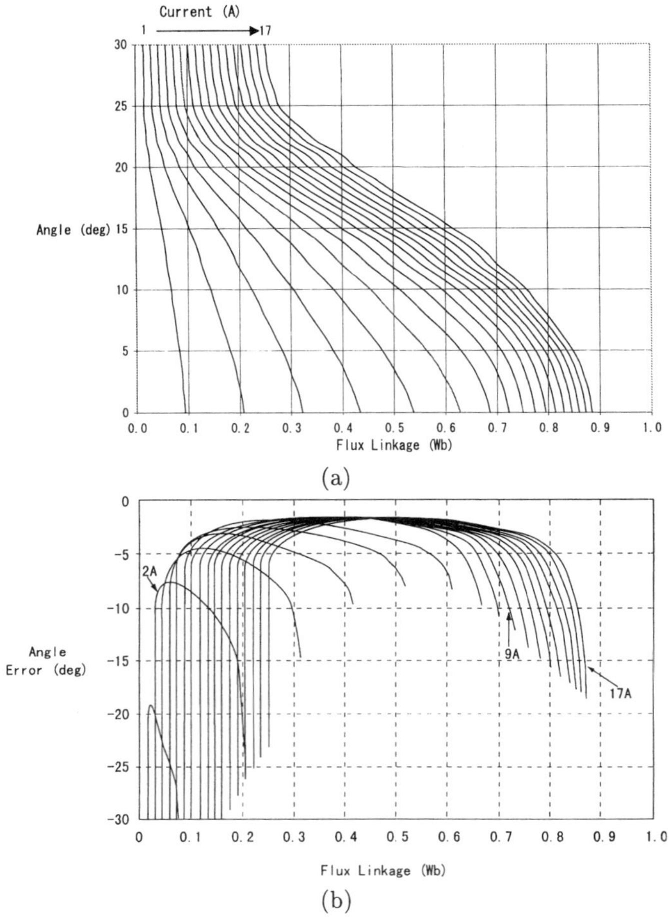

Fig. 3. (a) Rotor angle versus flux linkage, (b) angle estimation error for 10% flux linkage feedback error.

at other points in the magnetization region, the derivative of the functions, and the corresponding angle error can be much higher.

Effect of Current Measurement Error on Rotor Position Error To examine the effect of current feedback error on the rotor position estimation, the rotor position is defined as a function of current for constant values of flux linkage. The resultant curves, which are derived from the motor magnetization characteristics (Figure 1), are shown in Figure 4 (a).

From the resultant curves, the partial derivative function $\frac{\partial \theta}{\partial i}\big|_{(i,\psi)}$ can be calculated. Therefore using equation (10), the quantitative value of position error for a given value of current error may be found. Similarly to the previous example above, position error values are calculated for a current amplitude

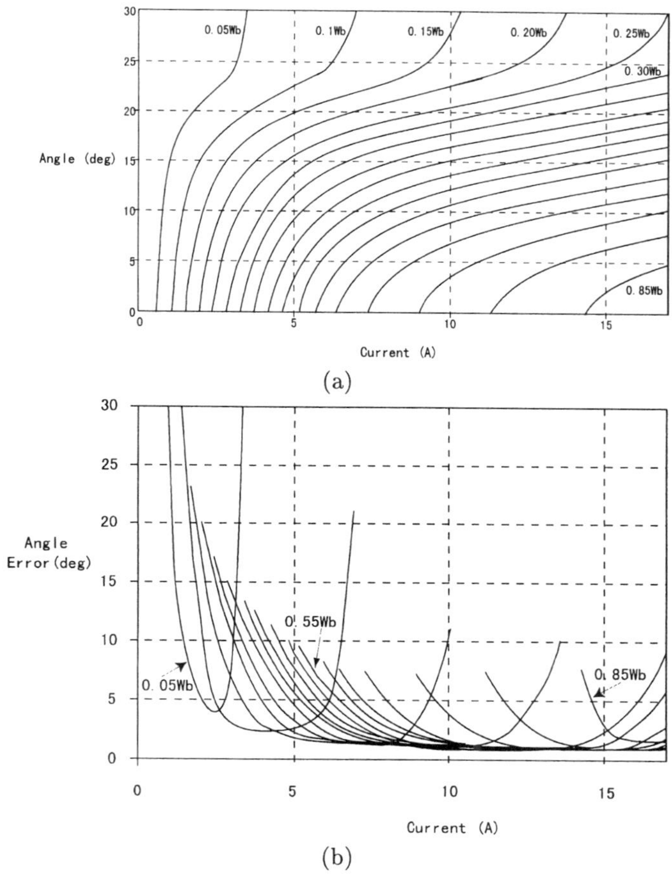

Fig. 4. (a) Rotor angle versus current, (b) angle error for 10% current feedback error.

error of +10% of the maximum operating motor current (17 A). The corresponding position error curves are plotted in Figure 4 (b). From the curves it can be confirmed that, as discussed above, there is more sensitivity to error when the current is low and the angle is close to the aligned or unaligned positions.

The *minimum* position error for each of the curves is detailed in Table 4. Once again it should be noted that these points correspond to the points when the value of derivative of the curves shown in Figure 4 (a) are minimum. However as can be seen from the resultant functions, the derivative and corresponding angle error at other points can be significantly higher.

It should be noted from the above results that for the same percentage error in the flux linkage and current feedback values, the resultant rotor position estimation error is generally lower for the case of current feedback error.

Table 3. Angle estimation error amplitude for +10% error in flux linkage feedback values (note that a local minimum of the error function occurs at approximately 16 A).

Current (A)	Error (deg)	Error (%)
1	19.16	63.87
2	7.40	24.65
3	4.59	15.29
4	3.34	11.14
5	2.57	8.57
6	2.17	7.24
7	1.99	6.56
8	1.84	6.13
9	1.78	5.93
10	1.74	5.81
11	1.73	5.79
12	1.73	5.79
13	1.72	5.74
14	1.72	5.74
15	1.71	5.72
16	1.70	5.67
17	1.74	5.81

Table 4. Angle estimation error amplitude for +10% error in current feedback values.

Flux (Wb)	Error (deg)	Error (%)
0.05	4.70	15.6
0.10	2.50	8.3
0.15	1.62	5.43
0.20	1.25	4.17
0.25	1.08	3.61
0.30	0.96	3.21
0.35	0.79	2.66
0.40	0.80	2.68
0.45	0.89	2.98
0.50	0.91	3.03
0.55	0.93	3.10
0.60	0.97	3.23
0.65	1.12	3.73
0.70	1.14	3.80
0.75	1.22	4.08
0.80	1.32	4.41
0.85	2.17	7.24

3 Fuzzy Logic Based Position Estimation Scheme

From the above discussion it was seen that the accuracy and reliability of sensorless position estimation is directly affected by feedback signal measurement noise and error. Therefore a new fuzzy logic (FL) based angle estimator has been developed with the major objective of having a low sensitivity to feedback signal error and a high robustness. This is achieved with a fuzzy rule based model of the motor which has the ability to be tolerant to input signal noise and error.

Before detailing the algorithm, firstly we will explain the rationale for using a fuzzy logic model, rather than a classical differential equation type model, by explaining the various advantages of fuzzy logic to this application:

(i) No Mathematical Model is Required: Fuzzy logic system identification provides a mathematical model free estimation and modeling approach. Hence, fuzzy logic is well suited to systems that are unknown, ill defined, or are structurally difficult to model due to naturally existing nonlinearity and other modeling difficulties. The switched reluctance motor is difficult to define in a precise and accurate manner with a conventional mathematical model. This is because the motor has non-linear magnetization characteristics and complex magnetic coupling interactions between the motor phases. Hence, fuzzy techniques are well qualified in providing a model of the switched reluctance motor. Neural networks can also be used to model the switched reluctance motor using a mathematical model free approach, but they often have a long learning time, and do not allow the examining of the internal structure of the model. Therefore, fuzzy models can be seen as favorable due to the fact that their behavior can be explained using linguistic rules and thus they can be easily adjusted by altering the rules [8].

(ii) Fuzzy Models are Universal Approximators: Although fuzzy modeling does not require any mathematical model, it has been proven that fuzzy models are universal approximators [9]. Therefore, a fuzzy model can approximate any continuous function to any degree of accuracy using arbitrary fuzzy sets. The switched reluctance motor model has a non-linear continuous function, and therefore, it is possible to model this function using fuzzy sets without the requirement for mathematical based modeling.

(iii) No Requirement for Large Look-up Table: In some previous SR motor position estimation schemes [1], a numerical table of measured static motor characteristics is used to describe the motor. The measured motor data provides a numerical model of the static angle versus flux linkage and current characteristics. These methods are commonly called table look-up methods, and although they have the advantage of being fast, they have the disadvantage of needing a large memory to store the table. The size of the table must be large for accurate position estimation, and this may be impractical for low cost or embedded applications. However in systems where numerical data is learnt by a fuzzy system, it is found that the memory requirement of the stored fuzzy model is much lower than that required by the equivalent look-up table [10].

(iv) Fuzzy Models Allow Fast Computation: An important criterion in the development of the angle estimation scheme, was the practical implementation in actual SR motor drives, where the algorithm must execute in real time. Thus, a major practical consideration is the complexity of computation of the angle estimation algorithm. A computationally demanding algorithm may have detrimental effects on the system performance, as during the delay time of the angle estimate calculation, the actual physical rotor position will change. In addition computationally demanding real time algorithms will demand a fast microprocessor, which may not be appropriate for some systems due to cost constrains. Approximate mathematical models (such as polynomials, splines or any other traditional approach) of the non-linear SR motor model could be used in principle to provide position estimation. However, the mathematical model will increase in complexity in proportion to its accuracy in modeling the non-linear motor. As in other non-linear mathematical models of power electronic systems and motor drives, their complexity can lead the model being computationally expensive with a relatively slow response. Hence previously developed model based sensorless schemes have assumed that the SR model was linear and time invariant in order for real time operation to be viable with a simplified model. However, fuzzy logic does not need a complex mathematical model, and thus has the advantage of relatively simple mathematical calculations used for the rule processing [11]. In addition, it has been shown that the fastest possible universal computation scheme corresponds exactly to the operations in fuzzy logic methods using Max-Min composition [12]. Therefore, in terms of real time sensorless SR motor operation it could be said that fuzzy logic techniques represent (in some reasonable sense) an ideal choice.

(v) Fuzzy Models Have Benefits for Practicing Industrial Engineers: Often in industry, a project will be driven primarily by time and monetary matters. Thus, there is often no time to find mathematically detailed models of a motor system. Fuzzy logic allows simple heuristic tools to be applied, and thus has found favor amongst practicing engineers [8]. Furthermore, if a different motor size or power rating is used for the system, it is normally much easier to reconfigure the system using fuzzy models. This is because the fuzzy linguistic rules can remain the same, and only the membership numerical values need to be changed [11].

3.1 SR Motor Model Training Scheme

To create a fuzzy rule based model of the SR motor characteristics, a training scheme is used which trains a fuzzy logic model based on measured numerical information about the SR motor. The information is derived from two main sources of measured data:

1. The static flux linkage versus current and position curves of the motor, which provide important information about the electromagnetic characteristics of the SR motor phases;
2. The dynamic real time operating waveforms of the motor, which can include mutual coupling between phases, temperature variations, eddy currents, and skin effects.

In both cases the measured data is defined as a two-input one-output input-output pair such that flux linkage ψ and current i are defined as the inputs, and the rotor position θ is defined as the output.

Each datum presented to the training system is defined as $(\psi(n), i(n), \theta(n))$ (where n is the nth data pair, ψ is the flux linkage, i is the current and θ is the position).

The datum is defined in this manner because the SR motor characteristics are defined as a two input-one output nonlinear function that relates flux linkage and current to rotor angle. Hence, the training task involves creating a fuzzy model of this function from the training data.

After the training is completed, the generated fuzzy logic rule base defines a function f mapping input values of flux linkage and current to output values of rotor position:

$$f : (\psi, i) \rightarrow \theta \tag{12}$$

Various methods can be used to train the fuzzy model from the measured data, such as using neural networks, genetic algorithms, and classical adaptive control techniques. However, due to its suitability to practical applications (fast, simple, and accurate), the table-look-up scheme [13] was used. This is because fast one pass learning and simple computation are features of the table lookup scheme. The training scheme consists of the following steps:

Step A - Dividing the Input and Output Domains into Fuzzy Regions: In this initialization step which is performed only once before the algorithm is executed, the range of intervals of the input and output variable domains are defined. In this case the input domains are flux ψ and current i, and are defined to have a range of 0 to 1 Wb and 0 to 20 A respectively (for the motor drive used in the research the actual values always lie within this range). Similarly the range of the angle θ is defined as 0 to 30 mechanical degrees, as this is the electrical cycle for the motor used.

Each domain is then divided up into fuzzy regions as shown in Figure 6. The variable spaces of flux linkage ψ, i, and θ were divided into 37, 39, and 30 regions respectively. The choice of the number of regions is a compromise between the number of resultant rules that are generated in the rule base (which as explained in [11] increases with the number of partitions) and the desired accuracy (which increases with the number of partitions). In this research it was found this partitioning gave an average error over all tests during steady state motor conditions of 0.40 degrees in simulation and 0.53 degrees in actual experimental results [14]. This was seen as a fair compromise between the system accuracy, and the number of sets and rules. However, to increase accuracy the number of fuzzy regions can be increased and optimized.

Each region is assigned a fuzzy membership function. In this work, the fuzzy sets were chosen to be isosceles triangular shapes, and maximum point of each triangle was chosen to lie at the center of its respective region. The other two vertices of the triangle are chosen to lie at the centers of the two neighboring fuzzy regions, and at these two points the membership values were made zero. Defining the membership functions in this relatively simple manner allowed a very fast microprocessor calculation time of the membership values.

Step B - Generating Fuzzy Rules from Input Data of Flux, Current, and Angle: After the fuzzy membership functions in each of the input and output domains are defined, the measured training data can be used to generate fuzzy rules in order to create the fuzzy rule base model. During the training phase, each input-output data pair, which consists of a crisp numerical value of measured flux linkage, current, and angle, is used to generate the fuzzy rules which model the system.

To determine a fuzzy rule from each input-output data pair, the first step is to find the degree of each data value (flux, current, and angle) in every membership region of its corresponding fuzzy domain. The variable is then assigned to the region with the maximum degree. As an example consider the input-output data pair n consisting of the values $\psi(n)$, $i(n)$, and $\theta(n)$. After assigning the data points to the fuzzy region in the input-output domain with the highest degree, the following fuzzy rule can be formed:

$$\textbf{If } \psi \textbf{ is } F^{M\psi(n)} \textbf{ and } i \textbf{ is } F^{Mi(n)} \textbf{ then } \theta \textbf{ is } F^{M\theta(n)} \quad (13)$$

where: $F^{M\psi(n)}$, $F^{Mi(n)}$, and $F^{M\theta(n)}$ is the fuzzy region of highest degree assigned to the data value $\psi(n)$, $i(n)$ and $\theta(n)$ respectively.

Step C - Assigning Rule Degrees: When each new rule is generated from the input-output data pairs, a rule degree or truth is assigned to that rule, where this rule degree is defined as the degree of confidence that the rule does in fact correlate to the function relating flux linkages and current to angle. In the developed method a degree is assigned which is the product of the membership function degree of each variable in its respective region. For example, the previous rule given in equation 14 would have degree:

$$\text{Degree(rule)} = \mu_{F^{M\psi(n)}} \times \mu_{F^{Mi(n)}} \times \mu_{F^{M\theta(n)}} \quad (14)$$

where: $\mu_{F^{M\psi(n)}}$, $\mu_{F^{Mi(n)}}$, and $\mu_{F^{M\theta(n)}}$ is the membership degree of the crisp data point $\psi(n)$, $i(n)$, and $\theta(n)$ in the sets $F^{M\psi(n)}$, $F^{Mi(n)}$, and $F^{M\theta(n)}$.

The purpose of this assignment is to choose between data sets that produce the same antecedents but different consequents. This would arise because when there is a large amount of measured data, some data pairs will produce rules that have the same antecedent but a different consequent (due to errors or noise in the measured data). This would mean that there are conflicting rules in the system. These are resolved by choosing the rule that has the highest degree to be placed in the fuzzy rule base.

Step D - Adaptive Rule Base Modification: As it can be seen from the above, every training data set produces a corresponding relating flux linkage and current to rotor position. Each new rule is stored on the fuzzy rule base unless there is a conflicting rule already existing that has a higher degree of truth.

3.2 Results of Training with Static Curves and Real Time Operating Data

As it was stated previously, the first stage in generating the fuzzy knowledge base is to use the static magnetization curves of the motor. Each point on the motor's magnetization curve represents an input data set of flux linkage and current and an output value of angle. The static flux linkage curves will characterize the motor to a good degree. However, as it was mentioned, real time operating effects such as mutual inductances between phases, eddy currents, temperature effects, and skin effects may be significant. Therefore, to model these effects, a further step of the training phase involved using the measured real time dynamic operating data of current, flux linkage, and angle of the motor phases.

After the learning tasks are performed, the generated fuzzy rules in the fuzzy rule base determine a fuzzy logic system mapping inputs flux and current to angle according to the modeled non-linear function. In Figure 5 results are shown for the generated fuzzy rule base of the motor used in the research after training the system with the measured static and dynamic motor data.

4 Position Sensing

After the training, the generated FL rule base defines a non-linear multi-input single-output function f which maps input values of flux linkage and current to output values of position:

$$f : (\psi, i) \rightarrow \theta \tag{15}$$

The model is then utilized during the operation of the sensorless rotor position estimation scheme to calculate the rotor position from measured values of flux linkage and current.

The sensorless system essentially operates as follows: while the motor is running, the phase currents and voltages in each of the four phases are measured and the flux linkages are estimated by trapezoidal integration using the motor voltage equation (1). Trapezoidal integration (2) was used due to its relatively high accuracy and simple calculations, which made it well suited for real time implementation.

The crisp or numeric flux linkage and current values of each phase are then input into the fuzzy logic rule based SR motor model. Normally, each crisp value of flux and current will be a member of more than one fuzzy set or membership function, and thus more than one rule from the rule base will normally be triggered by the numerical inputs. In the developed method, the rule implication used the Max-Product composition of the fuzzy rules:

$$\mu_{R^x(\theta)} = \max(\Psi, i)\min[\mu_{\Psi^x(\psi)}, \mu_{I^x(i)}] \times \mu_{\Theta^x(\theta)} \tag{16}$$

where: x = the xth rule that is triggered by the crisp inputs of flux and current, $\mu_{R^x(\theta)}$ = the membership function of the resultant output fuzzy set R^x in the angle domain, $\mu_{\Psi^x(\psi)}$ = the membership function of the input fuzzy set Ψ^x in the flux domain of input value ψ, $\mu_{I^x(i)}$ = the membership function of the input fuzzy set I^x in the current domain of input value i, and $\mu_{\Theta^x(\theta)}$ = the membership function of the output fuzzy set Θ^x in the angle domain.

If more than one rule is triggered by the crisp inputs, the triggered rules are aggregated using the union of the output of each of the individual rules.

ψ (columns) / i (rows)

	s19	s18	s17	s16	s15	s14	s13	s12	s11	s10	s9	s8	s7	s6	s5	s4	s3	s2	s1	m	b1	b2	b3	b4	b5	b6	b7	b8	b9	b10	b11	b12	b13	b14	b15	b16	b17	b18
s18	s15	b15																																				
s17	s15	m	b15	b15																																		
s16	s15	s6	m	b7	b15	b15																																
s15		s9	s4	m	b4	b9	b15	b15																														
s14		s15	s6	s3	m	b3	b6	b9	b14	b15																												
s13		s15	s7	s5	s2	m	b2	b4	b7	b9	b13	b15	b15																									
s12			s9	s6	s4	s2	s1	b1	b3	b5	b7	b9	b12	b15	b15																							
s11			s15	s7	s5	s4	s2	s1	b1	b2	b4	b6	b8	b9	b11	b14	b15																					
s10			s15	s8	s6	s4	s3	s2	s1	m	b2	b3	b5	b6	b8	b9	b11	b13	b15	b15																		
s9			s15	s9	s7	s5	s4	s3	s2	s1	m	b1	b3	b4	b5	b7	b8	b9	b11	b13	b15	b15																
s8				s13	s8	s6	s5	s4	s3	s2	s1	m	b1	b2	b4	b5	b6	b7	b8	b10	b11	b13	b15	b15														
s7				s15	s9	s7	s5	s4	s3	s3	s2	s1	m	b1	b2	b3	b4	b5	b6	b8	b9	b10	b12	b14	b15													
s6				s15	s10	s7	s6	s5	s4	s3	s2	s2	s1	m	b1	b2	b3	b4	b5	b6	b7	b8	b9	b11	b13	b15	b15											
s5					s12	s8	s7	s6	s5	s4	s3	s2	s1	s1	m	b1	b2	b3	b4	b5	b6	b7	b8	b9	b10	b12	b14	b15										
s4					s15	s9	s7	s6	s5	s4	s3	s3	s2	s1	m	m	b1	b2	b3	b4	b5	b6	b7	b8	b9	b10	b11	b14	b15									
s3					s15	s10	s8	s6	s5	s5	s4	s3	s2	s2	s1	m	m	b1	b2	b3	b4	b5	b6	b7	b8	b9	b10	b11	b14	b15								
s2						s11	s8	s7	s6	s5	s4	s3	s3	s2	s1	s1	m	b1	b1	b2	b3	b4	b5	b6	b7	b8	b9	b10	b12	b15								
s1						s15	s9	s7	s6	s5	s5	s4	s3	s2	s2	s1	m	m	b1	b2	b3	b3	b4	b5	b6	b7	b8	b9	b10	b13	b15							
m						s15	s10	s8	s7	s6	s5	s4	s3	s3	s2	s2	s1	m	b1	b1	**b2**	b3	b4	b4	b5	b6	b7	b8	b9	b11	b14							
b1						s15	s11	s8	s7	s6	s5	s5	s4	s3	s2	s2	s1	s1	m	b1	b2	b2	b3	b4	b5	b6	b7	b8	b9	b10	b12	b15						
b2							s13	s9	s8	s7	s6	s5	s4	s3	s3	s2	s2	s1	m	m	b1	b2	b3	b4	b4	b5	b6	b7	b8	b9	b11	b15						
b3							s15	s10	s8	s7	s6	s5	s4	s4	s3	s2	s2	s1	m	m	b1	b2	b2	b3	b4	b5	b6	b7	b8	b9	b10	b13	b15					
b4							s15	s11	s9	s7	s6	s6	s5	s4	s3	s3	s2	s2	s1	m	b1	b1	b2	b3	b4	b4	b5	b6	b7	b8	b10	b12	b15					
b5								s12	s9	s8	s7	s6	s5	s4	s4	s3	s2	s2	s1	m	m	b1	b2	b2	b3	b4	b5	b6	b7	b8	b9	b11	b14					
b6								s15	s10	s8	s7	s6	s6	s5	s4	s3	s3	s2	s1	s1	m	b1	b1	b2	b3	b4	b4	b5	b6	b7	b9	b10	b13	b15				
b7								s15	s11	s9	s8	s7	s6	s5	s4	s4	s3	s2	s2	s1	m	m	b1	b2	b3	b3	b4	b5	b6	b7	b8	b9	b11	b15				
b8								s15	s12	s9	s8	s7	s6	s5	s5	s4	s3	s3	s2	s1	s1	m	b1	b1	b2	b3	b4	b5	b5	b7	b8	b9	b11	b14				
b9									s14	s10	s8	s7	s7	s6	s5	s4	s4	s3	s2	s2	s1	m	m	b1	b2	b3	b4	b4	b5	b6	b7	b9	b10	b12	b15			
b10									s15	s10	s9	s8	s7	s6	s5	s5	s4	s3	s3	s2	s1	s1	m	b1	b1	b2	b3	b4	b5	b6	b7	b8	b9	b11	b15			
b11									s15	s12	s9	s8	s7	s7	s6	s5	s4	s3	s3	s2	s2	s1	m	m	b1	b2	b3	b4	b4	b5	b7	b8	b9	b11	b14			
b12										s13	s10	s9	s8	s7	s6	s5	s5	s4	s3	s3	s2	s1	s1	m	b1	b2	b2	b3	b4	b5	b6	b7	b9	b10	b13	b15		
b13										s15	s10	s9	s8	s7	s6	s6	s5	s4	s3	s3	s2	s2	s1	m	m	b1	b2	b3	b4	b5	b6	b7	b8	b10	b12	b15		
b14										s15	s11	s9	s8	s7	s7	s6	s5	s4	s4	s3	s2	s2	s1	s1	m	b1	b2	b3	b4	b4	b5	b7	b8	b9	b11	b15		
b15											s13	s10	s9	s8	s7	s6	s6	s5	s4	s3	s3	s2	s2	s1	m	b1	b1	b2	b3	b4	b5	b6	b8	b9	b11	b14		
b16											s15	s10	s9	s8	s7	s7	s6	s5	s4	s4	s3	s2	s2	s1	m	m	b1	b2	b3	b4	b5	b6	b7	b9	b10	b13		
b17											s15	s11	s9	s8	s8	s7	s6	s6	s5	s4	s3	s3	s2	s1	s1	m	b1	b1	b2	b3	b4	b5	b7	b8	b10	b12	b15	
b18											s15	s12	s10	s9	s8	s7	s6	s6	s5	s4	s4	s3	s2	s2	s1	m	b1	b1	b2	b3	b4	b5	b6	b8	b9	b11	b15	

Fig. 5. SR model fuzzy rule base (s = small, m = medium, b = big, empty = no rule (as seen in Figure 2 one can see there is no physical relationship between flux and current in these regions, and thus there can be no rules in these regions)). Highlighted cell is the fuzzy set shown in expanded form in Figure 7.

The aggregate of the fuzzy rules will thus produce an aggregated output fuzzy set in the angle domain. However, a single crisp numeric value of estimated rotor position is required and thus the output fuzzy set is then converted into a single crisp value of angle using defuzzification. In this method the center of average defuzzification formula was used to determine the crisp output angle θ_C:

$$\theta_C = \frac{\sum_{x=1}^{n} \mu_{R^x}(y^x) \times y^x}{\sum_{x=1}^{n} \mu_{R^x}(y^x)} \tag{17}$$

where: y^x is the point at which the fuzzy set R^x of the xth rule has peak membership value and $\sum_{x=1}^{n} \mu_{R^x}(y^x)$ is the membership value of the resultant output fuzzy set R^x in the angle domain for y^x.

Due to their low computational complexity the Max-Product rule implication and the center of average defuzzification methods have advantage for implementation of the estimation scheme in practical real time systems over other methods of rule processing.

It should be noted that all motor phases that are conducting current are used for position sensing, and thus there will be a crisp rotor position value found for each conducting phase. The method of combination of these rotor position estimates from each of the phases, in a way which takes into account the change in accuracy of the phases at different rotor angles, is described below.

5 Robustness of Fuzzy Model Based Method

To examine why there is robustness to noise and error in fuzzy rule based modeling, it should first be noted that each rule of the fuzzy rule base shown in Figure 5 expresses heuristically the motor characteristics in each region using fuzzy reasoning. Each rule will have the form:

$$R^{(n)} : \textbf{If } \psi \textbf{ is } A_\psi^{(n)} \textbf{ and } i \textbf{ is } A_i^{(n)} \textbf{ then } \theta \textbf{ is } B_\theta^{(n)} \tag{18}$$

where: $R^{(n)}$ = the nth rule, $A_\psi^{(n)}$ = antecedent fuzzy set in the flux linkage fuzzy set domain with membership function $\mu_{A_\psi^{(n)}}(\psi)$, $A_i^{(n)}$ = antecedent fuzzy set in the current fuzzy set domain with membership function $\mu_{A_i^{(n)}}(i)$, $B_\theta^{(n)}$ = consequent fuzzy set in the rotor angle fuzzy set domain with membership function $\mu_{B_\theta^{(n)}}(\theta)$.

Each of the antecedent and consequent fuzzy sets in the ψ, i, and θ variable domains represent a linguistic variable and are defined over a range of values or a region with membership value ranging from 0 to 1. Hence for any given rule, both the antecedents and the consequents in the input-output domains will be defined for a given range of values. For example consider a rule from the motor model that was shown in Figure 5:

$$\textbf{If } \psi \textbf{ is } b1 \textbf{ and } i \textbf{ is } m \textbf{ then } \theta \textbf{ is } b2 \tag{19}$$

Figure 7 shows an expanded form of Figure 5 where this rule is represented. The consequent of the rule is the fuzzy set $b2$ in the angle domain, and it is triggered by the antecedent fuzzy sets $b1$ in the flux linkage domain and m (medium) in the current domain.

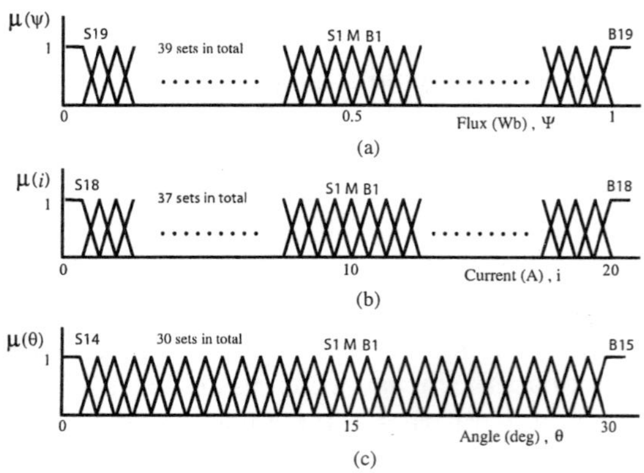

Fig. 6. Fuzzy domain regions for each variable (a) Flux Linkage. (b) Current. (c) Rotor position.

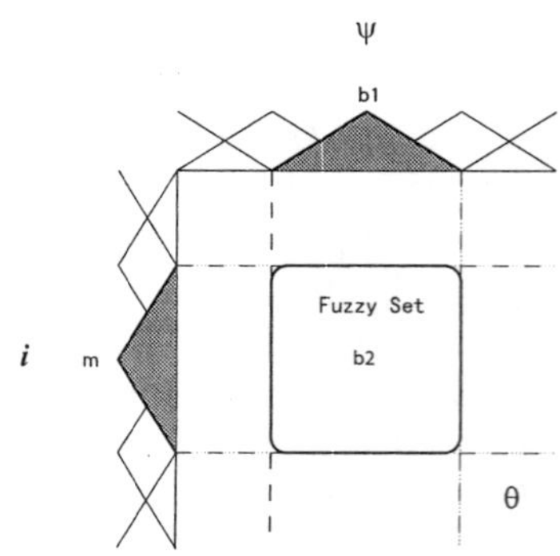

Fig. 7. Expanded sub-set of SR model fuzzy rule base.

The width of the membership functions of the fuzzy sets can be considered as an allowable level of noise [15], [16]. This means that an input data point with error or noise can still be placed in the same fuzzy set as the same input data with no error or noise. Thus by fuzzification of the input signals, input data which is corrupted by noise can be accepted in the same set as the same data without noise but with a different membership function [17]. The range of the fuzzy set membership function will determine the range of values that will be accepted as part of the same set. The range can be defined such that small deviations in the input data do not have a significant effect on the output position estimation.

However, it should also be noted that a disadvantage of using the fuzzy rule based modeling is that an inverse relationship exists between *robustness* and *accuracy*. As the robustness to input signal noise and error increases due to a widening of the fuzzy sets, the output resolution decreases due to the wider partitioning of the input and output signal domains. Therefore a balance must be found between error robustness and output accuracy. As discussed above, in this research work each variable domain of flux, current, and angle was divided up into the fuzzy regions as was shown in Figure 6. These resulted in a good compromise between accuracy and robustness to noise and error (as will be seen in the results sections, the error under normal conditions was approximately half a degree, which was comparable to the resolution of a physical position sensor).

However, as in all cases where finite width fuzzy sets are used, when there is zero or very small input signal error, the output will normally still contain some error or uncertainty due to the fuzzy set partitioning of the input and output domains. This entails that, strictly speaking, if robustness is not of concern then conventional schemes may result in higher accuracy at zero or no noise levels.

The above routine essentially encapsulates the algorithm used to estimate the rotor position from the fuzzy logic motor model using measurements of current and voltage only. However, as is discussed below, additional performance enhancement features have been added so that the sensorless angle estimation algorithm copes better with measurement errors and inaccuracies found in real motor drives. A block diagram of the complete algorithm including enhancement features is shown in Figure 8. This estimator directly replaces the position sensor required by the conventional SR motor controller.

6 Fuzzy Predictive Filters

A major concern is that in sensorless estimation schemes, measured motor feedback signals are used to calculate the motor position in real time. However motor drives are electrically noisy environments and practical measurement equipment is imperfect. Thus the feedback signals will be corrupted by some noise and error [5]. To improve the performance, some form of conventional filtering method could be used. However, in practice, the implementation of conventional filtering methods introduces some time delay. This may not be acceptable due to the time constrains of the real time position estimation and control of the motor. In addition problems emerge, due to the fact that the feedback signal noise in real motor drive environments is often impulsive or non-Gaussian in nature. Using conventional filtering methods to decrease this noise is difficult, due to the fact that non-Gaussian noise is difficult to mathematically model and predict, and therefore difficult to filter [18].

Hence, a novel method is shown here to decrease the problem of signal corruption in the feedback signals involved in SR position estimation. A fuzzy predictive filter is developed which combines both adaptive fuzzy logic based prediction, and a heuristic knowledge based algorithm implemented with fuzzy rules, to detect and discard noise on the feedback signals. The filter does not introduce any delay or phase shift in the feedback signals because it is predictive in nature. In addition, an important feature of the method is that the fuzzy predictive filter does not require any mathematical modeling of

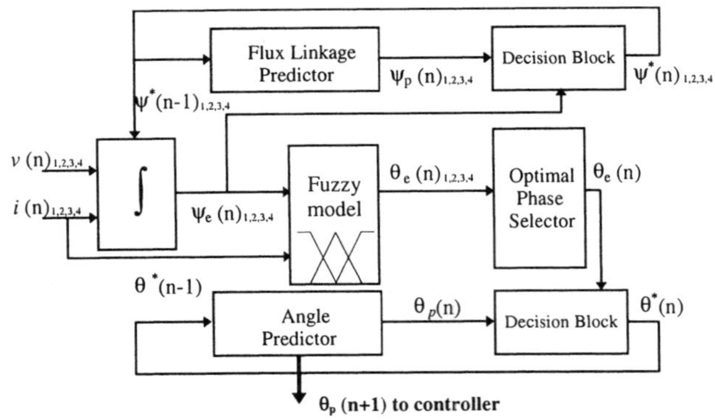

$v(n)_{1,2,3,4}$ = the four phase voltages of the SR motor

$i(n)_{1,2,3,4}$ = the four phase currents

$\psi_e(n)_{1,2,3,4}$ = the estimated four phase flux linkages (from integration)

$\psi^*(n-1)_{1,2,3,4}$ = the weighted flux of step (n - 1) used as integration constant

$\psi_p(n)_{1,2,3,4}$ = the predicted flux linkage

$\psi^*(n)_{1,2,3,4}$ = the weighted value of estimated and predicted flux linkage

$\theta_e(n)_{1,2,3,4}$ = the estimated position from the fuzzy model from all excited phases

$\theta_e(n)$ = the single weighted value of position after the phase selector

$\theta_p(n)$ = the predicted value of fuzzy position

$\theta^*(n)$ = the weighted value of estimated and predicted position

$\theta^*(n-1)$ = previous value of weighted position used for prediction

$\theta_p(n+1)$ = next step prediction of angle

Fig. 8. The complete block diagram of the fuzzy logic based position estimation algorithm.

the noise, and therefore can be used effectively with non-Gaussian impulsive noise [11].

6.1 Prediction of Flux Linkage and Position

The prediction algorithm used is implemented using fuzzy logic, and its purpose is to predict future values, so that if a measurement inaccuracy occurs, the predicted value may be used to lower the error, and therefore effectively filter the feedback signal using prediction. To achieve this, a comparison between estimated and predicted rotor position and flux values are made during each iteration. Then some combination of these is chosen in order to lessen the effect of errors.

The purpose of the predictor can be better understood if the ideal case is taken in the first instance. In the ideal case of accurate prediction, the estimated values of flux linkage (from integration) and position (from the output of the fuzzy motor model) could be predicted accurately using the prediction algorithm. Now, consider the case where for a short instant of time, the measured values of flux linkage and position are erroneous. At this time, the predicted values (which are predicted from the previous correct values of flux linkage and position) will be good approximations to the actual values of flux and position in the machine, if the prediction algorithm is accurate. In this case the estimated values can be ignored, and instead the predicted values can be used. However, this does not offer an ideal solution. The first question that arises is:

"If the estimated and predicted values are different, which is actually the more correct value?"

Human intuition is required to try to answer this question, and this will be discussed later.

The problem of predicting the flux linkage and angle in future steps of time is a problem of time-series prediction. It has been demonstrated that fuzzy logic systems are excellent solutions for time-series prediction [8], [13]. In addition, a wide array of fuzzy logic based prediction algorithms have been developed [19], [20]. In this work, real time motor drive implementation is an important issue, and thus the *table look-up learning scheme* [13] was used.

The fuzzy logic predictor using table look-up learning operates by generating fuzzy rules from the numerical input-output data pairs of previous data points, each separated by a constant step of time, and creating an adaptive predicting fuzzy rule base. The rules on this predicting fuzzy rule base define the relationship between the present value of data and past values of data. The fuzzy predictor can then make a prediction from this fuzzy rule base.

For example, consider $x(k), (k = 1, 2, 3, \ldots)$ as a time series of previous rotor position values (or previous flux linkage values) each separated with a constant time interval. If the future value of $x(k+l)$, at time step $k+l$ in the future, is to be determined from a window of n previous measurements $x(k-n+1), x(k-n+2), \ldots, x(k)$, then $k-n$ input-output pairs can be formed. These input-output pairs will be:

$$\begin{aligned}
[x(k-n+1), \ldots, x(k)] &\Rightarrow [x(k+l)] \\
[x(k-n), \ldots, x(k-1)] &\Rightarrow [x(k+l-1)] \\
&\cdots\cdots\cdots \\
[x(1), \ldots, x(n)] &\Rightarrow [x(n+l)] \qquad (20)
\end{aligned}$$

In the above equation, the symbol $\Rightarrow$ indicates some fuzzy logic based function relating a future output value with previous values. The prediction method consists of the following steps:

Step A: Dividing the Input and Output into Fuzzy Regions: In this initialization step which is performed only once before the algorithm is run, the range or intervals of the input and output variable domains are defined. In this case the input and output domains are the previous values and future values respectively of the flux linkage and angle. Each domain is then divided up into fuzzy regions. Finally, each fuzzy region is assigned a fuzzy membership function.

Step B: Fuzzy Rule Adaptation from Input-Output Data Pairs: Once the fuzzy membership functions in the input and output domains have been defined as in Step A, data can be used to determine fuzzy rules and create the fuzzy rule base. To determine a fuzzy rule from each input-output data pair the first step is to determine the degree of each crisp data point in every membership region of its corresponding fuzzy domain. The crisp data points are then all assigned to the fuzzy region with maximum degree. Then these fuzzy regions are combined to form a fuzzy rule relating the input values to the output values. Thus each new input-output data pair will produce a fuzzy rule.

As an example consider an input-output data pair $[x(k-n+1), \ldots, x(k)] \Rightarrow [x(k+l)]$. After assigning the data points to the fuzzy region in the input-output domain with the highest degree, the following fuzzy rule can be formed:

$$\begin{aligned} &\textbf{If } x(k-n+1) \textbf{ is } F^{M_{(k-n+1)}} \textbf{ and } \ldots \\ &\quad \textbf{and } x(k) \textbf{ is } F^{M_{(k)}} \textbf{ then } x(k+l) \textbf{ is } F^{M_{(k+l)}} \end{aligned} \tag{21}$$

where $F^{M_{(k-n+1)}} \ldots F^{M_{(k+l)}}$ is the fuzzy region of highest degree assigned to data values $x(k-n+1), \ldots, x(k+l)$ respectively.

Step C: Assigning Rule Degrees: In order to choose between data sets that produce the same antecedents but different consequent a rule *degree* is assigned to the rule. Then conflicting rules are resolved by choosing the conflicting rule with the highest degree. In the developed method, a degree is assigned which is the product of the membership degree of each variable in its respective region. For example, the rule in equation (21) would have degree:

$$\text{Degree(Rule)} = \mu_{F^{M_{(k-n+1)}}} \times \ldots \times \mu_{F^{M_{(k)}}} \times \mu_{F^{M_{(k+l)}}} \tag{22}$$

where $\mu_{F^{M_{(k)}}}$ is the membership degree of crisp data point $x(k)$ in fuzzy set $F^{M_{(k)}}$.

Step D: Adaptive Rule Base Modification: Every time a new measurement is made, a new input-output pair consisting of the previous and the present values is made, and thus a new rule can be formed relating previous values to future values. Each new fuzzy rule is stored on the fuzzy rule base unless there is a conflicting rule already existing which has a higher degree of truth. In this manner, the predicting fuzzy rule base is adaptive in nature, and will change during the actual operation according to the data input values.

Step E: Prediction from Rule Base: The adaptive fuzzy predicting rule base is then used to predict a future data value $x(k+l)$, from the previous n data points. Here the value l is a chosen integer that represents the value of $x(k)$ at the data point $k+l$ in the future. Essentially, the predicting fuzzy rule base is used as the non-linear mapping between the previous n data points and $x(k+l)$.

To make a prediction, the previous values $x(k-n+1), x(k-n+2), \ldots, x(k)$ are used as the crisp input to the fuzzy predictor. The inputs will trigger rules in the fuzzy predicting rule base and this will give an aggregated fuzzy output. This output is then defuzzified to give the crisp prediction of $x(k+l)$. The operation of the fuzzy predictors with the predicting fuzzy rule base uses max-min implication and center of average defuzzification. These methods use simple algorithms and thus have the advantage of fast computation.

In this application, values of $l = 1$ and $n = 4$ were used for the flux linkage predictor, and $l = 1$ and 2 and $n = 4$ are chosen for the angle predictor (where l and n are the parameters defined above). Therefore for any iteration k, the four previous and present values of flux linkage: $[\psi(k-n+1), \psi(k-n+2), \psi(k-n+3), \psi(k)] = [\psi(k-3), \psi(k-2), \psi(k-1), \psi(k)]$ are used to predict the next value of flux linkage $\psi(k+l) = \psi(k+1)$.

Similarly, the previous four values of angle are used to predict the next two values in time of angle $\theta(k+1)$ and $\theta(k+2)$. Two values are predicted because the next value of angle $\theta(k+1)$ can be used to compare the predicted and estimated value, and the next value in time after this angle $\theta(k+2)$ can also be used by the motor controller.

In each iteration, both training of the fuzzy rule base (Steps B, C, D) and prediction from the rule base (Step E) occurs. Specifically, when a new value of flux linkage $\psi(k)$ or rotor position $\theta(k)$ is estimated, it is used with the previous values to first modify the rule base by creating a new rule that maps past to present values. The updated rule base is used for prediction.

The second part of the fuzzy predictive filters, which is the fuzzy logic-based heuristic decision block, will be detailed in the next section.

6.2 Weighting of Predicted and Estimated Values Using Heuristic Rules

As can be seen in the block diagram of Figure 8, the predicted values of rotor position $\theta_p(n)$, and flux linkage $\psi_p(n)$, are used in conjunction with the estimated values of flux linkage $\psi_e(n)$ (from integration) and rotor position $\theta_e(n)$ (from the fuzzy model output). In the ideal case, the predicted and estimated values should be exactly the same. However, as mentioned above these may differ due to errors. In this case either the predicted values or the estimated values may be used, and a decision must be made as to which value should be chosen in order to increase the accuracy of the fuzzy predictive filter.

However, using fuzzy logic, instead of choosing either one of the values, a fuzzy combination of both can be used, to allow for a smoother decision transition. Thus, in this system a knowledge-based, heuristic decision maker was implemented. This decision making algorithm was termed a *Decision Block*. From Figure 8, it can be seen that the decision blocks of the flux linkage and angle produce a final weighted value $\psi^*(n)$ and $\theta^*(n)$ respectively.

The principal purpose of these decision blocks is to implement human intuition in selecting a good weighting between the estimated and predicted values of flux linkage and angle. For instance, it can be said that under steady state speeds, the rotor position and flux linkage trajectories will be regular and periodic functions. Under this condition, the fuzzy predictors will be able to forecast with good accuracy [21], [22]. Therefore, if the predicted and estimated values differ under steady speeds, then this would more likely be from errors in the estimated values and thus more weighting can be placed on the predicted value.

However, during transients the fuzzy predictors may not perform as well, due to the learning period always required with new events. In this case, if there is a difference between the estimated and predicted values, then the predicted value would more likely be in error, and thus more weighting will be placed on the estimated value.

Thus under *steady* speeds and conditions, it can be intuitively said that the confidence in predicted values will be *high*, whereas under transient speeds and conditions, this confidence will be *low*. In addition, it can be said that the confidence in the predicted values will be *higher* for *low acceleration* values than for *high acceleration* values.

From the above discussion it may seem that some conventional mathematical function relating confidence in the predicted values to the actual motor acceleration can easily be defined. However, some practical considerations make the use of a fuzzy system advantageous.

Firstly, it can be seen from the above discussion that *high*, *low* and *steady* are linguistic terms that contain a certain amount of fuzziness. With conventional mathematical logic functions it is difficult to adequately represent heuristic knowledge directly. However, fuzzy systems can deal with situations where sharp distinctions between the boundaries of application of rules do not occur.

Another advantage of using a fuzzy system in this case is that it allows flexibility to easily fine tune the relationship between acceleration and confidence by modifying the fuzzy sets and rules. During tuning, the general heuristic rules can remain unchanged.

Furthermore, a major advantage of using a fuzzy system is that it can cope with inherent uncertainty in the input signals, unlike traditional mathematical logic techniques which would require accurate inputs for accuracy in the outputs. In this system the input variable is acceleration. The acceleration cannot be directly measured by a mechanical sensor in this application because the system is sensorless. Another method is to calculate acceleration from speed values. Successful techniques have been recently developed to estimate acceleration by using predictive polynomial differentiators [23], [24] or model based state observation [25]. However, in this case no direct measurement of position or speed is possible. If the rotor angle estimates are used instead, the errors in the estimates may be too high for the exact calculation of acceleration.

Therefore, a fuzzy system was developed to relate prediction confidence to acceleration which eliminates the need for accurate inputs and can use heuristic knowledge. The fuzzy system only requires the acceleration feedback to be a fuzzy value to determine which predefined fuzzy domain the acceleration belongs to. Thus only imprecise (or fuzzy) knowledge is required about the motor acceleration.

Instead of acceleration, an *acceleration factor* is used which gives an approximation of the actual acceleration. As discussed above, this approximation can be used in this application because only the *relative* acceleration is important (e.g. *High* or *Low*), and not the actual numeric value. Therefore an acceleration factor A_n is defined as:

$$A_n = \frac{\theta_n - \theta_{n-k}}{k\Delta T} - \frac{\theta_{n-l} - \theta_{n-k-l}}{k\Delta T} \tag{23}$$

Here, θ_n is the rotor angle at step n and ΔT is the time between each iteration, and k and l were chosen to both equal 5. It should be noted that this factor is an imprecise representation of acceleration and contains a delay of $(k+1)\Delta T$ between the first measurement being available and when it applies.

The Decision Block uses this fuzzy system and gives the predicted values of flux linkage $\psi_p(n)$ and position $\theta_p(n)$ a confidence value based on the linguistic knowledge described above. It is a single input - single output fuzzy system where the input fuzzy domain is the acceleration factor, whilst the output domain is the confidence. The fuzzy rule base that was used is shown in Table 5. The rules on this table are created using the general heuristic knowledge discussed above. This demonstrates the ease in which fuzzy logic can encapsulate human intuition expressed in linguistic terms, to create a knowledge-based algorithm [26].

7 Optimal Phase Selection

In normal operation of the SR motor, the excitation region for each phase may overlap. Therefore, there will often be current flowing in two or more motor phase windings at a given time and hence, every excited phase can be used for estimating rotor position. Theoretically, the same position should be output by each of the excited phase rotor position estimations. However, each phase has rotor angle regions of optimal sensing, because in some rotor angle

Table 5. Fuzzy Rule Base for Confidence

Acceleration	*Confidence*
zero (Z)	very large (VL)
very small (VS)	medium (M)
small (S)	medium (M)
medium (M)	small (S)
large (L)	zero (Z)
very large (VL)	zero (Z)

regions the rotor position estimation will be more affected by errors than at other rotor angle regions. Therefore, for implementation in a practical drive, the estimated rotor position from the phase in the rotor angle that lies in the optimal region should be given the most weighting.

The optimal sensing region can be found from an analysis of the flux linkage curves. As was discussed above, the optimal sensing region is in the regions where the flux linkage curves are not tightly bunched up. Note that the region is difficult to exactly define in an exact manner. However, as detailed above, it can generally be said from the magnetization curves, that when the angle is near the unaligned position (or "small" relative angles) and the current is low, that the curves are very tightly bunched up. Therefore under this condition, small errors in the flux linkage estimation or current measurement will result in large errors in the position estimate. Additionally, when the angle is close to alignment (or "large" relative angles) the curves are also tightly bunched up, and therefore a small error in current measurement can produce large errors in rotor angle.

Therefore, the optimal sensing positions, where the best resolution is offered, is for medium angles between alignment and unalignment. However, the optimal sensing region in fact does not stop or start abruptly but has a transition region. In addition, the terms *small*, *medium*, and *large* used above to describe the regions, are also linguistic terms which can once again be naturally represented again by a fuzzy system. Therefore, to decide whether a motor phase measurement is in the optimal sensing region, a decision block was used. The decision block uses a two input - single output fuzzy system where the inputs are current and angle and the output is confidence. Based on the heuristic rules, it places a weight or confidence on each of the excited motor phase angle estimates and outputs one final weighted value $\theta_e(n)$.

The rule base that defines the linguistic rules that were used by the decision block is shown in Table 7. To determine the weighted angle value $\theta_e(n)$, each estimated phase angle is multiplied by its respective confidence value and the total is divided by the addition of all the confidence factors. Hence, if

Table 6. Fuzzy rule base for optimum-phase decision block relating current and angle to confidence

		Confidence	
Small i (S)	S	S	S
Medium i (M)	M	L	S
Large i (L)	M	L	S
	Small θ (S)	Medium θ (M)	Large θ (L)

there were two phases which produced an angle estimate, then the algorithm would compute the following:

$$\theta_e = \frac{\theta_1 C_1 + \theta_2 C_2}{C_1 + C_2} \tag{24}$$

where θ_e is the final angle estimate, θ_1 and θ_2 are the phase angle estimates of phase 1 and phase 2 respectively, and C_1 and C_2 are the confidence values of phase 1 and phase 2 respectively.

8 Simulation Results

The simulation tests allowed the method to be tested under operating conditions that could not be safely achieved using the real laboratory motor (without a high risk of motor or drive damage). For example, very high acceleration and sudden rotor stall from high speeds could be simulated, but would have been too risky to perform on the actual experimental motor and drive. However, the system can only be finally proven with actual experimental results which will be shown in the following section.

8.1 Transient Acceleration Test and Sudden Rotor Standstill (Zero Speed)

In this simulation test the motor is simulated to accelerate from 0 to 1000 r/min in 300 ms and then the rotor is brought to a sudden standstill. This can occur in practice in industrial factory drive operations where some object suddenly causes a jam in some equipment. For safe operation, the motor drive should not lose the position estimation tracking under this condition. If the position estimation tracking is not lost, then the system can detect a stall or jam situation and safely bring the system down.

The rotor angle, voltage, current, and the flux linkage waveforms for one phase of the motor are shown in Figure 9. In the waveform of the actual rotor position given in Figure 9 (a) it can be seen that the rotor position accelerates from standstill, until at $t = 0.3$ s when the speed is 1000 r/min, it is instantaneously stalled to zero speed. The rotor speed is then zero until $t =$ 3.5 s at which time the motor accelerates slightly. Therefore these simulation

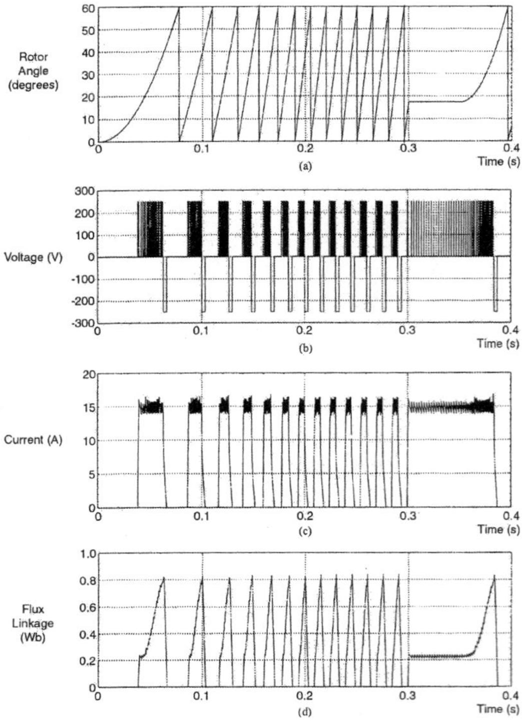

Fig. 9. Waveforms for 0-1000 r/min transient test (a) Actual rotor angle, (b) Phase voltage, (c) Phase current, (d) Phase flux linkage.

results test the performance of the sensorless algorithm under conditions of acceleration from start-up, sudden stall, and zero speed.

The voltage in this test, which can be seen in Figure 9 (b), is operating in the soft chopping mode. The corresponding phase current of the motor is detailed in Figure 9 (c). In this simulation, the chopping hysteresis band was set to have a high value of 16 A and a low value of 14 A. However it can be seen that at times the phase current falls outside these bands during the chopping period due to the fact that the simulated motor current controller has a minimum commutation frequency in order to more closely simulate a practical motor controller.

After 0.3 s when the motor is stalled and remains at zero speed for 0.5 s, one can see that the rotor angle remains constant, but that voltage and current are still applied to the phase. This is because the motor has stalled at an angle of 18 degrees, which means that the phase shown is still in its turn-on region. Therefore, the controller will apply current to the phase, which is maintained in the current hysteresis band by soft chopping. It should be noted that if the motor is stalled at zero speed there must be some current flowing in at least one phase (and hence finite torque production), otherwise position estimation cannot occur (as feedback current is required for the continuous position estimation). Thus, the scheme cannot operate at both zero speed and zero motor torque.

The phase flux linkage waveform is found from the numerical integration of the phase voltage and current. Figure 9 plots the flux linkage waveform corresponding to the currents and voltages of this test. One can see from the results in Figure 10 that the estimated angle is very close to the actual rotor position at all times of the test, including the acceleration and stall. The angle error between actual and estimated angle can also be seen in the figure and it has an average value of 0.4 degrees. This is more accurate than a typical industrial 8 bit position encoder which has a resolution of about 1.4 degrees, but less accurate than a more expensive 12 bit encoder which has a resolution of about 0.1 degrees. As it was mentioned above, the final accuracy is dependent on the number of rules or width of the membership functions in the fuzzy rule base.

As the estimation inputs are noise free in this simulation test (although there will be some error due to the numerical integration of flux linkage), these errors would be due to slightly erroneous outputs from the fuzzy rule based SR motor model (due to a not completely accurate model), or erroneous output of the fuzzy predictors (due to non perfect predictive rules). The average error due to erroneous outputs from the fuzzy rule based model is explained by the finite accuracy possible from the finite number of fuzzy regions that were chosen to divide the domains of rotor angle, current, and flux linkage.

As explained above, the position error could be reduced by increasing the amount of fuzzy sets in the variables domain, but an increase in the number of fuzzy sets would lead to more rules being required, and a corresponding increase in memory or storage requirements.

The predicted angles from the fuzzy predictor are also shown in Figure 10 (c). In this simulation test where there is no input signal noise it can be seen that the predictors have a fast learning time, and the predictions are accurate. There are a few points where no prediction can be made due to a lack of learnt rules (in this and other results in this Chapter, when a prediction cannot be made by the prediction algorithm due to lack of learnt rules, a zero value and vertical line is shown in the resultant plots). This would occur due to a particular combination of previous data points that had no corresponding learnt rules in the predictor's fuzzy rule base.

The operation of the flux linkage predictors is illustrated in Figure 11, where the actual phase flux linkage and the predicted flux linkage are shown for each motor. It can be seen that even with the highly non-linear chopping mode waveforms of the chopping mode flux during acceleration, that the predictor learns the motor behavior rapidly, with only a few points in the initial learning period where predictions cannot be made.

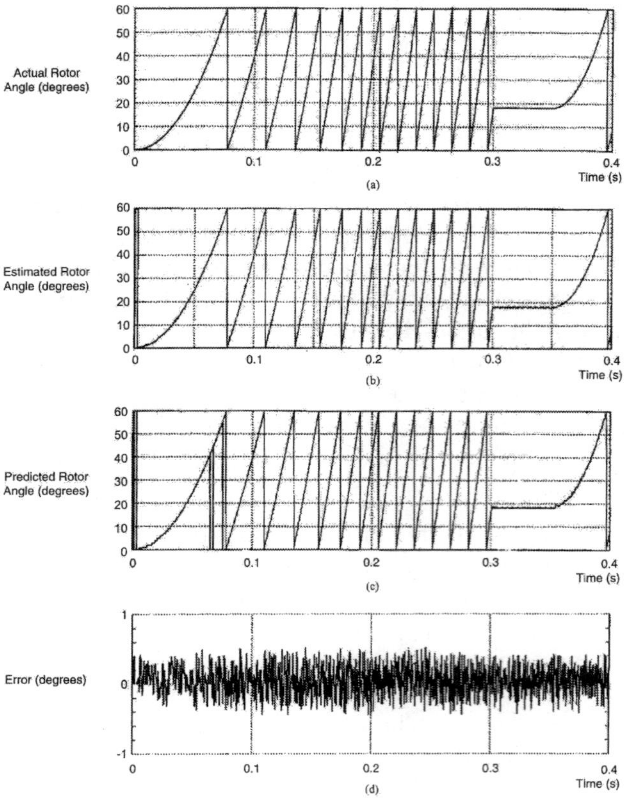

Fig. 10. Angle Waveforms for 0-1000 r/min transient test. (a) Actual rotor angle, (b) Estimated rotor angle, (c) Predicted rotor angle, (d) Angle error.

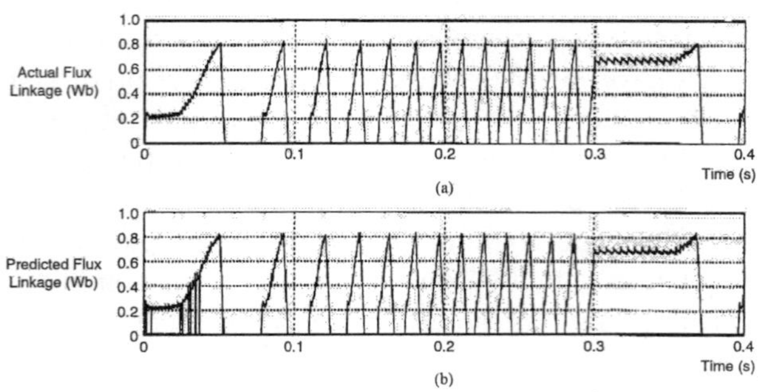

Fig. 11. Flux linkage waveforms for 0-1000 r/min transient test. (a) Actual Flux. (b) Predicted Flux.

9 Experimental Results

The actual operating effects of the SR motor drive includes measurement error, noise, calculation error, non ideal devices, electromagnetic coupling, controller delays, quantization error, mutual inductance between motor phases, parameter variation of motor inductances and resistances, asymmetrical inductance variation in the motor phases, variation in the magnetization curves in each of the phases, and effects on the motor waveforms of eddy currents. Therefore, experimental waveforms should be used to verify the ability of the sensorless scheme to operate with a real SR motor, as the above effects will affect the position estimation and prediction accuracy.

To obtain experimental results, a switched reluctance motor drive system was designed and constructed with a controller. The drive consists of several distinct sub systems as illustrated in Figure 8: 4-phase IGBT inverter (two switch per phase type), 4-phase (8 stator pole, 6 rotor pole) SR motor (4kW, 415V, 9A, 1500 r/min, four phase), an ADSP21020 digital signal processor board, 12 bit A/D converters, and the interfaces for signal input and output (currents, voltages, position and gate signals for IGBTs). Furthermore, a 12 bit shaft encoder was used to provide a reference for checking the estimated position.

In the actual motor drive implementation of the sensorless scheme, the start-up problem must be solved. This problem occurs because the magnetic characteristics of an SR motor phase are essentially symmetrical around the rotor - stator pole tip alignment point. Therefore, in essence there are two possible values of absolute rotor position for each point of flux linkage and current, that is, one on each side of the aligned position.

Hence during the start up of the sensorless motor drive, there are two problems. Firstly, the position is not known and therefore, the controller does not have knowledge of the required initial phase control strategy. Secondly, if only one phase is used initially, there will be two solutions to the estimate of rotor position for each flux linkage and current data pair.

However, this problem is well documented [27], and was easily solved using a simple method. In the initial starting instant, two phases of the motor are excited using a short pulse of current, to produce two sets of flux linkage and current pairs. This will produce four possible values of absolute position (two from each motor phase). Only one angle estimate of one phase will agree with one angle estimate of the other phase and this is the actual absolute rotor position. After this step is performed, the absolute value of rotor position has been found, and there will be no ambiguity in further measurements.

Using the experimental hardware, a wide range of operating modes and conditions were applied to the test motor. Some of the experimental results can be seen below.

9.1 Steady State Experimental Test

In Figure 12 (a) and (b), the measured current and voltage waveforms of one phase of the test SR motor are shown for the single pulse mode of the motor with a steady state speed of 810 r/min and 7 Nm torque.

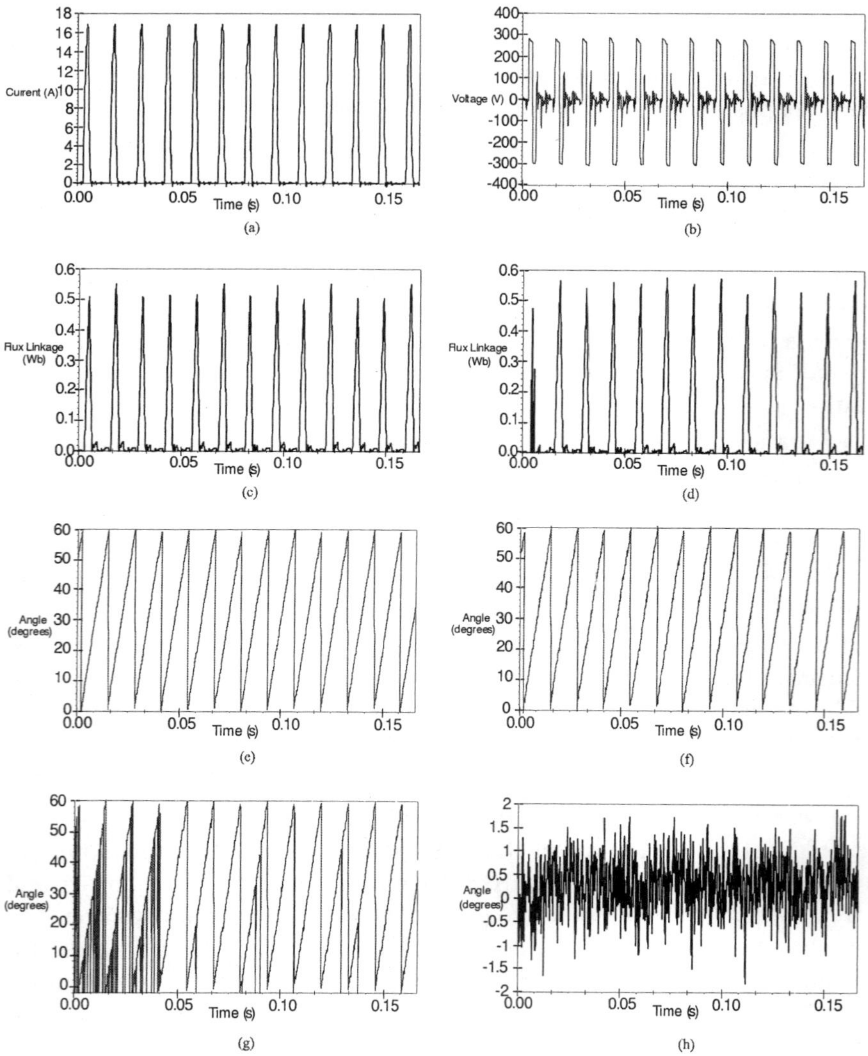

Fig. 12. Steady state test results. (a) Phase current, (b) Phase voltage, (c) Phase flux linkage, (d) Predicted flux linkage, (e) Measured rotor angle, (f) Estimated angle, (g) Predicted angle, (h) Error between estimated and measured angle.

The tests shown were taken with the A/D converter sampling frequency limited to 6000 Hz. This demonstrates the ability of the scheme to work with a relatively long iteration time of 166.7 μs. As was discussed above, due to time quantization effects, Low A/D frequencies act to increase the error in the flux linkage integration which is approximately inversely proportional to f^2, where f is the A/D sampling frequency (Table 2). It should be noted that due to this constraint, the limitation on the highest operating motor speed that can effectively be used with the sensorless scheme depends on the actual A/D converter frequency used with the motor drive. To operate at very high speeds without significant errors requires a corresponding high A/D converter frequency.

In Figure 12 (c) and (d), the flux linkage estimated by integration is detailed, together with the predicted flux linkage from the prediction algorithm. Note that there are variations in the maximum amplitudes of each pulse due to small variations in the on and off control angles and the dc supply voltage ripple and variation. It can be seen that the predicted values have a good accuracy and these can be used to detect and discard transient errors in the estimated flux. At time $t = 0$ there are no rules stored in the predictors fuzzy rule base and therefore there is an initial learning period (when there are many points where a prediction cannot be made). The learning period is less than 200 iterations which corresponds to 46% of one revolution.

The estimated and measured rotor positions for this test are shown in Figure 12 (e) and (f). The amplitude of error is also detailed and has an average of 0.52 degrees and a maximum of 1.92 degrees.

The results of the prediction of rotor angle are also shown in Figure 12 (g) where it is seen that approximately the first 240 iterations are needed for the initial learning period. This was equivalent to 52% of the one revolution of the rotor at constant speed. It was concluded after extensive testing that in the actual drive environment the developed prediction algorithm required approximately 200 - 250 numerical iterations to learn the single pulse mode waveforms.

9.2 Zero/Low Speed Operation

In this test the motor is initially stalled with a high inertial mechanical load that is thereafter slowly accelerated. The conditions of this test allow the sensorless position estimation algorithm to be tested for a number of important conditions. This includes zero speed operation (seen from $t = 0.3$ s to $t = 0.35$ s), slow acceleration, and low speed (seen from $t = 0.35$ s onwards).

The plot of the phase current, phase voltage, phase flux linkage, as well as measured and estimated rotor position, are shown together for one motor phase in Figure 13. At $t = 0.3$ s the rotor is stalled, and it can thus be seen that the rotor angle remains constant. However at $t = 0.35$ s the rotor is again accelerated with a low acceleration rate, and the rotor position again gradually increases. The average amplitude of error in this test is 0.59 degrees and a maximum of 2.37 degrees.

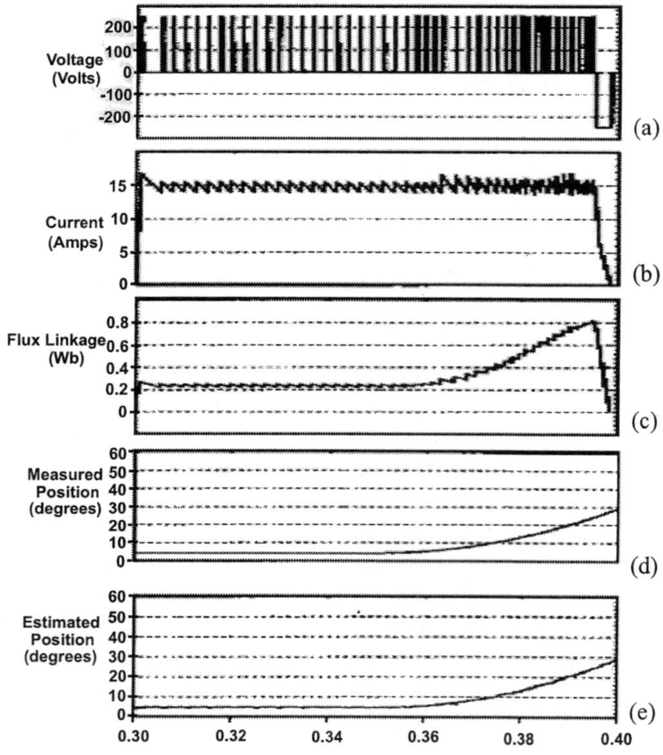

Fig. 13. Zero/low speed test. (a) Phase voltage, (b) Phase current, (c) Phase flux linkage, (d) Measured rotor angle, (e) Estimated angle

It can be seen in the phase current waveforms of Figure 13 that there are oscillations in the current hysteresis level even during zero and low speed. This is due to the maximum switching frequency limitation of the hysteresis current controller.

Furthermore it can be seen that the flux linkage changes when the rotor begins to accelerate from zero speed due to the change in incremental inductance with position. The estimated angle that is seen in these results can be seen to always have some error, even though the rotor position is constant. This can be explained by the fact that the current is not constant but has variations due to the hysteresis control. This high frequency variation will not be completely captured by the A/D converters, leading to measurement errors. Furthermore noise and modeling errors that are always present will lead to the error seen in the rotor position.

10 Experimental Robustness Tests

To examine the experimental robustness of the angle estimation algorithm, it is firstly instructive to examine the concept of robustness from the field of robust statistics. Huber has defined robust algorithms as having the following features [28]:

1. The algorithm should have a reasonably good accuracy in the assumed model of the underlying system.
2. Small deviations of the data should decrease the performance of the algorithm by only a small amount.
3. Large deviations in the data should not cause a catastrophe.

The first of these requirements essentially states that the output of the algorithm should be accurate when the input data is noise free. It was seen in [21], [22], and Figure 9 that the accuracy of the algorithm in predicting rotor position has a good accuracy for practically noise free feedback signals of current and voltage. Therefore, the sensorless position detection algorithm has been shown to satisfy the first requirement of a robust algorithm.

To satisfy the second requirement of a robust system entails that small deviations in the inputs of the sensorless algorithm should not have a significant adverse effect on the output rotor position estimate.

Furthermore, to satisfy the third requirement of algorithm robustness, the notion of a *breakdown point* should be considered [29]. The breakdown point is defined as a point where the algorithm fails by producing an unacceptably high output error due to a large deviation in the input data. A deviation in the input data can be defined as large when it has abnormally high deviation from the correct value, compared to the mean level of input data noise. For example in the SR motor drive there may be high amplitude but short duration periods of electromagnetic noise, due to fast turn off of the current in the inverter power devices. In the field of robust statistics these higher deviations in the data are termed *outliers*, as opposed to *inliers* which lie closer to the mean level of noise [30]. Although these robust statistic concepts are directly applicable to engineering tasks such as modeling and pattern recognition, the idea of a breakdown point and outlier and inlier data is applied in the FL based estimation algorithm by testing the algorithm with a very high level of signal noise, as detailed below.

10.1 Measurement Error Tests

In Figure 14 and 15 the waveforms of current and voltage for one phase of the SR motor are shown for the case when the motor is operating in single pulse mode at a steady speed of 660 r/min. There is an added random noise error with an amplitude of 10% of the maximum level of the measured signals. The added noise is Gaussian in nature. The expanded view of the current and voltage in Figures 14 (c) and 15 (c) clearly shows that the voltage and current with noise deviates significantly at various points in time from the measured values without added noise.

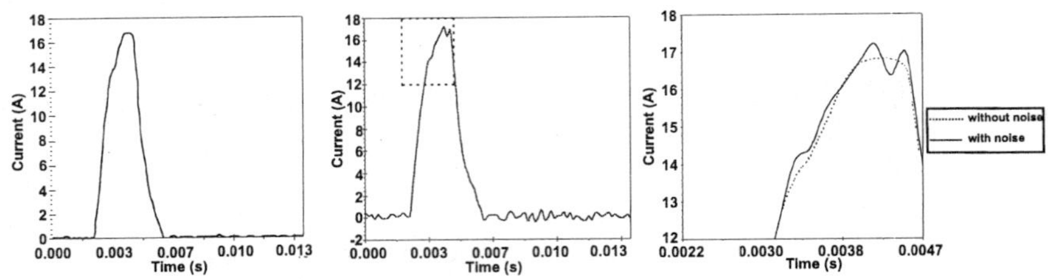

(a) (b) (c)

Fig. 14. SR motor phase current: (a) without noise, (b) with added noise (10% of maximum) and (c) expanded comparison of current waveforms, with and without noise.

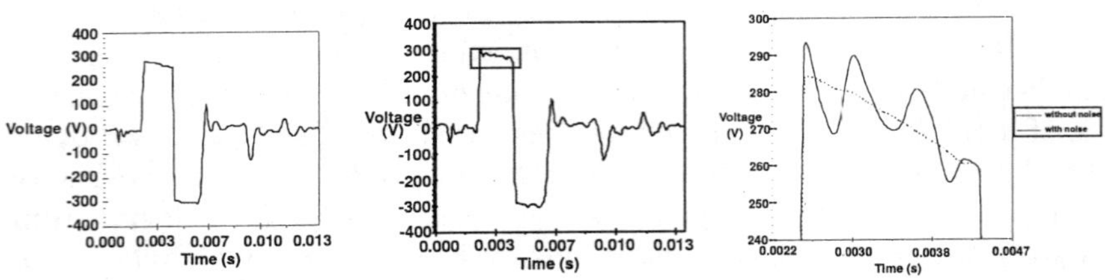

(a) (b) (c)

Fig. 15. SR motor phase voltage: (a) without noise, (b) with added noise (10% of maximum), (c) expanded comparison of voltage waveforms, with and without noise.

The measured and indirectly estimated rotor angle for this test is shown in Figure 16. The *average* amplitude of the error between the measured and estimated angle in this test is 1.24 degrees, whilst the *maximum* amplitude of error is 3.97 degrees. These values represent 2.06% and 6.6% of one electrical cycle of 60 degrees respectively. Note that the maximum angle estimation error is less than the 10% input signal error. Hence, the error in the estimated angle is not high relative to the input error, and operation of the motor is still possible. Therefore the second requirement of a robust algorithm, as described above, has been satisfied in this case.

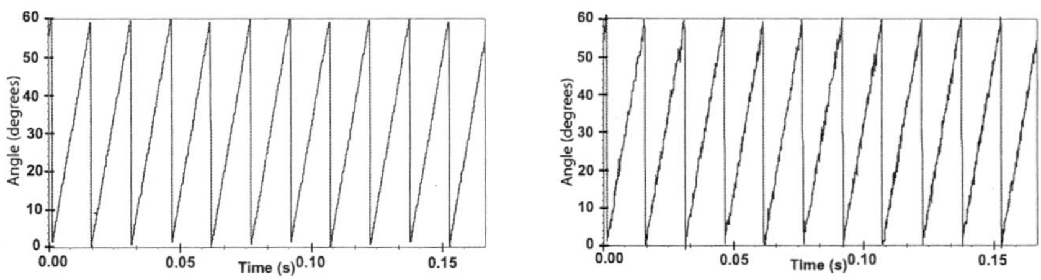

(a) (b)

Fig. 16. Ten percent of maximum amplitude noise: (a) measured rotor angle and (b) estimated angle.

For comparison, it can be recalled from the results shown in Section 2.1 for non-fuzzy methods of position estimation, that for a 10% error in the current feedback input the corresponding *minimum* position errors ranged from 2.8% to 15.6% depending on the flux linkage amplitude. Furthermore a 10% error in the flux linkage feedback led to *minimum* position estimation errors ranging from 5.67% to 63.87%. Additionally, it was seen that the error increased significantly in other non-optimum regions in the (ψ, i, θ) plane. As there are simultaneous errors in both the flux linkage and current feedback in this test, the total error will be a combination of the two errors as expressed in (11).

Therefore, it can be seen that the use of fuzzy sets improves the robustness of position estimation, due to the fact that the *average* position estimation error using the fuzzy model is of the same order as the *minimum* error using the non-fuzzy method, and the maximum error of the fuzzy method is significantly lower.

Further experiments were carried out to test the scheme at higher noise levels, and to see the operational limits of the current system. The level of noise imposed on the motor waveforms of current and voltage was increased to 20% of the input measured signals. In Figure 17 the estimated angle from the angle estimation algorithm is shown for this experiment (the measured angle was shown above in Figure 16). It can be seen that the general trajectory of the rotor angle has been retained in this test, even for this high level of noise. The average angle estimation error is 2.80 degrees (4.67%), whilst the peak error is 9.92 degrees, or 16.5% of the electrical angle cycle of 60 degrees, which is lower than the maximum input signal error amplitude. Furthermore, at no point in this test has the large deviation of the input current and voltage data caused a catastrophic breakdown of the algorithm, and thus the third requirement of a robust algorithm has also been satisfied.

10.2 Fuzzy Predictive Filters with Non-Gaussian Noise

To demonstrate the effectiveness of the fuzzy predictive filters, a test is shown for the case when the feedback signals are imposed with high amplitude im-

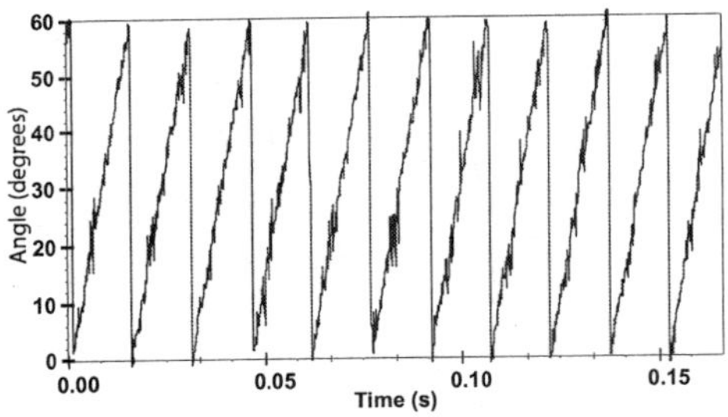

Fig. 17. Twenty percent noise: estimated angle.

pulsive noise. As mentioned above, the practical SR motor drive often has the problem of high amplitude impulse type noise caused by switching or commutation of high amplitude currents in the inverter circuit. The commutated current waveforms have short rise and fall times, and hence these waveforms contain significant amounts of energy at high frequencies. This radiated energy can be transmitted through parasitic stray capacitances to the control, interface, and measurement circuitry. Some parasitic capacitances will always exist between the high power inverter circuits, the high and low voltage side of the opto-isolation circuits, the low and high voltage side of the current and voltage measuring circuits, and the low voltage control and A/D converter circuits.

The characteristic feature of this generated noise is that it can have high amplitude during the switching of a power device. However, this noise is only seen during the switching instant. Therefore, the coupled noise in the control and current and voltage measurement circuits may have high amplitude, but be transient in nature. This type of high amplitude impulsive type noise is defined as non-Gaussian noise, and presents problems for practical systems because it is difficult to model mathematically and suppress efficiently. However, the fuzzy predictive filters of flux linkage and angle as described above were developed to lower the effect of non-Gaussian noise for the practical operation of the sensorless position estimation scheme.

In Figure 18, a demonstration of the fuzzy predictive filter's ability is shown. These figures show estimated flux linkage and angle derived from experimentally measured waveforms of current and voltage using the previously described motor drive running at 670 r/min. It can be seen in the figures that high amplitude error pulses occur in both the estimated flux linkage waveform and the estimated rotor position waveform. In order to obtain experimental consistency, these random pulses were added into the estimated signals.

In Figure 18 (a), the waveform of the flux linkage estimated from the measured current and voltages are shown, with a triangle representing each point where the flux is estimated from the current and voltage measurements. In the figure, a flux linkage waveform with high noise error can also be seen.

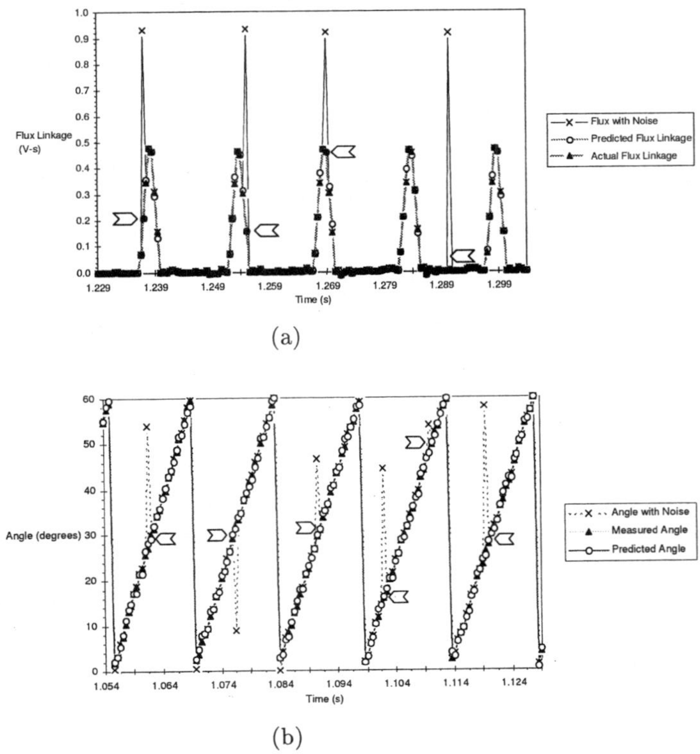

Fig. 18. Error elimination ability of FL-based predictors: a) Flux Linkage. b) Rotor Position. The points with arrows highlight the iterations where the predicted value can be used instead of the estimated angle to lessen the effect of switching noise.

In this test, the estimated flux linkage waveform with error is input to the flux linkage predictive filter instead of the actual estimated flux. When this waveform is input to the flux predictor it can be seen that the points with high level noise have effectively been replaced by predicted values. It should be noted that if the predictor could not remove the erroneous value, then due to the operation of integration, all future values of the estimated flux linkage would carry this error.

In Figure 18 (b), the waveforms are shown of the measured encoder angle and the estimated angle with impulsive type noise (at different test times, but with the same conditions as the flux linkage test). In the results, a triangular point shows the measured encoder positions at each sample time. In this test, the estimated angle that has been corrupted with high amplitude noise pulses is input to the angle predictor. It can be seen that the noise in the estimated angle is effectively reduced in the filtered angle value. The filtered value can thus be used instead of the estimated value to reduce the effect of switching noise.

The very high error pulses would normally lead to a controller error every time the error pulse occurred due to a high rotor position error. This could cause the controller to change the switching patterns of the motor phases and thus change the switching state of the inverter power switches. Hence, this would lead to erroneous switching of the motor voltage that may cause greater harmonics in the motor waveforms, increased torque ripple, and audible noise. In addition, there would be higher losses due to the motor phases being unnecessarily commutated, due to inverter losses and heating from the power devices being switched in error.

The results above have shown that when the fuzzy logic based predictive filters of flux linkage and angle are used, the high error pulses from sources such as switching noise are effectively eliminated, which leads to a more robust and stable motor drive operation.

11 Conclusion

In most applications where motor drives are used, the reliability of the drive is of utmost concern. This is particularly the case for some applications of the SR motor drive, such as in aerospace type applications [31], where the reliability and robustness, such as the ability to operate when one or more phases fail, are the main reasons for the choice of the motor type. Although it is commonly stated that position sensors may have the effect of reducing the reliability of the motor, position sensorless control methods may also reduce the system reliability unless their robustness in practical motor drive operation is proven, or at least thoroughly understood.

In this Chapter, it was shown using theoretical explanations and experimentally based results, that the fuzzy logic based angle estimation scheme has a high level of robustness and reliability, and is thus well suited for a wide range of practical systems. Results were detailed which showed that the scheme could successfully and reliably operate under erroneous operating conditions that are commonly found in motor drive environments such as:

1. Measurement errors due to noise.
2. Low bandwidth motor waveform measurements (e.g., due to slow A/D converters).
3. Flux Linkage estimation errors, which may be caused by:
   (a) Measurement errors.
   (b) Noise.

In order to compare the results of the fuzzy logic based method with non-fuzzy angle estimation, analytical expressions were given which described the effect of current and flux linkage input errors on the angle estimation accuracy of non-fuzzy schemes. Then the position estimation errors were calculated for non-fuzzy estimation. Hence it was confirmed that the average and maximum position estimation errors of the fuzzy logic based estimation scheme due to feedback signal error, were low compared to the non-fuzzy estimation methods.

The ability of the FL based method to cope with high levels of noise derives from the systems ability of fuzzifying the input signals and then processing with fuzzy linguistic rules. With this system, an input with noise will normally be able to be partially a member of the same membership function that the error free signal would be. Therefore, the same rules can be triggered by noisy and noise free signals, depending on the level of the noise. Hence, using the benefits of fuzzy reasoning rotor position can be successfully and reliably estimated under even high error and noise conditions in practical SR drives.

References

1. A. D. Cheok and N. Ertugrul, "Sensorless rotor position detection techniques in switched reluctance motor drives," in *Australasian Universities Power Engineering Conference*, Perth, Australia, pp. 84–89, 1995.
2. W. F. Ray and I. H. Al-bahadly, "Sensorless methods for determining the rotor position of switched reluctance motors," in *Proc. 5th European Conf. on Power Electronics and Applications*, pp. 7–13, 1993.
3. P. Acarnley, E. D. French, and I. H. Al-bahadly, "Position estimation in switched reluctance drives," in *Proc. 7th European Conf. on Power Electronics and Applications*, vol. 3, pp. 765–770, 1995.
4. I. Husain and M. Ehsani, "Error analysis in indirect rotor position sensing of switched reluctance motors," *IEEE Trans. Ind. Electron.*, vol. 41, no. 3, pp. 301–307, 1994.
5. S. Rehman and D. G. Taylor, "Issues in position estimation of sr motors," in *IEEE Power Electronics Specialists Conf. Rec.*, pp. 337–343, 1996.
6. T. Williams and R. Carter, "Measurement of machine inductances using an operational amplifier integrator," *International Journal of Electrical Engineering Education*, vol. 10, pp. 177–181, 1972/73.
7. A. D. Cheok and N. Ertugrul, "High robustness and reliability of a fuzzy logic based angle estimation algorithm for practical switched reluctance motor drives," in *IEEE Power Electronics Specialists Conf. Rec.*, Fukuoka, Japan, pp. 1302–1308, 1998.
8. J. Mendel, "Fuzzy logic systems for engieering: a tutorial," *Proceedings of the IEEE*, vol. 83, pp. 345–377, March 1995.
9. J. L. Castro, "Fuzzy logic controllers are universal approximators," *IEEE Trans. Systems Man and Cyber.*, vol. 25, pp. 629–635, April 1995.
10. G. C. D. Sousa and B. K. Bose, "A fuzzy set theory based control of a phase controlled converter dc machine drive," *IEEE Trans. Industry Applications*, vol. 30, no. 1, pp. 34–44, 1994.
11. B. Kosko, *Fuzzy Engineering.* Upper Saddle River: NJ: Prentice Hall, 1996.
12. R. Lea and V. Kreinovich, "Intelligent control makes sense even without expert knowledge: as explanation," *Int. Journal Reliable Computing*, 1995.
13. L. X. Wang and J. M. Mendel, "Generating fuzzy rules by learning from examples," in *Proc. 6th IEEE Int. Symp. on Intelligent Control*, New York, NY, pp. 263–268, 1991.

14. A. D. Cheok and N. Ertugrul, "High robustness and reliability of fuzzy logic based position estimation for sensorless switched reluctance motor drives," *IEEE Transactions on Power Electronics*, vol. 15, pp. 319–334, 2000.
15. P. J. C. Branco and J. A. Dente, "An experiment in automatic modeling an electrical drive system using fuzzy logic," *IEEE Trans. Syst. Man Cyber. Pt. C*, vol. 28, no. 2, pp. 254–262, 1998.
16. G. C. Mouzouris and J. M. Mendel, "Dynamic non-singleton fuzzy logic systems for nonlinear modeling," *IEEE Trans. Fuzzy Syst.*, vol. 5, no. 2, pp. 199–208, 1997.
17. J. van den Berg and D. Ettes, "Representation and learning capabilities of additive fuzzy systems," in *Proc. 1998 IEEE Int. Conf. Intelligent Engineering Systems*, pp. 121–126, 1998.
18. E. J. Wegman, S. G. Schwartz, and J. B. Thomas, *Topics in Non-Gaussian Signal Processing.* New York: Academic Press, 1989.
19. J. S. R. Jang, "ANFIS: adaptive-network-based fuzzy inference system," *IEEE Trans. Syst. Man Cyber.*, vol. 23, no. 3, pp. 665–684, 1993.
20. G. J. Deboeck, *Neural, Genetic, and Fuzzy System for Chaotic Finacial Markets.* New York: Wiley, 1994.
21. A. D. Cheok and N. Ertugrul, "A model free fuzzy logic based rotor position sensorless switched reluctance motor drive," in *IEEE Ind. Appl. Soc. Annual Meeting*, pp. 76–83, 1996.
22. N. Ertugrul and A. D. Cheok, "Indirect angle estimation in switched reluctance motor drives using fuzzy logic based predictor/corrector," in *IEEE Power Electronics Specialists Conf. Rec.*, pp. 845–851, 1998.
23. S. Valiviita and O. Vainio, "Delayless differentiation algorithm and its efficient implementation for motion control applications," in *IEEE Instrumentation and Measurement Technology Conference*, St. Paul, MN, pp. 881–886, 1998.
24. S. J. Ovaska and S. Valiviita, "Angular acceleration measurement: a review," *IEEE Transactions on Instrumentation and Measurement*, vol. 47, no. 5, pp. 1211–17, 1998.
25. P. R. Belanger, "Estimation of angular velocity and acceleration from shaft encoder measurements," in *Proc. 1992 IEEE Int. Conf. on Robotics and Automation*, Nice, France, pp. 585–92, 1992.
26. R. F. Sutton, *Modeling Human Operators.* New York: Wiley, 1991.
27. C. A. Ferreira, S. R. Jones, and B. T. Dragner, "Design and implementation of a five hp switched reluctance fuel lube pump motor drive for a gas turbine," *IEEE Transactions on Power Electronics*, vol. 10, pp. 55–60, 1995.
28. P. J. Huber, *Robust statistics.* New York: Wiley, 1981.
29. R. N. Dave and R. Krishnapuram, "Robust clustering methods: A unified view," *IEEE Trans. Fuzzy Syst.*, vol. 5, no. 2, pp. 270–293, 1997.
30. F. R. Hampel, E. M. Ponchotti, P. J. Rowsseuw, and W. A. Stahel, *Robust statistics: The approach based on influence functions.* New York: Wiley, 1986.
31. E. Richter, "Switched reluctance machines for high performance operation in a harsh environment," in *International Conference on Electrical Machines*, Cambridge, MA, pp. 18–24, 1990.

On the Use of Fuzzy Logic to Control Paralleled DC-DC Converters

Bogdan Tomescu* and Hugh F. VanLandingham†

* Space Systems/Loral

† Virginia Polytechnic Institute and State University

Keywords: *DC-DC converters, power electronics, fuzzy logic control, current sharing modules, master-slave structure, paralleled converters, boost mode converters, soft computing, Lyapunov stability, modern control formalism.*

1 Introduction

Paralleled DC-DC converter systems and, more generally, distributed power converter systems represent a major trend in today's power electronics research, [1]. Whereas at one time the stand-alone converter was the main feature, the single converter is now becoming an elementary cell for the next generation of power systems. The present focus on high-density integration and packaging will be naturally followed by a requirement for more advanced control methods. In this context, paralleled DC-DC converter systems are potential candidates for modern neuro-fuzzy control techniques due to their complex, nonlinear behavior, particularly for large load and line variations, [2]. Almost all existing control methods are based on simplified averaged linear models, [3] – [6]; and, even in the case of a single unit, it has been shown that the large-signal response deviates significantly from the predictions of a small-signal model, [7] – [9].

Several attempts have been made to apply modern control theory to the control of DC-DC converters, [10] – [14], including fuzzy logic control, [15] – [18]. A heuristic fuzzy logic approach becomes even more practical in the case of active control, e.g. master-slave control, for parallel operation of converters because of the difficulty in obtaining practical models for systems having a large number of feedback loops, [2]. The theme of this chapter is that the complexity of current sharing power converters fully justifies the use of the advanced control methods of soft computing (neural, fuzzy and evolutionary), whereas individual modules can, perhaps, be efficiently compensated with traditional techniques.

2 Improved Large Signal Performance of Paralleled DC-DC Converter Current Sharing Using Fuzzy Logic Control

This section demonstrates the implementation of a Master-Slave Control (MSC) fuzzy current sharing controller (CSC) with improved performance over a solidly developed classical design, such as in [6]. First, an improved PID compensator is introduced as an upgrade for the existing ones; then, the PID is employed as a "model" system in developing the rules for the fuzzy controller. In contrast, present control schemes make use of a PI or of a lead-lag with integrator controller for the active current sharing loop, [5] - [6], [19]-[23]. Fig. 2.1 shows the generic MSC scheme with converter #1 as master.

Because of the highly interconnected system, complex modeling does not provide a practical engineering way to consider both dynamic response and stability. Stability analysis is discussed later in the chapter. For improved transient performance, fuzzy control appears to have a strong potential. Present implementations can also be regarded as educated heuristic approaches since even after some analysis, they still use an iterative simulation-experimentation process to choose the final, stable, design, [19]-[21]. A first improvement that can be done is to upgrade the lead-lag plus integrator controller in Fig. 2.2 is to go to a PID, since an additional zero introduced in the low frequency range will shift up the phase thereby allowing higher bandwidth (BW) while retaining sufficient phase margin (PM). Considering eventual noise problems, a better design is found if using a trade-off between bandwidth and noise immunity, depending on the application.

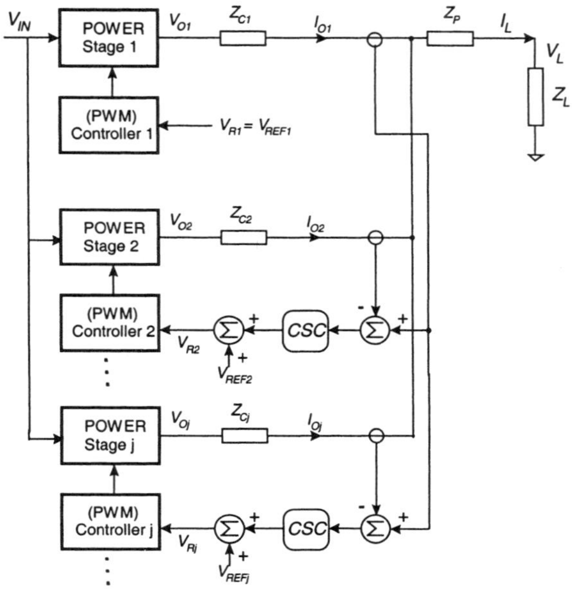

Fig. 2.1. Control scheme used for paralleling *N* converter modules with MSC.

A common current sharing controller (CSC) that uses a lead-lag cell plus an integrator, is shown in Fig. 2.2.

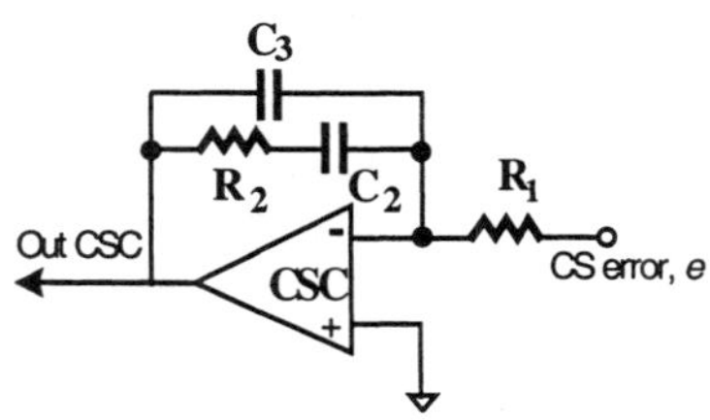

Fig. 2.2. Classical CSC architecture.

The small signal loop gain used in a three-voltage mode buck converter system, [2], [6], employing the MSC topology in Fig. 2.1 and CSC in Fig. 2.2, is reproduced here in Fig. 2.3 together with the PID improved loop gain. The system is supplied from 12 V input voltage, delivers a 5 V output voltage at 15 A load for each converter (45A nominal load for the whole system) and was simulated using PSpice. The bandwidth of the classical CSC is 29 KHz, with a PM of 47° and a Gain Margin (GM) of 14 dB. Its transfer function and parameters are:

$$H(s) = \omega_i \frac{1}{s} \cdot \frac{1 + \frac{s}{\omega_z}}{1 + \frac{s}{\omega_p}},$$

$$\omega_i = \frac{1}{R_1 \cdot (C_2 + C_3)} = 1200 \sec^{-1}, \tag{1}$$

$$\omega_z = \frac{1}{R_2 \cdot C_2} = 2 \cdot \pi \cdot 50 \sec^{-1},$$

$$\omega_p = \frac{C_2 + C_3}{R_2 \cdot C_2 \cdot C_3} = 2 \cdot \pi \cdot 5 \cdot 10^4 \sec^{-1}.$$

The PID design has been adjusted for a BW of 100 kHz, a reasonable trade-off between noise immunity and speed, with a PM also of 47° and infinite GM.

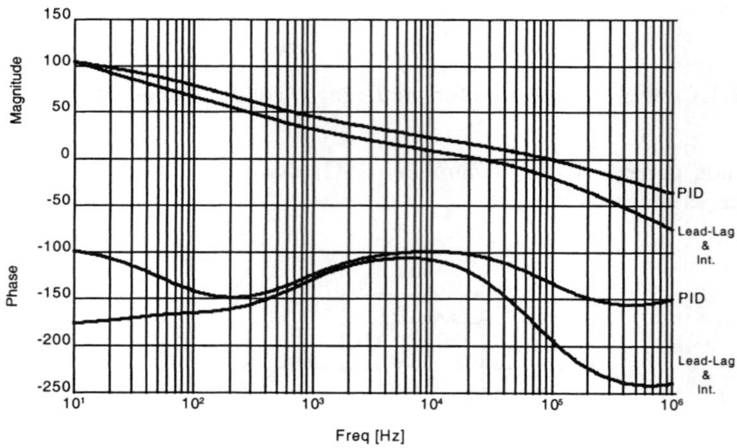

Fig. 2.3. Small signal loop gains for a classical design and improved PID design.

The PID controller is described by a transfer function of the form

$$H(s) = k_p + \frac{k_i}{s} + k_d \cdot \frac{s}{s + \omega_p} \tag{2}$$

with k_p=20, k_i=1 and k_d=0.1. The implementation of the derivative part includes the high frequency pole for realizability, chosen to be the same value as in (1); a considerable improvement in the dynamic performance can be seen.

The design methodology used for the fuzzy logic controller (FLC) is to use the PID improved results as the "expert" in developing the inference rules. Figure 2.4 shows a generic controller that uses the current share error, *e*, into a PD fuzzy scheme combined with an integral component needed for the zero steady state error.

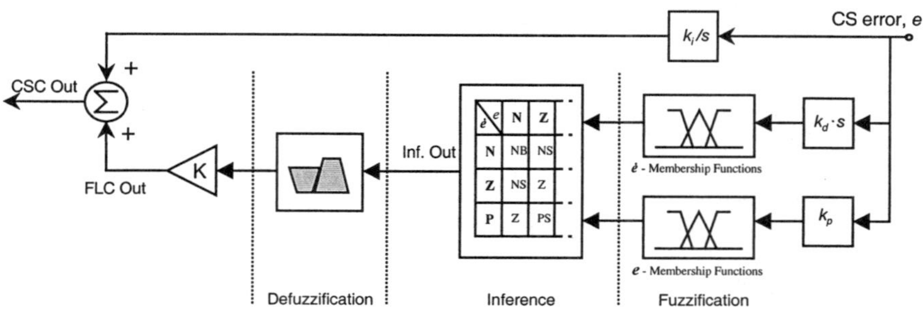

Fig. 2.4. Fuzzy logic implementation of CSC.

A three range partition, negative (N), zero (Z) and positive (P), can be chosen for the inputs, *e* and $\dot{e}$. A more refined five-range partition, negative-big (NB), negative-small (NS), zero (Z), positive-small (PS) and positive-big (PB) is used for the output. Based on the usual action of a PD cell, by looking at its output range as a function of the input range from (2), a rule-based inference is chosen for each consequent; the membership functions and respective rules are shown in Fig. 2.5.

We employ a product inference engine, singleton fuzzifier and center average defuzzifier, [25]. An example of establishing the fuzzy rule values based on the PID expert in (2) is as follows. Since the ratio between the gains of the proportional and derivative part is 200 (k_p/k_d), for a range of *e* between -1 and 1, we would need to consider a range between -200 and 200 for $\dot{e}$. Therefore, if *e* is N (-1) and $\dot{e}$ is N (-200), then the PID would roughly give a control value of -40, hence the value NB of -40 for the consequent. The other rules are inferred similarly. Further optimization in partitioning the input spaces as well as using neuro-fuzzy techniques in an adaptive grid of controllers, [24]-[25], can be considered. Some simulation results will now be presented.

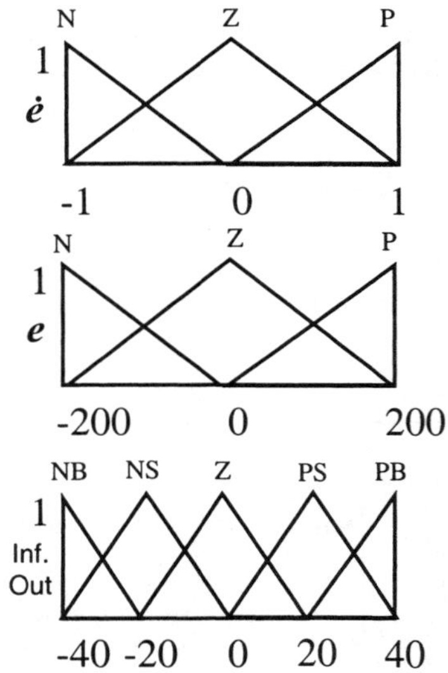

e \ e	N	Z	P
N	NB (-40)	NS (-20)	Z (0)
Z	NS (-20)	Z (0)	PS (20)
P	Z (0)	PS (20)	PB (40)

Rule Table of the Fuzzy Controller
(Output Singletons)

Fig. 2.5. Membership functions and rule-based inference for the fuzzy logic controller in Fig. 2.4.

Time domain simulations have been done on the previous mentioned system where one module has the nominal parameters - reference voltage, output filter values, internal compensator R and C values - and the parameters of the other two deviate from nominal within usual fabrication tolerances for these components, i.e., 1% to 5%. Comparative 50% and 90% nominal load step responses are given in Fig. 2.6 for the classical design in [6], Fig. 2.6 (a), the PID design, Fig. 2.6 (b) and the fuzzy controller, Fig. 2.6 (c).

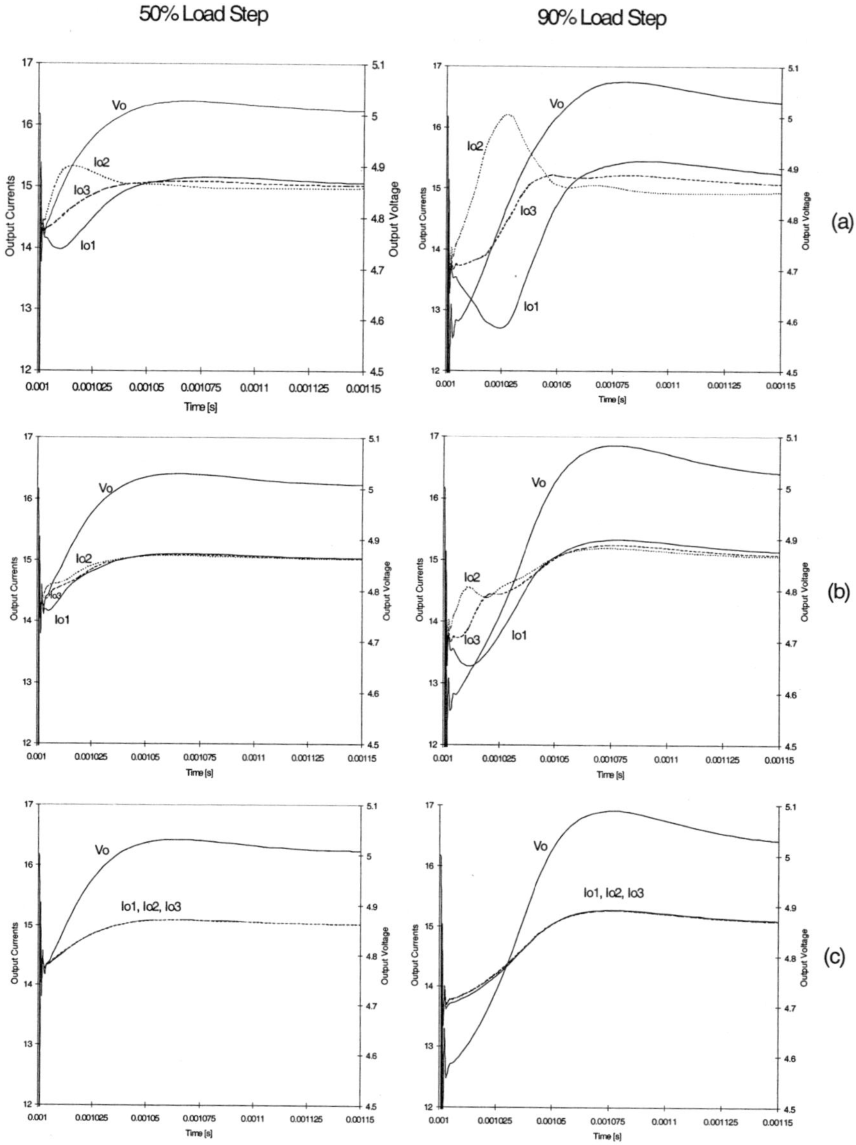

Fig. 2.6. Step load responses for 50% and 90% of the nominal load: (a)-System with the classical CSC in Fig. 2.2; (b)-System employing a PID CSC; (c)-System using a Fuzzy Logic CSC.

A substantial improvement in the currents overshoot/settling time as well as their equalization is noted. Almost identical responses for the two step load cases when using the fuzzy control approach, shows the validity of fuzzy control in dealing naturally with the nonlinear character of the system. The low performance index in the traditional cases is due to the fact that a conventional linear controller, designed only on a small signal model, is valid only near the operating point and not for the entire plant trajectory. The fuzzy approach gives also a simple practical way of overcoming modeling complexity by using heuristic judgment. It can also be noticed that for each step load, the voltage waveforms for the three controllers are very close since, in general, the output voltage dynamics are influenced only by the internal loops of the master converter, the slave converters exhibiting practically no voltage gain, [19]-[20].

The drawbacks of the fuzzy control could be a lower noise immunity as well as the need to ensure stability. The first problem is usually handled on an application specific basis either through conventional, analog or digital, filtering or by building filtering capabilities in the fuzzy controller itself, e.g. by using triangular or Gaussian fuzzifiers instead of the singleton fuzzifiers, [25]. For the second issue there are several usual paths, one being to use a hierarchical structure in which the fuzzy controller is allowed to take care of the dynamics and a supervisory stable Lyapunov or PID controller intervenes only to maintain stability. Another possibility would be to try to keep the trajectories on a Lyapunov descent path that ensures stability for the network, using the network energy as the Lyapunov function, [10]. Consider now the control effort involved. Figure 2.7 depicts the averaged duty cycle plots for the same cases shown in Fig. 2.6.

One can again see that the fuzzy controller gives the minimal needed effort to control the system current sharing. The duty cycle of the two slave converters, d_2 and d_3, never saturate in this case, Fig. 2.7 (c). In comparison, the use of the compensator in Fig. 2.2, Fig. 2.7 (a), is the slowest and requires the most effort. In all cases the master duty cycle saturates for a certain period since the master does not get back any information about current sharing and hence tries to use all its control range for output voltage regulation, as mentioned above.

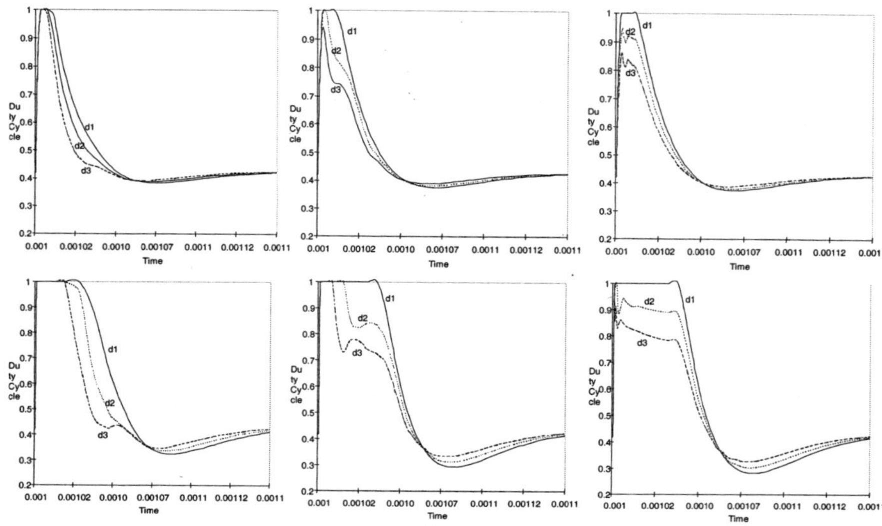

(a) (b) (c)

Fig. 2.7. CSC Control Effort (averaged duty cycles):

(a)-With the classical CSC in Fig. 2.2;

(b)-Employing a PID CSC; (c)-Using a Fuzzy Logic CSC

(upper plots)-50% Load Step; (lower plots)-90% Load Step.

Summarizing, a fuzzy logic controller for the master-slave current sharing loop of a paralleled DC-DC system has been introduced with considerably improved large signal performance over the presently employed (small signal) compensators, both in system response and control effort. The design uses a PID expert to derive the fuzzy inference rules, and simulation results show a good parameter insensitive transient response in load step responses of 50% and 90% of the nominal load. The technique benefits also from the heuristic approach to the problem that overcomes the complexity in modeling such systems and hence offers a practical engineering tool, amenable to both analog and digital implementations.

3 Large Signal, Disturbance and Robustness Formulation for DC-DC Converters

The basic DC-to-DC power conversion function of switching converters is achieved by repetitive switching between two linear networks consisting of storage elements, inductors and capacitors, [36]. Time averaging allows the switched, discontinuous, system to be approximated as a continuous, nonlinear, "large signal" model. Then linearization around the DC, steady state, operating point is applied in order to obtain an approximate, linear model. Thus, in the small signal analysis, all the results of the linear control systems can be applied, provided that the perturbation in the system is small enough. For large signal perturbations, so important for time domain performance, the linear models prove less accurate, especially for paralleled systems. In all present state space approaches, both the input voltage and the load are treated more as parameters than as disturbances. The real circuit, however, functions as a regulator and the loop dynamics have to adapt to a load step and a line impulse or step disturbance. Therefore, the control problem is one of disturbance rejection and robustness of a regulator or a system of paralleled regulators.

Related to this, and a source of nonlinearity, is the way the load comes into the system model. All previous state space approaches consider a resistive load, $i_o = v_o/R_{ld}$ and write v_o in terms of the states $(i_l \;\; v_c)'$ – inductor current and capacitor voltage. Hence, they obtain a model that has R_{ld} in the associated matrices and no i_o terms. But, in many cases, especially for distributed systems, the loads are other converters that are constant power loads, or when used in RF applications, the load is often a secondary, linear regulator. Therefore, working with the "unterminated" model, [14], [26], in which the output current is another variable, is a good way to approach the problem, but one should pose the control problem in a robustness and disturbance rejection formalism. So far, the unterminated model has only been detailed mainly in a transfer functions, *s* domain, analysis. In the next paragraphs, such a formalism is introduced using state space modeling and considering the large signal problem, since it is at that juncture that improved control strategies are needed, in order to deal with the nonlinear response under large load and line disturbances.

Without loss of generality, considering PWM CCM, any switching DC-DC converter, terminated to the resistive load R_{ld}, can be described by the state space equations for the two linear networks corresponding to the two switching stages. For example, considering the boost converter in Fig. 3.1, we have

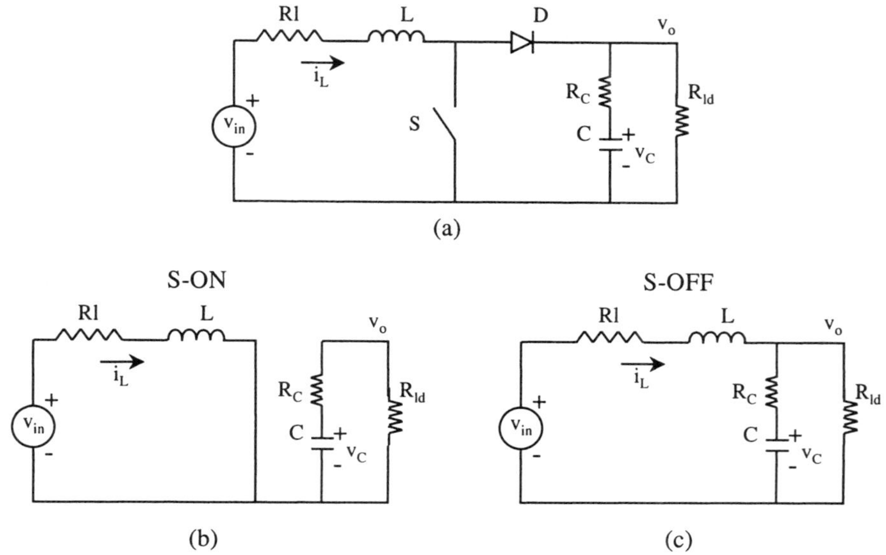

Fig. 3.1. Boost Converter in CCM using the resistive load terminated model.

When the switch is ON, during the interval $d{\cdot}T_s$, Fig. 3.1- (b):

$$\dot{\mathbf{x}} = \begin{bmatrix} \dot{i}_L \\ \dot{v}_c \end{bmatrix} = \begin{bmatrix} -\frac{R_L}{L} & 0 \\ 0 & -\frac{1}{C}\cdot\frac{1}{R_{ld}+R_C} \end{bmatrix} \cdot \begin{bmatrix} i_L \\ v_c \end{bmatrix} + \begin{bmatrix} \frac{1}{L} \\ 0 \end{bmatrix} \cdot v_{in} = \mathbf{A_1} \cdot \mathbf{x} + \mathbf{E}_{\mathbf{1x}}^{\mathbf{v_{in}}} \cdot v_{in}$$

$$v_o \equiv y = \begin{bmatrix} 0 & \frac{R_{ld}}{R_{ld}+R_C} \end{bmatrix} \cdot \begin{bmatrix} i_L \\ v_c \end{bmatrix} = \mathbf{C_1} \cdot \mathbf{x} \tag{3}$$

and when the switch is OFF, during the interval $(1\text{-}d){\cdot}T_s = d'{\cdot}T_s$, Fig. 3.1 - (c):

$$\dot{\mathbf{x}} = \begin{bmatrix} \dot{i}_L \\ \dot{v}_c \end{bmatrix} = \begin{bmatrix} -\frac{1}{L}\cdot\left(R_L + \frac{R_{ld}\cdot R_C}{R_{ld}+R_C}\right) & -\frac{1}{L}\cdot\frac{R_{ld}}{R_{ld}+R_C} \\ \frac{1}{C}\cdot\frac{R_{ld}}{R_{ld}+R_C} & -\frac{1}{C}\cdot\frac{1}{R_{ld}+R_C} \end{bmatrix} \cdot \begin{bmatrix} i_L \\ v_c \end{bmatrix} +$$

$$+ \begin{bmatrix} \frac{1}{L} \\ 0 \end{bmatrix} \cdot v_{in} = \mathbf{A_2} \cdot \mathbf{x} + \mathbf{E}_{\mathbf{2x}}^{\mathbf{v_{in}}} \cdot v_{in} \tag{4}$$

$$v_o \equiv y = \begin{bmatrix} \frac{R_{ld}\cdot R_C}{R_{ld}+R_C} & \frac{R_{ld}}{R_{ld}+R_C} \end{bmatrix} \cdot \begin{bmatrix} i_L \\ v_c \end{bmatrix} = \mathbf{C_2} \cdot \mathbf{x}$$

Almost all the modeling used today is based on the classical small signal analysis, [3], using the "straight line" approximation of the fundamental matrix, $\mathbf{e}^{\mathbf{A}\cdot t} \cong \mathbf{I} + \mathbf{A}\cdot t$, to obtain a unique, linear, averaged model, for the entire switching cycle. The same averaging process is employed to derive the discrete averaged model, [27], where it is still needed to have a single model for one whole switching cycle. The matrices of this model are obtained to be simply the average of each switching half-cycle model matrix, weighed with the duty cycle, D, at the DC operating point, $\mathbf{X_0}$, Y, V_{in}. This linear approximation works well as long as the perturbation is small so that one stays in the vicinity of the steady state operation point, correlated with the ratio between the equivalent corner frequency of the output filter and the switching frequency being much smaller than unity, $f_c/f_s << 1$.

One obtains thus first the continuous, nonlinear (bilinear) model

$$\begin{aligned} \dot{\mathbf{x}} &= \mathbf{A}\cdot\mathbf{x} + \mathbf{B}\cdot u + \mathbf{E}_{\mathbf{x}}^{\mathbf{v_{in}}}\cdot v_{in} + \Delta\mathbf{A}\cdot\mathbf{x}\cdot u + \Delta\mathbf{E}_{\mathbf{x}}^{\mathbf{v_{in}}}\cdot v_{in}\cdot u \\ v_o &\equiv y = \mathbf{C}\cdot\mathbf{x} + \mathbf{D}\cdot u + \Delta\mathbf{C}\cdot\mathbf{x}\cdot u \end{aligned} \tag{5}$$

with

$$\begin{aligned} &\mathbf{A} = \mathbf{A_1}\cdot D + \mathbf{A_2}\cdot(1-D), \quad \mathbf{B} = \Delta\mathbf{A}\cdot\mathbf{X}_{\mathbf{0}r} + \\ &+ \Delta\mathbf{E}_{\mathbf{x}}^{\mathbf{v_{in}}}\cdot V_{in}, \quad \mathbf{C} = \mathbf{C_1}\cdot D + \mathbf{C_2}\cdot(1-D) \\ &\mathbf{D} = \Delta\mathbf{C}\cdot\mathbf{X_0}, \quad \mathbf{E}_{\mathbf{x}}^{\mathbf{v_{in}}} = \mathbf{E}_{\mathbf{1x}}^{\mathbf{v_{in}}}\cdot D + \mathbf{E}_{\mathbf{2x}}^{\mathbf{v_{in}}}\cdot(1-D), \\ &\mathbf{X}_{\mathbf{0}r} = -\mathbf{A}^{-1}\cdot\mathbf{E}_{\mathbf{x}}^{\mathbf{v_{in}}}\cdot V_{in} \end{aligned} \tag{6}$$

and Δ denoting the difference between the two quantities for the first and respectively second switching cycle. For the boost converter above, the detailed model resulting from (5) would be

$$\mathbf{X}_{\mathbf{0}r} = \begin{bmatrix} \frac{R_{ld}}{R'^2}\cdot V_{in} \\ \frac{1}{1-D}\cdot V_{in}\cdot\frac{(1-D)^2\cdot R_{ld}}{R'} \end{bmatrix} \tag{7}$$

$$\dot{\mathbf{x}} = \begin{bmatrix} -\frac{R_L + (R_{ld} \| R_C)(1-D)}{L} & -\frac{R_{ld} \cdot (1-D)}{(R_{ld}+R_C) \cdot L} \\ \frac{R_{ld} \cdot (1-D)}{(R_{ld}+R_C) \cdot C} & -\frac{1}{(R_{ld}+R_C) \cdot C} \end{bmatrix} \cdot \mathbf{x} +$$

$$+ \begin{bmatrix} \frac{1}{L} \cdot \frac{R_{ld}}{R'} \cdot \frac{(1-D) \cdot R_{ld} + R_C}{R_{ld}+R_C} \\ -\frac{R_{ld}}{R'} \cdot \frac{1}{(R_{ld}+R_C) \cdot C} \end{bmatrix} \cdot V_{in} \cdot u + \begin{bmatrix} \frac{1}{L} \\ 0 \end{bmatrix} \cdot v_{in} +$$

$$+ \begin{bmatrix} \frac{R_{ld} \| R_C}{L} & \frac{R_{ld}}{(R_{ld}+R_C) \cdot L} \\ -\frac{R_{ld}}{(R_{ld}+R_C) \cdot C} & 0 \end{bmatrix} \cdot \mathbf{x} \cdot u \qquad (8)$$

$$y = \begin{bmatrix} (R_{ld} \| R_C) \cdot (1-D) & \frac{R_{ld}}{(R_{ld}+R_C) \cdot C} \end{bmatrix} \cdot \mathbf{x} - V_{in} \cdot \frac{R_{ld} \| R_C}{R'} \cdot u +$$

$$+ \begin{bmatrix} -(R_{ld} \| R_C) & 0 \end{bmatrix} \cdot \mathbf{x} \cdot u$$

with

$$R' = R_L + (R_{ld} \| R_C) \cdot D \cdot (1-D) + (1-D)^2 \cdot R_{ld} . \qquad (9)$$

By keeping the linear terms, the small signal model can be obtained, and one can calculate the salient Laplace transfer functions and use Bode plot techniques for stability analysis. This is useful and motivated by the practical requirements in the industry; in addition, the vast majority of commercially available network analyzers offer linear gain and impedance measurements on logarithmic scales. When the nonlinear, large signal model was first considered [7], it was shown that the converter can substantially deviate from the predictions of the small signal model. The averaged converter model comes out as bilinear and, depending on the design, may have multiple equilibrium points, both real and virtual. Therefore it is very important to consider the nonlinear, large signal performance as a strong design goal. The next paragraph offers a proof using another approach that employs series expansion and links the model to the ideas proposed in this study.

Consider now the same boost converter above, but using the unterminated model, where the output current is the load disturbance variable, Fig. 3.2.

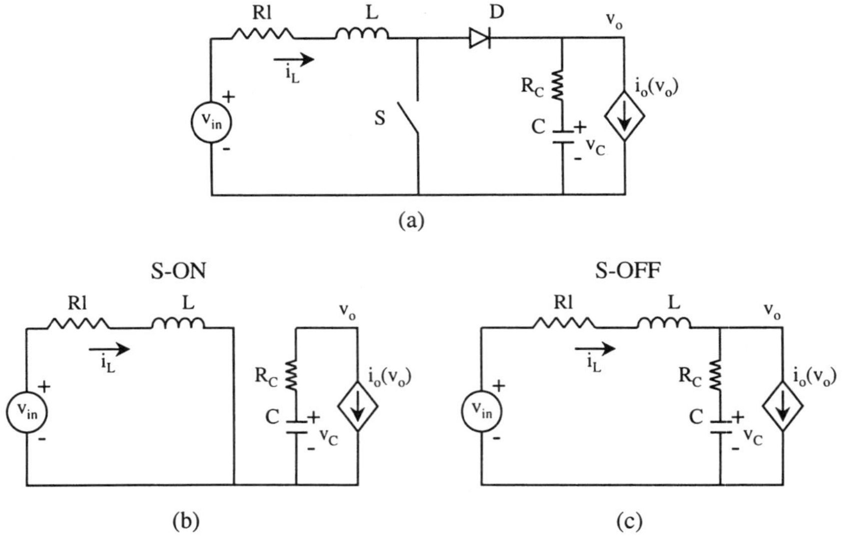

Fig. 3.2. Boost Converter in CCM using the Unterminated Model.

Continuing the case in the previous paragraph, the model equations are

$$\begin{bmatrix} \dot{i}_L \\ \dot{v}_C \end{bmatrix} = \begin{bmatrix} -\frac{R_l}{L} & 0 \\ 0 & 0 \end{bmatrix} \cdot \begin{bmatrix} i_L \\ v_C \end{bmatrix} + \begin{bmatrix} \frac{1}{L} \\ 0 \end{bmatrix} \cdot v_{in} + \begin{bmatrix} 0 \\ -\frac{1}{C} \end{bmatrix} \cdot i_o = \mathbf{A_1} \cdot \mathbf{x} + \mathbf{E_{1x}^{v_{in}}} \cdot v_{in} + \mathbf{E_{1x}^{i_o}} \cdot i_o$$

$$v_o = \begin{bmatrix} 0 & 1 \end{bmatrix} \cdot \begin{bmatrix} i_L \\ v_C \end{bmatrix} - R_C \cdot i_o = \mathbf{C_1} \cdot \mathbf{x} + \mathbf{E_{1y}^{i_o}} \cdot i_o$$

(10)

for the ON time and

$$\begin{bmatrix} \dot{i}_L \\ \dot{v}_C \end{bmatrix} = \begin{bmatrix} -\frac{R_l + R_C}{L} & -\frac{1}{L} \\ \frac{1}{C} & 0 \end{bmatrix} \cdot \begin{bmatrix} i_L \\ v_C \end{bmatrix} + \begin{bmatrix} \frac{1}{L} \\ 0 \end{bmatrix} \cdot v_{in} + \begin{bmatrix} \frac{R_C}{L} \\ -\frac{1}{C} \end{bmatrix} \cdot i_o = \mathbf{A_2} \cdot \mathbf{x} + \mathbf{E_{2x}^{v_{in}}} \cdot v_{in} + \mathbf{E_{2x}^{i_o}} \cdot i_o$$

$$v_o = \begin{bmatrix} R_C & 1 \end{bmatrix} \cdot \begin{bmatrix} i_L \\ v_C \end{bmatrix} - R_C \cdot i_o = \mathbf{C_2} \cdot \mathbf{x} + \mathbf{E_{2y}^{i_o}} \cdot i_o$$

(11)

for the OFF time. The same equations can be written in a compact form,

$$\begin{aligned}\dot{\mathbf{x}} &= \mathbf{A}_1 \cdot \mathbf{x} + \begin{bmatrix} \mathbf{E}_{1x}^{v_{in}} & \mathbf{E}_{1x}^{i_o} \end{bmatrix} \cdot \begin{bmatrix} v_{in} \\ i_o \end{bmatrix} = \mathbf{A}_1 \cdot \mathbf{x} + \mathbf{E}_{1x} \cdot \mathbf{w} \\ v_o \equiv y_o &= \mathbf{C}_1 \cdot \mathbf{x} + \begin{bmatrix} \mathbf{0} & \mathbf{E}_{1y}^{i_o} \end{bmatrix} \cdot \begin{bmatrix} v_{in} \\ i_o \end{bmatrix} = \mathbf{C}_1 \cdot \mathbf{x} + \mathbf{E}_{1y} \cdot \mathbf{w}\end{aligned} \tag{12}$$

and respectively

$$\begin{aligned}\dot{\mathbf{x}} &= \mathbf{A}_2 \cdot \mathbf{x} + \begin{bmatrix} \mathbf{E}_{2x}^{v_{in}} & \mathbf{E}_{2x}^{i_o} \end{bmatrix} \cdot \begin{bmatrix} v_{in} \\ i_o \end{bmatrix} = \mathbf{A}_2 \cdot \mathbf{x} + \mathbf{E}_{2x} \cdot \mathbf{w} \\ v_o \equiv y_o &= \mathbf{C}_2 \cdot \mathbf{x} + \begin{bmatrix} \mathbf{0} & \mathbf{E}_{2y}^{i_o} \end{bmatrix} \cdot \begin{bmatrix} v_{in} \\ i_o \end{bmatrix} = \mathbf{C}_2 \cdot \mathbf{x} + \mathbf{E}_{2y} \cdot \mathbf{w}\end{aligned} \tag{13}$$

Here the disturbance vector, $\mathbf{w} = [v_{in}\ i_o]$, contains both the input voltage and the output (load) current, which is assumed constant over one switching cycle, a very good practical approximation. Since the output filter requirement, $f_c/f_s<<1$, for using the straight line approximatione mentioned above holds for most present designs, we obtain the solutions

$$\mathbf{x}(t) = (\mathbf{I} + \mathbf{A}_1 \cdot t) \cdot \mathbf{x}(0) + \mathbf{E}_{1x} \cdot \mathbf{w} \cdot t \tag{14}$$

$$y(t) = (\mathbf{I} + \mathbf{C}_1 \cdot t) \cdot \mathbf{x}(0) + \mathbf{E}_{1y} \cdot \mathbf{w} \cdot t \tag{15}$$

for $0 \le t < d \cdot T_s$ and

$$\mathbf{x}(t) = [\mathbf{I} + \mathbf{A}_2 \cdot (t - d \cdot T_s)] \cdot \mathbf{x}(d \cdot T_s) + \mathbf{E}_{2x} \cdot \mathbf{w} \cdot (t - d \cdot T_s) \tag{16}$$

$$y(t) = [\mathbf{I} + \mathbf{C}_2 \cdot (t - d \cdot T_s)] \cdot \mathbf{x}(d \cdot T_s) + \mathbf{E}_{2y} \cdot \mathbf{w} \cdot (t - d \cdot T_s) \tag{17}$$

for $d \cdot T_s \le t < T_s$. The combination of equations (14) – (17) yields the following expression for $x(T_s)$ and $y(T_s)$, respectively:

$$\mathbf{x}(T_s) = [\mathbf{I} + T_s \cdot (d \cdot \mathbf{A}_1 + d' \cdot \mathbf{A}_2)] \cdot \mathbf{x}(0) + T_s \cdot (d \cdot \mathbf{E}_{1x} + d' \cdot \mathbf{E}_{2x}) \cdot \mathbf{w} \tag{18}$$

$$y(T_s)=[\mathbf{I}+T_s\cdot(d\cdot\mathbf{C}_1+d'\cdot\mathbf{C}_2)]\cdot\mathbf{x}(T_s)+T_s\cdot(d\cdot\mathbf{E}_{1\mathbf{y}}+d'\cdot\mathbf{E}_{2\mathbf{y}})\cdot\mathbf{w} \tag{19}$$

After $n+1$ switching intervals, we have

$$\begin{aligned}\mathbf{x}((n+1)\cdot T_s)=\mathbf{x}_{\mathbf{n+1}}=[\mathbf{I}+T_s\cdot(d_n\cdot\mathbf{A}_1+d_n'\cdot\mathbf{A}_2)]\cdot\mathbf{x}_n+\\+T_s\cdot(d_n\cdot\mathbf{E}_{1\mathbf{x}}+d_n'\cdot\mathbf{E}_{2\mathbf{x}})\cdot\mathbf{w}_{\mathbf{n}}\end{aligned} \tag{20}$$

$$\begin{aligned}y((n+1)\cdot T_s)=y_{\mathbf{n+1}}=[\mathbf{I}+T_s\cdot(d\cdot\mathbf{C}_1+d_n'\cdot\mathbf{C}_2)]\cdot\mathbf{x}_{\mathbf{n}}+\\+T_s\cdot(d_n\cdot\mathbf{E}_{1\mathbf{y}}+d_n'\cdot\mathbf{E}_{2\mathbf{y}})\cdot\mathbf{w}_{\mathbf{n}}\end{aligned} \tag{21}$$

which constitutes the discrete, nonlinear, averaged model for the system.

Since the DC-DC regulator is trying to regulate the output voltage at some constant value, e.g., +5 V, we need to shift the axes around the nominal, DC operating point, which occurs at

$$\begin{aligned}\mathbf{x}_{\mathbf{n+1}}=\mathbf{x}_{\mathbf{n}}=\mathbf{X}_0,\quad y_{\mathbf{n+1}}=y_{\mathbf{n}}=Y_0,\\d_{\mathbf{n+1}}=d_{\mathbf{n}}=D_0=D,\quad \mathbf{w}_{\mathbf{n+1}}=\mathbf{w}_{\mathbf{n}}=\mathbf{W}_0\end{aligned} \tag{22}$$

The quiescent operating point of the regulator is at the origin. In this way, the development of the discrete, averaged model would be complete. Although the discrete equations are well suited for digital simulation, they are not fully relevant for in-depth, analytical calculations. It is then preferred to work with the continuous time model that can easily model the OpAmp compensator and offers physical insight into the problem. Moreover, due to the lowpass filter at the output, with a cut-off frequency much lower than the switching frequency, this is an energy processing system and not an informational system. Therefore, because the switching frequency is much higher than the natural frequencies of the circuit, the switching period is short enough to approximate the derivatives of the state vector using the Euler forward-differentiation approximation, $\frac{d\mathbf{x}(t)}{dt}=\frac{\mathbf{x}_{\mathbf{n+1}}-\mathbf{x}_{\mathbf{n}}}{T_s}$. We obtain thus a continuous, nonlinear system

$$\dot{\mathbf{x}} = \left[\mathbf{A}_1 \cdot u + \mathbf{A}_2 \cdot (1-u)\right] \cdot \mathbf{x} + \left[\mathbf{E}_{1\mathbf{x}} \cdot u + \mathbf{E}_{2\mathbf{x}} \cdot (1-u)\right] \cdot \mathbf{w}$$
$$y = \left[\mathbf{C}_1 \cdot u + \mathbf{C}_2 \cdot (1-u)\right] \cdot \mathbf{x} + \mathbf{E}_{\mathbf{y}} \cdot \mathbf{w} \tag{23}$$

where the duty cycle, d, was replaced using the modern control formalism notation with u and we used the fact that $\mathbf{E_{1y}} = \mathbf{E_{2y}} = [0 \ \text{-}R_C]$.

Since we are interested in the dynamic performance of the DC-DC regulator, the continuous model needs to be also shifted around the DC operating point, as mentioned above. If only a linear model were desired, one would keep only the first order terms of the Taylor expansion. In order to develop a complete, nonlinear, model the higher order terms in the three variables x, u, w, are needed.

$$\begin{aligned}
f(x,u,w) &= [f(X_0,U_0,W_0)]_{=OP} + \left.\frac{\partial f}{\partial x}\right|_{OP} \cdot \underbrace{(x-X_0)}_{new\ x} + \left.\frac{\partial f}{\partial w}\right|_{OP} \cdot \underbrace{(w-W_0)}_{new\ w} + \left.\frac{\partial f}{\partial d}\right|_{OP} \cdot \underbrace{(u-U_0)}_{new\ u} + \\
&\left.\frac{\partial^2 f}{\partial x \partial u}\right|_{OP} \cdot (x-X_0)\cdot(u-U_0) + \left.\frac{\partial^2 f}{\partial x \partial w}\right|_{OP} \cdot (x-X_0)\cdot(w-W_0) + \left.\frac{\partial^2 f}{\partial x \partial u}\right|_{OP} \cdot (x-X_0)\cdot(u-U_0) + \\
&\frac{1}{2!}\cdot\left.\frac{\partial^2 f}{\partial x^2}\right|_{OP} \cdot (x-X_0)^2 + \frac{1}{2!}\cdot\left.\frac{\partial^2 f}{\partial w^2}\right|_{OP} \cdot (w-W_0)^2 + \frac{1}{2!}\cdot\left.\frac{\partial^2 f}{\partial u^2}\right|_{OP} \cdot (u-U_0)^2 + \\
&\left.\frac{\partial^3 f}{\partial x \partial u \partial w}\right|_{OP} \cdot (x-X_0)\cdot(u-U_0)\cdot(w-W_0) + \ldots
\end{aligned} \tag{24}$$

The differences in parentheses are the dynamics of the perturbation from the nominal Operating Point (OP), $[V_{IN}\ ,\ I_O]=\mathbf{W_0}$, $D=U_0$, and $\mathbf{X_0}$ and will be our default new variables, i.e., x, u, w; the same applies to the output, y, around its DC value $V_0=Y_0$. As it can be seen from detailing (24), only the terms in the DC OP, the first derivatives and the second derivatives *not* in the same variable are non-zero. Hence, we obtain a model in the bilinear form,

$$\dot{\mathbf{x}} + \left(\dot{\bar{\mathbf{X}}}_0\right)_{"OP=0"} = \left\{\mathbf{A}\cdot\mathbf{X}_0 + \mathbf{E}_x\cdot\mathbf{W}_0\right\}_{"OP=0"} + \left\{\mathbf{A}\cdot\mathbf{x} + \mathbf{B}\cdot u + \left(\mathbf{E}_x\cdot\mathbf{w}\right)_{"Dist"}\right\}_{Lin} +$$
$$\left\{\underbrace{\left(\mathbf{A}_1 - \mathbf{A}_2\right)}_{\Delta\mathbf{A}}\cdot\mathbf{x}\cdot u + \underbrace{\left[\mathbf{E}_{1x}^{\mathbf{v}_{in}} - \mathbf{E}_{2x}^{\mathbf{v}_{in}} \quad \mathbf{E}_{1x}^{\mathbf{i}_o} - \mathbf{E}_{2x}^{\mathbf{i}_o}\right]}_{\Delta\mathbf{E}_x}\cdot\mathbf{w}\cdot u\right\}_{"Nonlin"} \tag{25}$$

$$y + \left(Y_O\right)_{"OP=V_o"} = \left\{\mathbf{C}\cdot\mathbf{X}_0 + \mathbf{E}_y\cdot\mathbf{W}_0\right\}_{"OP=V_o"} + \left\{\mathbf{C}\cdot\mathbf{x} + \mathbf{D}\cdot u + \left(\mathbf{E}_y\cdot\mathbf{w}\right)_{"Dist"}\right\}_{"Lin"} +$$
$$\left\{\underbrace{\left(\mathbf{C}_1 - \mathbf{C}_2\right)}_{\Delta\mathbf{C}}\cdot\mathbf{x}\cdot u + \left(\underbrace{\left[\mathbf{0} \quad \mathbf{E}_{1y}^{\mathbf{i}_o} - \mathbf{E}_{2y}^{\mathbf{i}_o}\right]}_{\Delta\mathbf{E}_y}\cdot\mathbf{w}\cdot u\right)_{=0}\right\}_{"Nonlin"} \tag{26}$$

with

$$\begin{aligned}
&\mathbf{A} = \mathbf{A}_1\cdot D + \mathbf{A}_2\cdot(1-D),\ \mathbf{B} = \Delta\mathbf{A}\cdot\mathbf{X}_0 + \Delta\mathbf{E}_x\cdot\mathbf{W}_0,\ \mathbf{C} = \mathbf{C}_1\cdot D + \mathbf{C}_2\cdot(1-D)\\
&\mathbf{D} = \Delta\mathbf{C}\cdot\mathbf{X}_0 + \left(\mathbf{E}_{1y}^{\mathbf{i}_o} - \mathbf{E}_{2y}^{\mathbf{i}_o}\right)\cdot I_O = \Delta\mathbf{C}\cdot\mathbf{X}_0,\ \text{since } \mathbf{E}_{1y}^{\mathbf{i}_o} = \mathbf{E}_{2y}^{\mathbf{i}_o} = -R_C \ \text{(Capacitor ESR)}\\
&\mathbf{E}_x = \mathbf{E}_{1x}\cdot D + \mathbf{E}_{1x}\cdot(1-D),\ \mathbf{E}_y = \mathbf{E}_{1y}\cdot D + \mathbf{E}_{1y}\cdot(1-D) = -R_C,\\
&\text{where } \mathbf{X}_0 = -\mathbf{A}^{-1}\cdot\mathbf{E}_x\cdot\mathbf{W}_0
\end{aligned} \tag{27}$$

In brief,

$$\dot{\mathbf{x}} = \mathbf{A}\cdot\mathbf{x} + \left(\Delta\mathbf{A}\cdot\mathbf{x} + \mathbf{B}\right)\cdot u + \left(\Delta\mathbf{E}_x\cdot u + \mathbf{E}_x\right)\cdot\mathbf{w} = f_1(\mathbf{x}) + g_1(\mathbf{x})\cdot u + p_1(u)\cdot\mathbf{w} \tag{28}$$

$$y = \mathbf{C}\cdot\mathbf{x} + \left(\Delta\mathbf{C}\cdot\mathbf{x} + \mathbf{D}\right)\cdot u + \mathbf{E}_y\cdot\mathbf{w} = f_2(\mathbf{x}) + g_2(\mathbf{x})\cdot u + \mathbf{E}_y\cdot\mathbf{w}\,. \tag{29}$$

This result is the same as the one obtained by employing the elementary method of writing the equations using the perturbations around the DC operating point shifted axes, **x+X₀, w+W₀,** etc. and maintaining the second order terms; the linear model would only keep the first order terms. However, this approach does not offer the necessary insight to properly formalize the real control issues noted above. By using the unterminated model in this new form, one has now a unified, straightforward approach to model the system in a modern control formalism. The disturbance vector is highlighted and separated, and the system can be implemented in Matlab for further developments.

As can easily be inferred and has also been shown before, the worst case for stability and peak ratings response occurs for the loss-less network. Therefore, the design oriented analysis will continue for models with no parasitics while the simulated results will be done using the real, complete model system. In addition, this technique will also allow us to properly identify the plant structure in the light of our control problem and, in general, to better describe this type of systems.

In this study, the boost converter in Figs. 3.1 and 3.2 will be used as a baseline example since it offers both the nonlinear terms and a non-minimal phase (RHPZ) behavior making it a good case to verify the validity and performance of the new control strategy. If we reconsider the converter in Fig. 3.2, without parasitics, average the system over one switching cycle and perturb it around the DC operating point, we obtain

$$\begin{aligned}
\dot{\mathbf{x}} &= \begin{bmatrix} 0 & -\frac{1-D}{L} \\ \frac{1-D}{C} & 0 \end{bmatrix} \cdot \mathbf{x} + \begin{bmatrix} \frac{V_{in}}{L\cdot(1-D)} \\ -\frac{I_o}{C\cdot(1-D)} \end{bmatrix} \cdot u + \begin{bmatrix} \frac{1}{L} \\ 0 \end{bmatrix} \cdot v_{in} + \begin{bmatrix} 0 \\ -\frac{1}{C} \end{bmatrix} \cdot i_o + \begin{bmatrix} 0 & \frac{1}{L} \\ -\frac{1}{C} & 0 \end{bmatrix} \cdot \mathbf{x} \cdot u = \\
&= \begin{bmatrix} 0 & -\frac{1-D}{L} \\ \frac{1-D}{C} & 0 \end{bmatrix} \cdot \mathbf{x} + \begin{bmatrix} \frac{V_{in}}{L\cdot(1-D)} \\ -\frac{I_o}{C\cdot(1-D)} \end{bmatrix} \cdot u + \begin{bmatrix} \frac{1}{L} & 0 \\ 0 & -\frac{1}{C} \end{bmatrix} \cdot \mathbf{w} + \begin{bmatrix} 0 & \frac{1}{L} \\ -\frac{1}{C} & 0 \end{bmatrix} \cdot \mathbf{x} \cdot u \\
y &= \begin{bmatrix} 0 & 1 \end{bmatrix} \cdot \mathbf{x} = \mathbf{C} \cdot \mathbf{x}
\end{aligned} \tag{30}$$

A resistive load (RL), R_{ld}, Fig. 3.1, becomes part of the passive structure of the electrical network during each cycle and therefore the model for each half cycle changes by replacing $i_o=v_o/R_{ld}=x_2/R_{ld}$, in the loss-less case, as a function of the states and the rest of the passive elements; for the model including parasitics, the specific i_o expression in each half cycle would need to be considered. We obtain,

$$\begin{aligned}
\dot{\mathbf{x}} &= \begin{bmatrix} 0 & -\frac{(1-D)}{L} \\ \frac{(1-D)}{C} & -\frac{1}{R_{ld}\cdot C} \end{bmatrix} \cdot \mathbf{x} + \begin{bmatrix} \frac{V_{in}}{L}\frac{1}{1-D} \\ -V_{in}\cdot\frac{1}{(1-D)^2}\cdot\frac{1}{R_{ld}\cdot C} \end{bmatrix} \cdot u + \begin{bmatrix} \frac{1}{L} \\ 0 \end{bmatrix} \cdot v_{in} + \begin{bmatrix} 0 & \frac{1}{L} \\ -\frac{1}{C} & 0 \end{bmatrix} \cdot \mathbf{x} \cdot u \\
y &= \begin{bmatrix} 0 & 1 \end{bmatrix} \cdot \mathbf{x}
\end{aligned} \tag{31}$$

3.1 Resistive Load Variations

All existing models, consider the load somehow only as a parameter in the linear transfer functions of the converter. However, Eq. (31) shows that any change in the load, e.g., load step, will appear as a change in the internal terms of the plant matrices, hence a robustness problem. For very low perturbations, the modification in the **A** and **B** matrix terms can be omitted but for large perturbations the existing models should consider not only the state variations but also the new load resistance, which is the parallel combination of the nominal and step resistances,

$$\frac{1}{R_{ld}} = \frac{1}{R_{ld\_nom}} + \frac{1}{R_{ld\_step}} \tag{32}$$

No present modeling addresses it and the clarification should also be made for the small signal model, for consistency. Eq. (31) also shows that the load resistance only impacts the regulator significantly through the a_{22} and the b_2 terms, of the respective **A** and **B** matrices. Therefore, by using (32) in (31) and factoring the terms, the system's state model becomes

$$\begin{aligned}\dot{\mathbf{x}} = &\left(\begin{bmatrix} 0 & -\frac{(1-D)}{L} \\ \frac{(1-D)}{C} & -\frac{1}{C}\cdot\frac{1}{R_{ld\_nom}} \end{bmatrix} + \begin{bmatrix} 0 & 0 \\ 0 & -\frac{1}{C}\cdot\frac{1}{R_{ld\_nom}} \end{bmatrix}\cdot\frac{1}{xtp}\cdot\mathrm{Step}(t-t_{step})\right)\cdot\mathbf{x} + \\ &V_{in}\cdot M\cdot\left(\begin{bmatrix} \frac{1}{L} \\ -\frac{1}{C}\cdot\frac{1}{(1-D)\cdot R_{ld\_nom}} \end{bmatrix} + \right. \\ &\left. + \begin{bmatrix} 0 \\ -\frac{1}{C}\cdot\frac{1}{(1-D)\cdot R_{ld\_nom}} \end{bmatrix}\cdot\frac{1}{xtp}\cdot\mathrm{Step}(t-t_{step})\right)\cdot u + \\ &\begin{bmatrix} \frac{1}{L} \\ 0 \end{bmatrix}\cdot v_{in} + \begin{bmatrix} 0 & \frac{1}{L} \\ -\frac{1}{C} & 0 \end{bmatrix}\cdot\mathbf{x}\cdot u\end{aligned} \tag{33}$$

with $xtp = R_{ld\_step}/R_{ld\_nom}$, the percentage of the load step occurring at t_{step} and M the DC voltage conversion ratio. Eq. (33) shows more clearly the resistive load robustness (and disturbance rejection) issues. By re-writing Eq. (33) in a compact form we obtain,

$$\dot{\mathbf{x}} = (\mathbf{A}+\mathbf{L})\cdot\mathbf{x} + (\mathbf{B}+\mathbf{L})\cdot u + \mathbf{E}^{v_{in}}\cdot v_{in} + \Delta\mathbf{A}\cdot\mathbf{x}\cdot u \tag{34}$$

Eq. (34) formalizes how the load is affecting the plant's structure, through the variation/uncertainty term **L**. After the step load has been applied, the term **L** is absorbed again in each corresponding matrix. In addition, in the current sharing problem for a system of paralleled converters, only the load disturbance is of interest and therefore the CSC design will consider a zero line disturbance. We obtain thus the bilinear form

$$\dot{\mathbf{x}} = \mathbf{A} \cdot \mathbf{x} + \mathbf{B} \cdot u + \Delta\mathbf{A} \cdot \mathbf{x} \cdot u\,. \qquad (35)$$

The forms (34) and (35) will be employed for analysis and stability design while forms (28), (29) will be used for controller design and simulated results, which are run with a full converter model that includes parasitics, saturations, etc.

All basic DC-DC converter topologies can be factored in this manner, by containing the $1/R_{ld}$ term, which in the loss-less case appears only in the last, main diagonal, element of the **A** matrix and, sometimes, in the **B** vector. This shows the importance of separating the real converter structure from the case including the parasitics in order to properly highlight the control problem.

3.2 Constant Power Load

For a constant power load, $i_o = -p_{ld}/v_o$ AC-wise independently of the network passive elements and for either position of the switch, since the constant power source is an exogenous source. We obtain

$$\begin{aligned}
\dot{\mathbf{x}} = &\begin{bmatrix} 0 & -\frac{1-D}{L} \\ \frac{1-D}{C} & 0 \end{bmatrix} \cdot \mathbf{x} + \begin{bmatrix} \frac{V_{in}}{L\cdot(1-D)} \\ -\frac{1}{C\cdot(1-D)} \cdot \frac{P_{ld}}{Y_o} \end{bmatrix} \cdot u + \begin{bmatrix} \frac{1}{L} \\ 0 \end{bmatrix} \cdot v_{in} + \\
&\begin{bmatrix} 0 \\ \frac{1}{C} \end{bmatrix} \cdot \frac{p_{ld}}{y} + \begin{bmatrix} 0 & \frac{1}{L} \\ -\frac{1}{C} & 0 \end{bmatrix} \cdot \mathbf{x} \cdot u = \\
&\mathbf{A}_{pl} \cdot \mathbf{x} + \mathbf{B}_{pl} \cdot u + \mathbf{E}_{\mathbf{x}pl}^{\mathbf{v}_{in}} \cdot v_{in} + \\
&\mathbf{E}_{\mathbf{x}pl}^{\mathbf{i}_o} \cdot \frac{p_{ld}}{y} + \Delta\mathbf{A}_{pl} \cdot \mathbf{x} \cdot u \\
&y = \begin{bmatrix} 0 & 1 \end{bmatrix} \cdot \mathbf{x} = \mathbf{C} \cdot \mathbf{x}
\end{aligned} \qquad (36)$$

It can be seen that besides the nonlinearity, the constant power load (CPL) has the potential to destabilize the system by introducing right hand plane poles. In order to see that, it is useful to also write the converter's model with a negative load .resistance equivalent to the constant power load. The negative incremental resistance behavior can be easily inferred by writing the expression of the load power as the product of current and voltage, acquaint for the signs (when the voltage increases the current decreases and vice-versa) and then express one of the quantities as a function of the equivalent load resistance. Now, one can use (31) with R_{ld} replaced with $-R_{CPL}$ and similarly with (33), we have that

$$\dot{\mathbf{x}} = \left(\begin{bmatrix} 0 & -\frac{(1-D)}{L} \\ \frac{(1-D)}{C} & \frac{1}{C}\cdot\frac{1}{R_{CPL\_nom}} \end{bmatrix} + \begin{bmatrix} 0 & 0 \\ 0 & \frac{1}{C}\cdot\frac{1}{R_{CPL\_nom}} \end{bmatrix} \cdot \frac{1}{xtp}\cdot \text{Step}(t - t_{step}) \right) \cdot \mathbf{x} +$$
$$V_{in}\cdot M \cdot \left(\begin{bmatrix} \frac{1}{L}\cdot \\ \frac{1}{C}\cdot\frac{1}{(1-D)\cdot R_{CPL\_nom}} \end{bmatrix} + \begin{bmatrix} 0 \\ \frac{1}{C}\cdot\frac{1}{(1-D)\cdot R_{CPL\_nom}} \end{bmatrix} \cdot \frac{1}{xtp}\cdot \text{Step}(t - t_{step}) \right) \cdot u +$$
$$\begin{bmatrix} \frac{1}{L} \\ 0 \end{bmatrix} \cdot v_{in} + \begin{bmatrix} 0 & \frac{1}{L} \\ -\frac{1}{C} & 0 \end{bmatrix} \cdot \mathbf{x}\cdot u \tag{37}$$

Several studies, [26], [28], have been carried out to deal with the potential instability introduced by the negative resistance equivalent with the constant power load, but they are not the direct objective of this study. In practice, a constant power load has a resistive behavior for low voltages and thus, a positive load resistance of very high value exists in parallel with it. Around the steady state operating point, the negative resistance behavior takes place. Theoretically, for any electrical network that gets energized, if a perfectly equivalent but negative impedance is placed in tandem with that network's positive output impedance, then oscillations will occur. Practically, depending of how large the perturbation is, the system may still be stable provided that there is enough damping (positive resistance) at the respective interface for that DC operating point. In those cases, using the unterminated model and applying the Nyquist criterion, one can still have a stable system, even though RHP poles exist. Each example is analyzed on an application specific basis and several means of compensating the current sharing loop have been used.

3.3 Closed-Loop Model

By replacing the control input, *u*, with its expression as a function of the state feedback variables, a closed loop model is obtained. As an illustration, the most used control strategy today for PWM converters will be used i.e., peak current mode control (PCMC). It offers good generality and the expressions for voltage mode control (VMC) and average current mode control (ACMC) can be written in the same form for the feedback function, less and respectively more complicated, but without any major difference with respect to the concerns and goal of this study. PCMC is widely employed in the industry due to its inherent pulse by pulse current limiting capability combined with offering a lower-order system for the power stage. In addition, its overcurrent protection feature enables a PCMC converter to be easier integrated in distributed, parallel systems.

Small signal models including the sample and hold effect, "$H_e(s)$" have been currently developed, [30] and also in a modern, state space formalism, [31]. Since in this study we are addressing the nonlinear, fuzzy logic, control of the output current sharing of master – slave paralleled converter systems under large load disturbance, we are interested in the system's behavior mainly from a large signal point of view. Moreover, in dealing with a paralleled system, one assumes that the individual DC/DC regulator modules are already properly designed in terms of their own small signal stability and hence $H_e(s)$ was employed, before they are integrated in the respective distributed system. Therefore, in order to concentrate on the new control approach, without restricting any generality, we will assume that the module design has the appropriate ramp compensation and the effects of $H_e(s)$ are negligible. The equations can be adapted to include the detailed sample-and-hold model, but they would only become cumbersome to read in state space form, with no real contribution to the newly proposed current sharing control strategy.

In addition, especially for large signal paralleled DC-DC systems, there is no real advantage in separating the converter circuit at the PWM modulator output in order to consider the real modulator circuit and use the modulator's gain expression that comes out from the signal theory for that circuit. That is what the existing modeling techniques do so that they can use in their calculations the quantities that are available in the converter circuitry. One of the main reasons behind it is to be able to measure for each individual module, on the real hardware, the associated gains, in this case proportional with the inverse of the PWM reference ramp amplitude. However, the integrated box level results are practically identical if the mathematical averaged model is used as a whole, for the overall converter module. That is how all the models start, but they are then split and get modified at the PWM modulator output. Reference [31] comes the closest to addressing this issue, highlighting both forms, but with a different goal, i.e., trying to include the effects of "$H_e(s)$" in a state space form. Those forms are useful in the single converter small signal, Bode analysis, but they would just hide the real control issues revealed here, when used for the paralleled system analysis.

Based on the above findings, considering the overall averaged model and neglecting the high frequency effects - $H_e(s)$, the PCMC control input can be written as, [7], [31], [32],

$$u = F_m \cdot v_{ea} - \mathbf{F_{ix}} \cdot \mathbf{x} \tag{38}$$

or, respectively

$$u = F_m \cdot v_{ea} - \mathbf{F}_{\mathbf{ix}} \cdot \mathbf{x} - \mathbf{F}_{\mathbf{iw}} \cdot \mathbf{w}\,, \tag{39}$$

if the original unterminated model were used, with

$$\mathbf{F}_{\mathbf{ix}} = \mathbf{C}_{\mathbf{i}} \cdot \left(\mathbf{I} + \frac{D \cdot T_s}{2} \cdot \mathbf{A}_1 \right) \tag{40}$$

and

$$\mathbf{F}_{\mathbf{iw}} = \frac{D \cdot T_s}{2} \cdot \mathbf{C}_{\mathbf{i}} \cdot \mathbf{E}_{\mathbf{x}1} = \left[\mathbf{F}_{\mathbf{i}}^{\mathbf{v_{in}}} \quad \mathbf{F}_{\mathbf{i}}^{\mathbf{i_o}} \right] = \left[\mathbf{F}_{\mathbf{i}}^{\mathbf{v_{in}}} \quad \mathbf{0} \right], \tag{41}$$

since $\mathbf{C}_\mathrm{i} = [\mathrm{R}_\mathrm{i} \quad 0]$, R_i being the equivalent resistance and ratio of the current sensing network.

The voltage loop compensator is described by

$$\begin{aligned} \dot{\mathbf{x}}_{\mathbf{f}} &= \mathbf{A}_{\mathbf{f}} \cdot \mathbf{x}_{\mathbf{f}} + \mathbf{B}_{\mathbf{f}} \cdot (v_{ref} - x_2) = \mathbf{A}_{\mathbf{f}} \cdot \mathbf{x}_{\mathbf{f}} + [0 \quad -\mathbf{B}_{\mathbf{f}}] \cdot \mathbf{x} + \mathbf{B}_{\mathbf{f}} \cdot v_{ref} \\ v_{ea} &= \mathbf{C}_{\mathbf{f}} \cdot \mathbf{x}_{\mathbf{f}} + \mathbf{D}_{\mathbf{f}} \cdot (v_{ref} - x_2) + v_{ref} \\ &= \mathbf{C}_{\mathbf{f}} \cdot \mathbf{x}_{\mathbf{f}} + [0 \quad -\mathbf{D}_{\mathbf{f}}] \cdot \mathbf{x} + (\mathbf{D}_{\mathbf{f}} + \mathbf{I}) \cdot v_{ref} \end{aligned} \tag{42}$$

This comes from the TF description of the OpAmp, having v_{ref} at its non-inverting input and x_2 at its inverting one. For example, for the compensator employed in PCMC and given in Fig. 3.3 below,

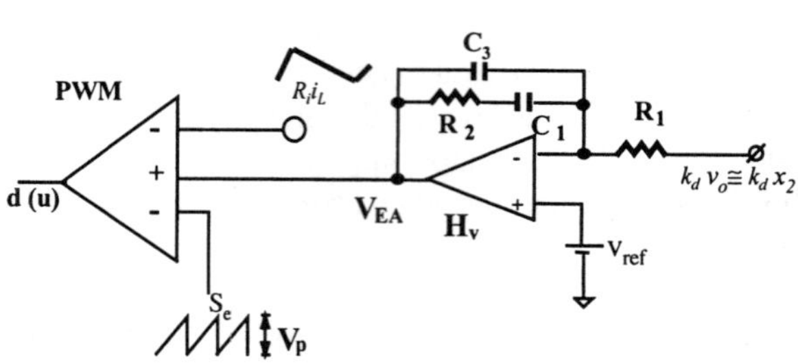

Fig. 3.3. PCMC Control Scheme.

we have that

$$H_v(s) = \frac{V_{ea}(s)}{-k_d \cdot X_2(s)} = \frac{1}{s \cdot R_1 \cdot C_3} \cdot \frac{\left(s + \frac{1}{R_2 \cdot C_1}\right)}{\left(s + \frac{1}{R_2 \cdot \frac{C_1 \cdot C_3}{C_1 + C_3}}\right)}$$
$$= \frac{\omega_i}{s} \cdot \frac{(s + \omega_{z1})}{(s + \omega_{p1})} = \frac{\omega_i}{s} \cdot \left(1 + \frac{\omega_{z1} - \omega_{p1}}{s + \omega_{p1}}\right) \tag{43}$$

and hence,

$$\mathbf{A_f} = \begin{bmatrix} -\omega_{p1} & 0 \\ \omega_{z1} - \omega_{p1} & 0 \end{bmatrix}, \quad \mathbf{B_f} = \begin{bmatrix} 1 \\ 1 \end{bmatrix}, \quad \mathbf{C_f} = \omega_i \cdot [0 \quad 1], \quad \mathbf{D_f} = [0] \tag{44}$$

For the compensator used in VMC, the structure would be similar, but of third order.

Based on the unterminated model in the formalism introduced above, a new representation of the large signal model for the PCMC DC-DC regulator has thus been obtained. Its block diagram is given in Fig. 3.4 below

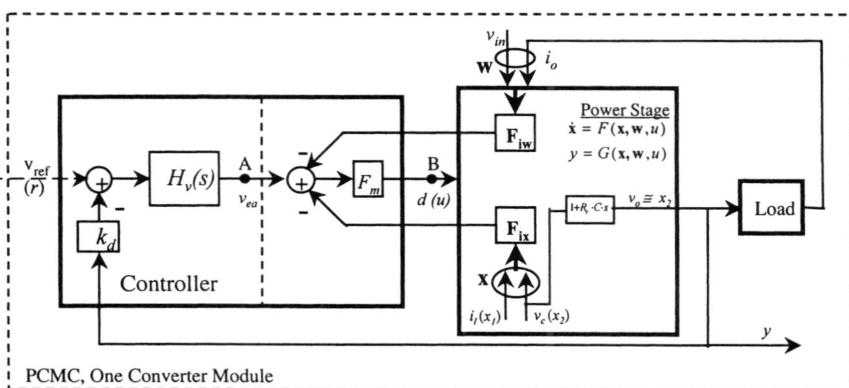

Fig. 3.4. PCMC DC-DC Regulator Control Block diagram.

It can be seen now that following this approach, the converter can be modeled in a straightforward, explicit way in simulators like Simulink, for subsequent design developments. Replacing (38) in (35) gives

$$\dot{\mathbf{x}} = \left(\mathbf{A} - F_m \cdot \mathbf{B} \cdot \mathbf{F}_{\mathbf{ix}}\right) \cdot \mathbf{x} - F_m \cdot \Delta\mathbf{A} \cdot \mathbf{x} \cdot \mathbf{F}_{\mathbf{ix}} \cdot \mathbf{x} + \left(\mathbf{B} + \Delta\mathbf{A} \cdot \mathbf{x}\right) \cdot (F_m \cdot v_{ea}) \tag{45}$$

or, in abbreviated form,

$$\dot{\mathbf{x}} = f_1(\mathbf{x}) + g_1(\mathbf{x}) \cdot (F_m \cdot v_{ea}). \tag{46}$$

Replacing (42) in (45) gives the augmented plant, i.e., closed loop model

$$\begin{bmatrix} \dot{\mathbf{x}} \\ \dot{\mathbf{x}}_{\mathbf{f}} \end{bmatrix} = \begin{bmatrix} \mathbf{A} - F_m \cdot \mathbf{B} \cdot \mathbf{C}_{\mathbf{i}} \cdot \mathbf{F}_{\mathbf{ix}} - F_m \cdot \Delta\mathbf{A} \cdot \mathbf{x} \cdot \mathbf{C}_{\mathbf{i}} \cdot \mathbf{F}_{\mathbf{ix}} & F_m \cdot \left(\mathbf{B} + \Delta\mathbf{A} \cdot \mathbf{x}\right) \cdot \mathbf{C}_{\mathbf{f}} \\ \left[\mathbf{0} \quad -\mathbf{B}_{\mathbf{f}}\right] & \mathbf{A}_{\mathbf{f}} \end{bmatrix} \cdot \begin{bmatrix} \mathbf{x} \\ \mathbf{x}_{\mathbf{f}} \end{bmatrix} + \begin{bmatrix} F_m \cdot \left(\mathbf{B} + \Delta\mathbf{A} \cdot \mathbf{x}\right) \\ \mathbf{B}_{\mathbf{f}} \end{bmatrix} \cdot v_{ref} \tag{47}$$

or, in brief,

$$\dot{\mathbf{x}}_{\mathbf{cl}} = \mathbf{A}_{\mathbf{cl}} \cdot \mathbf{x}_{\mathbf{cl}} + \mathbf{B}_{\mathbf{cl}} \cdot v_{ref} = f_{cl}(\mathbf{x}_{\mathbf{cl}}) + g_{cl}(\mathbf{x}_{\mathbf{cl}}) \cdot v_{ref} \tag{48}$$

Expanding the terms for the boost and respective buck converters, we obtain the large signal models in the newly introduced formalism,

$$\begin{bmatrix} \dot{x}_1 \\ \dot{x}_2 \\ \dot{x}_{f1} \\ \dot{x}_{f2} \end{bmatrix} = \begin{bmatrix} -F_m \cdot \frac{R_i}{L} \cdot \left(V_{in} \cdot M + x_2\right) & -\frac{D'}{L} & 0 & F_m \cdot \frac{\omega_i}{L} \cdot \left(V_{in} \cdot M + x_2\right) \\ \frac{D'}{C} + F_m \cdot \frac{R_i}{C} \cdot \left(\frac{V_{in} \cdot M}{D' \cdot R_{ld}} + x_1\right) & -\frac{1}{R_{ld} \cdot C} & 0 & -F_m \cdot \frac{\omega_i}{C} \cdot \left(\frac{V_{in} \cdot M}{D' \cdot R_{ld}} + x_1\right) \\ 0 & -1 & -\omega_{p1} & 0 \\ 0 & -1 & \omega_{z1} - \omega_{p1} & 1 \end{bmatrix} \cdot \begin{bmatrix} x_1 \\ x_2 \\ x_{f1} \\ x_{f2} \end{bmatrix} + \begin{bmatrix} F_m \cdot \frac{1}{L} \cdot \left(V_{in} \cdot M + x_2\right) \\ -F_m \cdot \frac{1}{C} \cdot \left(\frac{V_{in} \cdot M}{D' \cdot R_{ld}} + x_1\right) \\ 1 \\ 1 \end{bmatrix} \cdot v_{ref} \tag{49}$$

for the boost converter and, respectively,

$$\begin{bmatrix} \dot{x}_1 \\ \dot{x}_2 \\ \dot{x}_{f1} \\ \dot{x}_{f2} \end{bmatrix} = \left[\begin{array}{cc|cc} -F_m \cdot \frac{R_i}{L} \cdot V_{in} & \frac{1}{L} \cdot \left(F_m \cdot \frac{R_i}{L} \cdot V_{in} \cdot \frac{D \cdot T_s}{2} - 1 \right) & 0 & F_m \cdot \frac{\omega_i}{L} \cdot V_{in} \\ \frac{1}{C} & -\frac{1}{R_{ld} \cdot C} & 0 & 0 \\ \hline 0 & -1 & -\omega_{p1} & 0 \\ 0 & -1 & \omega_{z1} - \omega_{p1} & 1 \end{array}\right] \cdot \begin{bmatrix} x_1 \\ x_2 \\ x_{f1} \\ x_{f2} \end{bmatrix} + \begin{bmatrix} F_m \cdot \frac{1}{L} \cdot V_{in} \\ 0 \\ 1 \\ 1 \end{bmatrix} \cdot v_{ref} \tag{50}$$

for the buck converter. As known, [4], [36], [39], the buck converter's model is linear, if no input voltage perturbation is considered.

Reference-to-Output transfer function block diagrams can be derived for a linear, small signal verification, [37], similar to the approach in [29], [30], which consider only a stand alone module, with zero reference input, dynamics wise. They verify the normally used salient loop gains, e.g., the overall and outer loop gain but contain additional terms due to the direct feedthrough term in the reference voltage given by the action of the OpAmp compensator for any variable present at its non-inverting input. As shown in [29], [30], except for individual blocks, the detailed closed form expressions are very lengthy and, depending on the PCMC gains modeling approach, require more or less simplifications in order to highlight the main quantities. However, block level Matlab and Simulink linearization can be employed in order to use and validate the above models, the results being shown in Section 7 both for small and large signal simulations.

For a complete record, if one would like to keep the unterminated model form and also include the input voltage perturbation and parasitics, we obtain by using (28), (29) and (39)

$$\begin{aligned} \dot{\mathbf{x}} = {} & \underbrace{\left(\mathbf{A} - F_m \cdot \mathbf{B} \cdot \mathbf{F_{ix}}\right) \cdot \mathbf{x} - F_m \cdot \Delta\mathbf{A} \cdot \mathbf{x} \cdot \mathbf{F_{ix}} \cdot \mathbf{x}}_{f_1(\mathbf{x})} + \\ & + \underbrace{F_m \cdot \left(\mathbf{B} + \Delta\mathbf{A} \cdot \mathbf{x}\right)}_{g_1(\mathbf{x})} \cdot v_{ea} + \underbrace{F_m \cdot \Delta\mathbf{E_x} \cdot \mathbf{w}}_{q_1(\mathbf{w})} \cdot v_{ea} + \\ & + \underbrace{\left(\mathbf{E_x} - F_m \cdot \Delta\mathbf{A} \cdot \mathbf{x} \cdot \mathbf{F_{iw}} - F_m \cdot \mathbf{B} \cdot \mathbf{F_{iw}}\right)}_{p_1(\mathbf{x})} \cdot \mathbf{w} - \\ & - \underbrace{F_m \cdot \left(\Delta\mathbf{E_x} \cdot \mathbf{w} \cdot \mathbf{F_{iw}} \cdot \mathbf{w} + \Delta\mathbf{E_x} \cdot \mathbf{w} \cdot \mathbf{F_{ix}} \cdot \mathbf{x}\right)}_{\mathbf{Q_{rest}}} \end{aligned} \tag{51}$$

$$y = \underbrace{\left(\mathbf{C} - F_m \cdot \mathbf{D} \cdot \mathbf{F_{ix}}\right) \cdot \mathbf{x} - F_m \cdot \Delta\mathbf{C} \cdot \mathbf{x} \cdot \mathbf{F_{ix}} \cdot \mathbf{x}}_{f_2(\mathbf{x})} + F_m \cdot \underbrace{\left(\mathbf{D} + \Delta\mathbf{C} \cdot \mathbf{x}\right)}_{g_2(\mathbf{x})} \cdot v_{ea} + \underbrace{\left(\mathbf{E_y} - F_m \cdot \mathbf{D} \cdot \mathbf{F_{iw}} - F_m \cdot \Delta\mathbf{C} \cdot \mathbf{x} \cdot \mathbf{F_{iw}}\right)}_{p_2(\mathbf{x})} \cdot \mathbf{w} \tag{52}$$

or, in a compact form,

$$\dot{\mathbf{x}} = f_1(\mathbf{x}) + \left(g_1(\mathbf{x}) + q_1(\mathbf{w})\right) \cdot v_{ea} + p_1(x) \cdot \mathbf{w} + \mathbf{Q_{rest}} \tag{53}$$

$$y = f_2(\mathbf{x}) + g_2(\mathbf{x}) \cdot v_{ea} + p_2(x) \cdot \mathbf{w} \, . \tag{54}$$

For the loss-less, zero input voltage perturbation model, we obtain the augmented plant similarly to (47),

$$\begin{bmatrix} \dot{\mathbf{x}} \\ \dot{\mathbf{x}}_\mathbf{f} \end{bmatrix} = \left[\begin{array}{c:c} \left(\mathbf{A} - F_m \cdot \mathbf{B} \cdot \mathbf{F_{ix}}\right) - F_m \cdot \Delta\mathbf{A} \cdot \mathbf{x} \cdot \mathbf{F_{ix}} & F_m \cdot \left(\mathbf{B} + \Delta\mathbf{A} \cdot \mathbf{x}\right) \cdot \mathbf{C_f} \\ \hdashline \left[\mathbf{0} \quad -\mathbf{B_f}\right] & \mathbf{A_f} \end{array}\right] \cdot \begin{bmatrix} \mathbf{x} \\ \mathbf{x_f} \end{bmatrix} + \begin{bmatrix} \mathbf{E_x^{i_o}} \\ \mathbf{0} \end{bmatrix} \cdot i_o + \begin{bmatrix} F_m \cdot \left(\mathbf{B} + \Delta\mathbf{A} \cdot \mathbf{x}\right) \\ \mathbf{B_f} \end{bmatrix} \cdot v_{ref} \tag{55}$$

Analogous expressions to (49) and (50) can also be obtained. Eq. (55) can be written in a compact form similarly to (48),

$$\dot{\mathbf{x}}_\mathbf{cl} = \mathbf{A_{cl}} \cdot \mathbf{x_{cl}} + \mathbf{E_{cl}} \cdot w + \mathbf{B_{cl}} \cdot r = f_{cl}(\mathbf{x_{cl}}) + p_{cl}(\mathbf{x}) \cdot w + g_{cl}(\mathbf{x_{cl}}) \cdot r \tag{56}$$

These forms will be used in the next sections to design the fuzzy logic control for the current sharing loop in the master-slave paralleled system. The use of the unterminated model combined with the modern control formalism was able not only to reveal the disturbance rejection and robustness issues but also to offer a direct, explicit way of mathematically simulating the system in the time domain, e.g., in Matlab and Simulink. Circuit simulators, like Spice, which are excellent tools for verifying the real circuit, solve the electrical network implicitly and do not necessarily offer insight into the control systems aspect of the problem.

4 Paralleled Converters System Modeling

4.1 The Master – Slave Current Sharing Control as a Tracking Problem

Consider now a paralleled system comprised of two converters, of which converter #1 is master. This covers most practical cases since even in a "maximum current master", [33], type of control, where the master role rotates, there is only one master converter at a time. Denoting with "sub-1" the quantities related to that converter and considering for start the condensed notation, we have, using, for example, (54)

$$\begin{aligned} \dot{\mathbf{x}}_1 &= f_1(\mathbf{x}_1) + p_1(x) \cdot w_1 \\ y_1 &= \mathbf{x}_1(2) = C \cdot \mathbf{x}_1 \end{aligned} \tag{57}$$

$$\begin{aligned} \dot{\mathbf{x}}_2 &= f_2(\mathbf{x}_2) + p_2(x) \cdot w_2 + g_2(\mathbf{x}_2) \cdot r_2 \\ y_2 &= \mathbf{x}_2(2) = C \cdot \mathbf{x}_2 \end{aligned} \tag{58}$$

The master's model does not have the term in the reference input since it does not receive any current sharing information. Electrically, the two converter's connection looks as in Fig. 4.1.

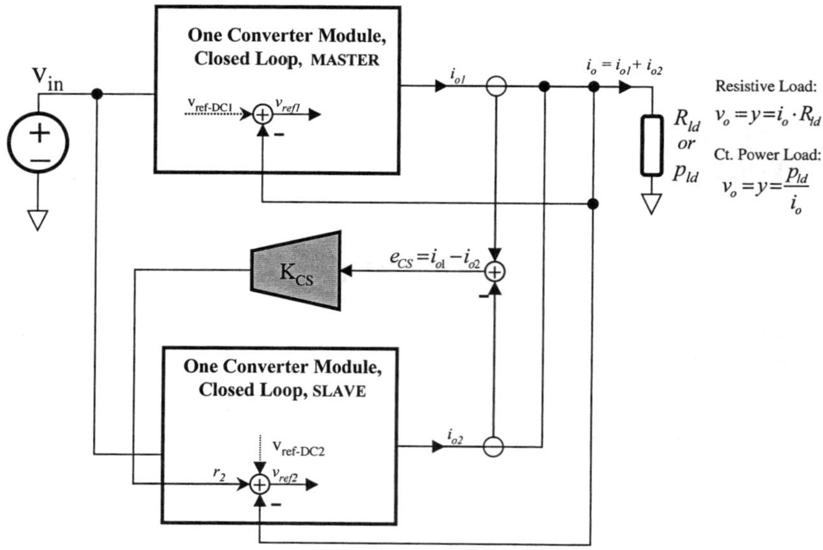

Fig. 4.1. Electrical diagram for two M-S control paralleled converters.

Considering the unterminated model and writing the expression of the load current for Resistive Load (RL) and respectively Constant Power Load (CPL) for n paralleled plants, we obtain the expression for each total, parallel system's output as

$$v_o = R_{load} \cdot i_{oLoad} = \underbrace{R_{load} \cdot \sum_{i=1}^{n} i_{oi}}_{\sum_{v_{oi}}^{n} v_{oi}}, \text{ hence } y = \sum_{i=1}^{n} y_i, \quad \text{for RL}$$

$$v_o = \frac{P_{load}}{i_{oLoad}} = \frac{P_{load}}{\sum_{i=1}^{n} i_{oi}} = \frac{1}{\underbrace{\left(\sum_{i=1}^{n} \frac{i_{oi}}{P_{load}}\right)}_{=\sum_{i=1}^{n} \frac{1}{v_{oi}}}}, \text{ hence } y = \frac{1}{\sum_{i=1}^{n} \frac{1}{y_i}}, \quad \text{for CPL} \tag{59}$$

Hence, from a control systems standpoint, the diagram in Fig. 4.1 can be drawn as in Fig. 4.2 for RL and respectively Fig. 4.3 for CPL.

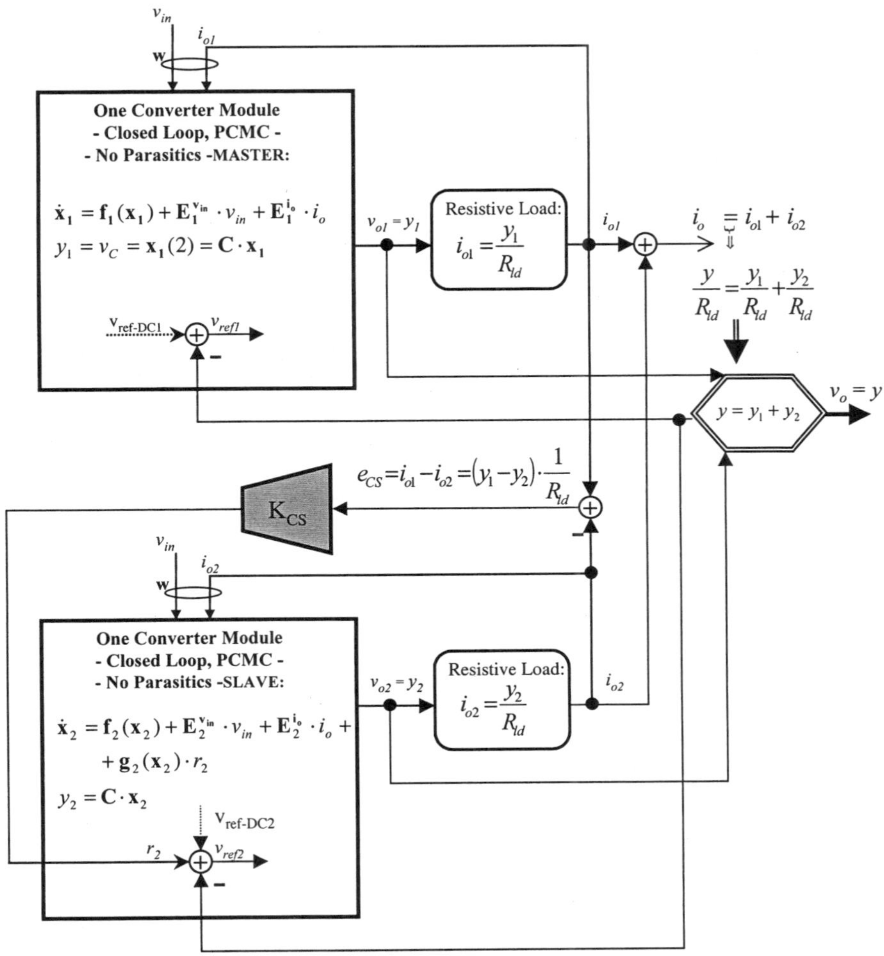

Fig. 4.2. Control systems diagram-for two M-S control paralleled converters - resistive load.

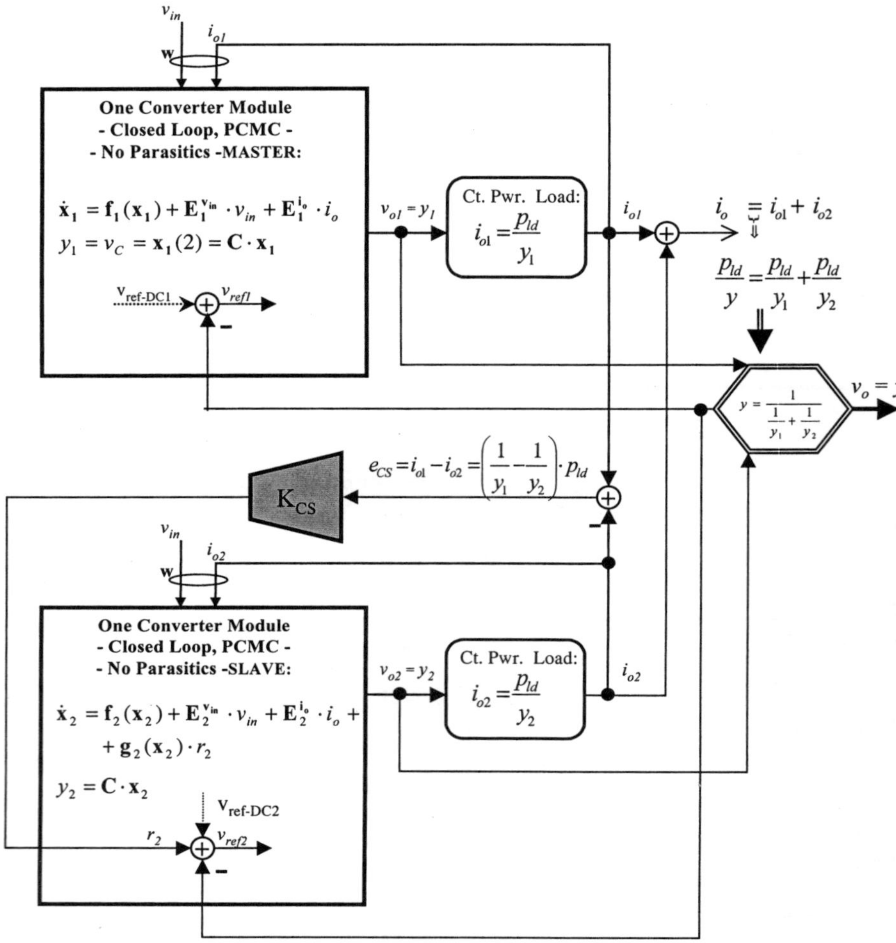

Fig. 4.3. Control system diagram-for two M-S control paralleled converters – constant power load.

The importance of having both the electrical and the modern control system interfaces can now be seen, as they bring insight into the control problem. From examining the control scheme for both types of loads, one can see that in a master – slave setup the error fed to the current sharing controller (CSC) is directly proportional to the state's second component error, ideally equal to the output voltage. Therefore, the control problem is that of the slave tracking the master, once an output load disturbance has occurred. The difference between the master and slave currents is altering the reference of the slave converter - normally on zero for the individual regulator - so it tracks the current of the master. The slave is practically transformed into a controlled-current source while the output voltage dynamics are regulated by the master converter, the only converter having only a voltage feedback and its reference unaltered.

The current sharing (CS) error for the two load types disturbances can be written

$$e_{cs} = \frac{1}{R_{ld}} \cdot e_y = \frac{1}{R_{ld}} \cdot \mathbf{C} \cdot \mathbf{e_x} \tag{60}$$

for a resistive load and

$$e_{cs} = p_{ld} \cdot \frac{e_y}{y_1 \cdot y_2} = p_{ld} \cdot \frac{\mathbf{C} \cdot \mathbf{e_x}}{\mathbf{C} \cdot \mathbf{x_1} \cdot \mathbf{C} \cdot \mathbf{x_2}} = p_{ld} \cdot \frac{\mathbf{C} \cdot \mathbf{e_x}}{\mathbf{x_1}(2) \cdot \mathbf{x_2}(2)}. \tag{61}$$

for a constant power load, with $\mathbf{e_x}=\mathbf{x_1}-\mathbf{x_2}$.

The subsequent study will address the resistive load case as a starting point for developing the new fuzzy control algorithm, which is the main focus of this study. Constant power loads require additional developments, but once the design approach for the fuzzy logic compensator is introduced in the simpler case of a resistive load they can be treated similarly and will be the subject of future research.

Using Eq. (49) we obtain for the error equation in the case of the boost converter

$$
\dot{\mathbf{e}}_{\mathbf{x}} = \begin{bmatrix}
\begin{array}{l}
-V_{in}\cdot M\cdot\left(F_{m1}\cdot\frac{R_{i1}}{L_1}\cdot x_{11}-F_{m2}\cdot\frac{R_{i2}}{L_2}\cdot x_{12}\right)- \\
\left(F_{m1}\cdot\frac{R_{i1}}{L_1}\cdot x_{21}\cdot x_{11}-F_{m2}\cdot\frac{R_{i2}}{L_2}\cdot x_{22}\cdot x_{12}\right)- \\
D'\cdot\left(\frac{x_{21}}{L_1}-\frac{x_{22}}{L_2}\right)+ \\
V_{in}\cdot M\cdot\left(F_{m1}\cdot\frac{\omega_{i1}}{L_1}\cdot x_{f21}-F_{m2}\cdot\frac{\omega_{i2}}{L_2}\cdot x_{f22}\right)+ \\
\left(F_{m1}\cdot\frac{\omega_{i1}}{L_1}\cdot x_{21}\cdot x_{f21}-F_{m2}\cdot\frac{\omega_{i2}}{L_2}\cdot x_{22}\cdot x_{f22}\right)
\end{array} \\[2ex]
\begin{array}{l}
D'\cdot\left(\frac{x_{11}}{C_1}-\frac{x_{12}}{C_2}\right)+\frac{V_{in}\cdot M}{D'\cdot R_{ld}}\cdot\left(F_{m1}\cdot\frac{R_{i1}}{C_1}\cdot x_{11}-F_{m2}\cdot\frac{R_{i2}}{C_2}\cdot x_{12}\right)+ \\
\left(F_{m1}\cdot\frac{R_{i1}}{C_1}\cdot x_{11}^2-F_{m2}\cdot\frac{R_{i2}}{C_2}\cdot x_{12}^2\right)-\frac{1}{R_{ld}}\cdot\left(\frac{x_{21}}{C_1}-\frac{x_{22}}{C_2}\right)- \\
\frac{V_{in}\cdot M}{D'\cdot R_{ld}}\cdot\left(F_{m1}\cdot\frac{\omega_{i1}}{C_1}\cdot x_{f21}-F_{m2}\cdot\frac{\omega_{i2}}{C_2}\cdot x_{f22}\right)- \\
\left(F_{m1}\cdot\frac{\omega_{i1}}{C_1}\cdot x_{11}\cdot x_{f21}-F_{m2}\cdot\frac{\omega_{i2}}{C_2}\cdot x_{11}\cdot x_{f21}\right)
\end{array} \\[2ex]
-\left(x_{21}-x_{22}\right)-\left(\omega_{p11}\cdot x_{f11}-\omega_{p12}\cdot x_{f12}\right) \\[2ex]
\begin{array}{l}
-\left(x_{21}-x_{22}\right)+\left(\omega_{z11}-\omega_{p11}\right)\cdot x_{f11}- \\
\left(\omega_{z12}-\omega_{p12}\right)\cdot x_{f12}+\left(x_{f21}-x_{f22}\right)
\end{array}
\end{bmatrix} -
\begin{bmatrix}
F_{m2}\cdot\frac{1}{L_2}\cdot\left(V_{in}\cdot M+x_{22}\right) \\[2ex]
-F_{m2}\cdot\frac{1}{C_2}\cdot\left(\frac{V_{in}\cdot M}{D'\cdot R_{ld}}+x_{12}\right) \\[2ex]
1 \\[2ex]
1
\end{bmatrix}\cdot r_2 \qquad (62)
$$

where the first index for states denotes the state number and the second index denotes the converter number. The buck converter can be addressed similarly. From equation (62) we can see that:

a) The FLC has to be achieved through r_2 and to be stable.

b) The paralleled system state, and hence current sharing, error has a nonlinear, complicated expression, not directly amenable for obtaining a closed-form control law, which again justifies the fuzzy logic approach.

For ideally identical converters, using the closed loop, loaded, converter model, (47) and (40), the error equation can be written,

$$\dot{\mathbf{e}} = \begin{bmatrix} \mathbf{e}_{\mathbf{x}} \\ \mathbf{e}_{\mathbf{x}_{\mathbf{f}}} \end{bmatrix} = \begin{bmatrix} \mathbf{A} - F_m \cdot \mathbf{B} \cdot \mathbf{C}_{\mathbf{i}} \cdot \left(\mathbf{I} + \frac{D \cdot T_s}{2} \cdot \mathbf{A}_1 \right) & F_m \cdot \mathbf{B} \cdot \mathbf{C}_{\mathbf{f}} \\ [\mathbf{0} \quad -\mathbf{B}_{\mathbf{f}}] & \mathbf{A}_{\mathbf{f}} \end{bmatrix} \cdot \begin{bmatrix} \mathbf{e}_{\mathbf{x}} \\ \mathbf{e}_{\mathbf{x}_{\mathbf{f}}} \end{bmatrix} - \begin{bmatrix} F_m \cdot (\mathbf{B} + \Delta\mathbf{A} \cdot \mathbf{x}_2) \\ \mathbf{B}_{\mathbf{f}} \end{bmatrix} \cdot r +$$

$$\begin{bmatrix} F_m \cdot \Delta\mathbf{A} \cdot \left(\begin{array}{c} \mathbf{x}_2 \cdot \mathbf{C}_{\mathbf{i}} \cdot \left(\mathbf{I} + \frac{D \cdot T_s}{2} \cdot \mathbf{A}_1 \right) \cdot \mathbf{x}_2 - \mathbf{x}_1 \cdot \mathbf{C}_{\mathbf{i}} \cdot \left(\mathbf{I} + \frac{D \cdot T_s}{2} \cdot \mathbf{A}_1 \right) \cdot \mathbf{x}_1 + \\ + \mathbf{x}_1 \cdot \mathbf{C}_{\mathbf{f}} \cdot \mathbf{x}_{\mathbf{f1}} - \mathbf{x}_2 \cdot \mathbf{C}_{\mathbf{f}} \cdot \mathbf{x}_{\mathbf{f2}} \end{array} \right) \\ \mathbf{0} \end{bmatrix} \tag{63}$$

Eq. (63) highlights now better the paralleled system's structure in terms of the current sharing error. The disadvantage of this approach is that we do not have all the problem's details. As can be seen in (63) for the linearized model, the two-converter response is identical with the response of one converter (same linear equation for the states and for the states error). Therefore, the converters would share the current almost evenly, with the difference being in the nonlinear terms. On the other hand, using the exact error equation, (62), makes it practically impossible to separate the state error for deriving a control law as well as keeping it stable. The analysis and design of the CSC can be thus done employing this equation. The validity of the design can then be checked using the real model, (51), (52), which includes the parameters variation among modules and parasitics.

5 Lyapunov Stability Analysis for DC-DC Regulators

This section discusses the stability issues and properties that DC-DC regulators have from a Lyapunov energy function perspective. The main reason for the approach is that if available, even though sometimes too stringent, Lyapunov functions offer stable trajectories for any control method desired. This is particularly important for fuzzy logic control methods of complex systems, e.g., the present master – slave paralleled converters current sharing problem, as will be shown below.

5.1 Open Loop Stability

It has been shown, [10], that an open loop, resistive load, loss-less DC-DC regulator is an incrementally passive system and it is open-loop Lyapunov stable. If the damping from the parasitics is considered, then the open loop converter is asymptotically stable. Using similar calculations, in the case of a constant power load, e.g., another DC-DC regulator, the converter may or may not be stable depending on the amount of damping, load power and perturbation size. Different control strategies are then needed to achieve the stability task.

Let the switching converter be composed of ideal DC sources, ideal switches, incrementally passive resistors and linear passive reactive elements. Consider the currents and voltages in each network branch with respect to the nominal OP trajectory solution and their increments (perturbations), *di, dv*. Tellegen's theorem states that they are orthogonal, both as full quantities and as increments. We obtain thus,

$$\begin{aligned}&\sum_{\text{DC Sources}} di \cdot dv + \sum_{\text{Switches}} di \cdot dv + \\ &+ \sum_{\text{"Positive" Resistors}} di \cdot dv + \sum_{\text{Inductors}} di \cdot dv + \sum_{\text{Capacitors}} di \cdot dv = 0\end{aligned} \tag{64}$$

The summations in the terms involving DC sources and switches are zero as the increment in either voltage or current of each term is necessarily zero. The summation for "Positive Resistors" is always non-negative as each term is individually zero or positive, as a result of incremental passivity of resistors. This summation can be thought of as the dissipated power in the increment with respect to the given nominal trajectory. The terms concerning the reactive, passive inductors and capacitors, represent the time rate of change, $\dot{V}(d\mathbf{x})$, of the stored energy in the increment, $V(d\mathbf{x})$, [10],

$$\mathrm{V}(d\mathbf{x}) = \sum_{\text{Ind.}} \tfrac{1}{2} \cdot di \cdot dv + \sum_{\text{Cap.}} \tfrac{1}{2} \cdot di \cdot dv \cdot \tag{65}$$

which is a positive definite quadratic function of the perturbations of the state variables. Because (64) holds identically for any of the possible circuit topologies, the energy in the increment is a Lyapunov function for the given nominal trajectory. From identifying the energy rate-of-change, $\dot{V}(d\mathbf{x})$ and Eqs. (64) and (65) we have that

$$\dot{\mathbf{V}}(d\mathbf{x}) = - \sum_{\text{"Positive" Resistors}} di \cdot dv \leq 0 \tag{66}$$

Hence the nominal trajectory is Lyapunov stable. Moreover, asymptotic stability can be concluded as well as at least some parasitic loss is always associated with each inductor and capacitor and hence the inequality in (66) becomes strict. Because the nominal trajectory above can be taken as any solution trajectory, we have that any two solutions trajectories do not diverge. It is also depicted in [10] that the case when the switches are operated with a periodic switching pattern of finite frequency can be extended to the case where the switching frequency becomes infinite. This is the case naturally analyzed with the state-space averaged model of the converter, since it would idealize the assumptions of having the time constants of the circuit much smaller than the switching frequency, which are the modeling premises discussed previously. In this way an open-loop stability result can be obtained for state-space averaged models. The main difference from the case of finite switching frequency is that only the stability of an equilibrium point for an averaged model needs to be considered rather than a limit cycle. Therefore, we are able now to include nonlinear reactive and resistive elements that are of interest for the practical cases in power electronics circuits.

Recognizing that the vector $d\mathbf{x}$ and its components di, dv above are actually the dynamic perturbations from nominal employed throughout this study, i.e., $\mathbf{x}$ and respectively i_L and v_C, we can now use our previous models in the subsequent Laypunov analysis. Consider again the system's model for the states, as given in Eq. (31),

$$\begin{aligned}\dot{\mathbf{x}} &= \begin{bmatrix} 0 & -\frac{(1-D)}{L} \\ \frac{(1-D)}{C} & -\frac{1}{R_{ld}\cdot C} \end{bmatrix} \cdot \mathbf{x} + V_{in} \cdot M \cdot \begin{bmatrix} \frac{1}{L} \\ -\cdot\frac{1}{(1-D)}\cdot\frac{1}{R_{ld}\cdot C} \end{bmatrix} \cdot u + \\ &+ \begin{bmatrix} \frac{1}{L} \\ 0 \end{bmatrix} \cdot v_{in} + \begin{bmatrix} 0 & \frac{1}{L} \\ -\frac{1}{C} & 0 \end{bmatrix} \cdot \mathbf{x} \cdot u = \\ &= \mathbf{A}\cdot\mathbf{x} + \mathbf{B}\cdot u + \mathbf{E}\cdot v_{in} + \Delta\mathbf{A}\cdot\mathbf{x}\cdot u \end{aligned} \tag{67}$$

where the **B** matrix has been re-arranged and only the state equation is considered since the output is just one of the states in this case. Also, without loss of generality, the load resistance, R_{ld}, will be considering as including the step load one in order to simplify the notations. In addition, we will consider that the load resistance includes the sign, i.e., positive for resistive load and negative for constant power load. If one picks now the simple Lyapunov function, given by the filter components energy function, we have

$$V(\mathbf{x}) = \frac{1}{2} \cdot \mathbf{x}^{\mathrm{T}} \cdot \begin{bmatrix} L & 0 \\ 0 & C \end{bmatrix} \cdot \mathbf{x} = \frac{1}{2} \cdot \mathbf{x}^{\mathrm{T}} \cdot \mathbf{Q} \cdot \mathbf{x}, \tag{68}$$

and its derivative becomes

$$\begin{aligned} \dot{V}(\mathbf{x}) &= \frac{1}{2} \cdot \mathbf{x}^{\mathrm{T}} \cdot \left(\mathbf{Q} \cdot \mathbf{A} + \mathbf{A}^{\mathrm{T}} \cdot \mathbf{Q}\right) \cdot \mathbf{x} + \mathbf{x}^{\mathrm{T}} \cdot \mathbf{Q} \cdot \mathbf{B} \cdot u + \\ &+ \mathbf{x}^{\mathrm{T}} \cdot \mathbf{Q} \cdot \mathbf{E} \cdot v_{in} + \frac{1}{2} \cdot \mathbf{x}^{\mathrm{T}} \cdot \left(\mathbf{Q} \cdot \Delta\mathbf{A} + \Delta\mathbf{A}^{\mathrm{T}} \cdot \mathbf{Q}\right) \cdot \mathbf{x} \cdot u \end{aligned} \tag{69}$$

or, using (67),

$$\dot{V}(\mathbf{x}) = -\frac{1}{R_{ld}} \cdot x_2^2 + V_{in} \cdot M \cdot \left(x_1 - \frac{x_2}{(1-D)} \cdot \frac{1}{R_{ld}} \right) \cdot u + x_1 \cdot v_{in} + 0 \cdot \tag{70}$$

Eq. (70) reveals several important facts about the open loop system and its energy function. First, it verifies, as in [10], that the worst case for stability under a passive resistor load is for infinite load, i.e., the first term on the right side equal to zero; in that case, the **A** matrix itself is skew symmetric. It can be seen however that for a constant power load, equivalent to an active (negative) resistor, the first term actually hinders substantially this energy function's derivative from being negative definite. In addition, for a constant power load, the linear system's eigenvalues have a positive real part, which makes the energy function candidate to fail the Lyapunov test even after a similarity transformation that would employ those eigenvalues. Furthermore, this idealized plant would be open loop unstable. The reasons for which most converters are still stable for small to medium constant power load variations are: **1)** making the constant power load equivalent with a negative resistance for any variation and throughout the entire converter bandwidth is a severely over-conservative approach; **2)** the parasitics also play an important role, helping the stability through the damping that they introduce, here a worst case, i.e. no damping system is considered; **3)** in closed loop, some control strategies introduce additional negative terms. As mentioned before however, there is no real universal approach, to accommodate for all type of variations for both resistive and constant power loads, unless the controllers and/or the systems become very complicated. All the above conclusions hold for all six basic DC-DC converter topologies established in the existing literature, [3], [36], [40] and therefore they prove the generality of the new approach.

Considering, again, the term in the control, u, in equation (70), one of the first approaches to a Lyapunov stable control was for an infinite load, passive circuit, where the control is determined so it guarantees that the derivative of the energy function is negative, [10]. That approach addressed single converter modules though, already well established in the industry in terms of control implementation and size. A complex digital controller, because of its large size and cost, does not offer a better performance for an individual DC-DC regulator model than the present control methods, almost all with good performances. However, in the case of distributed systems, due to their complexity and size allowance, modern digital controllers are fully justified and needed as the next step for improved DC-DC power conversion. It is therefore important to investigate the Lyapunov stability properties of the complete, closed loop, DC-DC regulator plant. The design-oriented analysis will continue for the worst-case stability models, with no parasitics, and the simulated results will be done using the complete, real system.

6 On a Lyapunov-Stable Fuzzy Logic Control

This section proves the existence of a stable Lyapunov adaptive control algorithm for the case of employing a fuzzy logic controller for paralleled converter systems. First, the theoretical aspect is detailed and then a controller is proposed based on the above findings.

6.1 Input – Output Linearization for Fuzzy Logic Control

Recalling Eqs. (47) and (40), we have the closed loop model

$$\begin{bmatrix} \dot{\mathbf{x}} \\ \dot{\mathbf{x}}_\mathbf{f} \end{bmatrix} = \left[\begin{array}{c|c} \mathbf{A} - F_m \cdot \mathbf{B} \cdot \mathbf{C}_\mathbf{i} \cdot \left(\mathbf{I} + \frac{D \cdot T_s}{2} \cdot \mathbf{A}_1 \right) - & \\ F_m \cdot \Delta\mathbf{A} \cdot \mathbf{x} \cdot \mathbf{C}_\mathbf{i} \cdot \left(\mathbf{I} + \frac{D \cdot T_s}{2} \cdot \mathbf{A}_1 \right) & F_m \cdot (\mathbf{B} + \Delta\mathbf{A} \cdot \mathbf{x}) \cdot \mathbf{C}_\mathbf{f} \\ \hline [\mathbf{0} \quad -\mathbf{B}_\mathbf{f}] & \mathbf{A}_\mathbf{f} \end{array} \right] \cdot \begin{bmatrix} \mathbf{x} \\ \mathbf{x}_\mathbf{f} \end{bmatrix} + \begin{bmatrix} F_m \cdot (\mathbf{B} + \Delta\mathbf{A} \cdot \mathbf{x}) \\ \mathbf{B}_\mathbf{f} \end{bmatrix} \cdot r = \mathbf{A}_\mathbf{cl} \cdot \mathbf{x}_\mathbf{cl} + \mathbf{B}_\mathbf{cl} \cdot r \tag{71}$$

A good way to highlight the CS error is to employ an input-output linearization, [25], [34], [35], so the reference input is linearly present in the expression of the output, which is the second component of the state vector, x_2, as detailed in the previous sections. Regardless of each state component index, the general model is of the form

$$x^{(n)} = f(x) + g(x) \cdot u \,, \tag{72}$$

Eq. (71) can be then put in the form (72) for the desired output and we would obtain systems of different orders and relative degrees, depending on each specific DC-DC topology that is analyzed. For example, from (49) and (50) we obtain first and respectively second order equations. For the generality of the proof, we consider a case when n derivations are needed.

6.2 Adaptation Law

From equation (72), if we could easily implement the ideal control,

$$u_{id} = \frac{1}{g(\mathbf{x})} \cdot \left[-f(\mathbf{x}) + y_r^{(n)} + \mathbf{k}^T \cdot \mathbf{e} \right] \tag{73}$$

for tracking the reference y_r and hence canceling the error $e = y_r - y$, one can take $\mathbf{k} = [k_n, \ldots, k_1]^\mathrm{T}$ such that all roots of the polynomial $s^n + k_1 s^{n-1} + \ldots + k_n$. are in the open left-half complex plane, then the closed-loop system becomes

$$e^{(n)} + k_1 \cdot e^{(n-1)} + \cdots + k_n \cdot e = 0 \tag{74}$$

Because of the way we chose **k**, we would have that $e(t) \rightarrow 0$ as $t \rightarrow \infty$, i.e., the plant output y tracks the reference y_r asymptotically and the poles can be placed so that we get the desired transient response. Since $f(\mathbf{x})$ and $g(\mathbf{x})$ are too complicated to use for deriving the ideal control in (73), or not known in detail, one can use the averaged model behavior to heuristically come up with a fuzzy controller. Also, we assume, as in this case, that we can determine state dependent bounds, which can be fairly loose for this type of applications, such that $|f(\mathbf{x})| \leq f^{U}(\mathbf{x})$ and $0 < g(\mathbf{x}) \leq g^{U}(\mathbf{x}) < b$.

We use an easy-to-implement fuzzy logic control using product inference engine, ξ, singleton fuzzifier, θ and center average defuzzifier, $u_{fz}(e \mid \theta) = \theta^T \cdot \xi(e)$, [25]. Denoting with θ^* the optimal singletons after the Lyapunov stability adaptation, and defining the minimum approximation error $w = u_{fz}(e \mid \theta^*) - u_{id}$, we obtain after some algebra,

$$\dot{\mathbf{e}} = \mathbf{\Lambda} \cdot \mathbf{e} + \mathbf{b} \cdot (\theta^* - \theta) \cdot \xi(\mathbf{x}) - \mathbf{b} \cdot w \tag{75}$$

where $\mathbf{e} = [e, e', \ldots, e^{(n-1)}]^T$, Λ has the controllable (companion) canonical form with the elements of **k** as coefficients and **b** is a column matrix with the last element b and the rest zero.

Consider now the Lyapunov function candidate

$$V = \frac{1}{2} \cdot \mathbf{e}^T \cdot \mathbf{P} \cdot \mathbf{e} + \frac{b}{2 \cdot \gamma} \cdot (\theta^* - \theta)^T \cdot (\theta^* - \theta) \tag{76}$$

with **P** satisfying the Lyapunov equation $\mathbf{\Lambda}^T \cdot \mathbf{P} + \mathbf{P} \cdot \mathbf{\Lambda} = -\mathbf{Q}$, where **Q** and **P** are positive definite. We obtain

$$\dot{V} = -\frac{1}{2} \cdot \mathbf{e}^T \cdot \mathbf{Q} \cdot \mathbf{e} - \mathbf{e}^T \cdot \mathbf{p_n} \cdot b \cdot w + \frac{b}{\gamma} \cdot (\theta^* - \theta)^T \cdot \left(\gamma \cdot \mathbf{e}^T \cdot \mathbf{p_n} \cdot \xi(\mathbf{x}) - \dot{\theta}\right) \tag{77}$$

where p_n is the last column of the **P** matrix. Therefore, if we choose the generic adaptation law

$$\dot{\theta} = \gamma \cdot \mathbf{e}^T \cdot \mathbf{p_n} \cdot \xi(\mathbf{x}) \tag{78}$$

then the last term in (77) is canceled and if we design the fuzzy controller with proper rules from the knowledge we have about the system, then the error w will be small enough so the first term in (77) prevails and V' will be negative.

The previous algorithm has the merit of proving the existence of a stable fuzzy control. It is very general, but it does not suggest a direct implementation since a high degree of trial and error gain adjustments is needed in order to see any results. However, based on its proof, a Lyapunov supervisory control can be initiated. The results are given next.

6.3 Closed Loop Lyapunov Stability Analysis

Using the same type of Lyapunov function as in (68), but accounting also for the OpAmp compensator states, Fig. 3.3, we take

$$\mathbf{Q}_{\mathrm{cl}} = \begin{bmatrix} L & 0 & 0 & 0 \\ 0 & C & 0 & 0 \\ 0 & 0 & C_1 & 0 \\ 0 & 0 & 0 & C_3 \end{bmatrix} = \begin{bmatrix} \mathbf{Q}_{\mathrm{x}} & \mathbf{0} \\ \mathbf{0} & \mathbf{Q}_{\mathrm{x_f}} \end{bmatrix} \tag{79}$$

Using the last equality in (71), the derivative of the energy function can be written now

$$\dot{V}_{cl}(\mathbf{x}) = \frac{1}{2}\cdot\mathbf{x}_{\mathrm{cl}}^{\mathrm{T}}\cdot\left(\mathbf{Q}_{\mathrm{cl}}\cdot\mathbf{A}_{\mathrm{cl}} + \mathbf{A}_{\mathrm{cl}}^{\mathrm{T}}\cdot\mathbf{Q}_{\mathrm{cl}}\right)\cdot\mathbf{x}_{\mathrm{cl}} + \mathbf{x}_{\mathrm{cl}}^{\mathrm{T}}\cdot\mathbf{Q}\cdot\mathbf{B}_{\mathrm{cl}}\cdot r \tag{80}$$

or, after expanding the terms,

$$\dot{V}_{cl}(\mathbf{x}) = \frac{1}{2}\cdot\left[\mathbf{x}^{\mathrm{T}} \quad \mathbf{x}_{\mathrm{f}}^{\mathrm{T}}\right]\cdot\left\{\begin{array}{c|c} \mathbf{a}_{11} & F_m\cdot\begin{bmatrix} \mathbf{Q}_{\mathrm{x}}\cdot(\mathbf{B}+\Delta\mathbf{A}\cdot\mathbf{x})\cdot\mathbf{C}_{\mathrm{f}} + \\ +\mathbf{C}_{\mathrm{f}}^{\mathrm{T}}\cdot(\mathbf{B}+\Delta\mathbf{A}\cdot\mathbf{x})^{\mathrm{T}}\cdot\mathbf{Q}_{\mathrm{x}} \end{bmatrix} \\ \hline \begin{array}{l} \mathbf{Q}_{\mathrm{x_f}}\cdot\left[\mathbf{0} \quad -\mathbf{B}_{\mathrm{f}}\right] + \\ +\left[\mathbf{0} \quad -\mathbf{B}_{\mathrm{f}}\right]^{\mathrm{T}}\cdot\mathbf{Q}_{\mathrm{x_f}} \end{array} & \mathbf{Q}_{\mathrm{x_f}}\cdot\mathbf{A}_{\mathrm{f}} + \mathbf{A}_{\mathrm{f}}^{\mathrm{T}}\cdot\mathbf{Q}_{\mathrm{x_f}} \end{array}\right\}\cdot\begin{bmatrix} \mathbf{x} \\ \mathbf{x}_{\mathrm{f}} \end{bmatrix} + \\ + \left[\mathbf{x}^{\mathrm{T}} \quad \mathbf{x}_{\mathrm{f}}^{\mathrm{T}}\right]\cdot\begin{bmatrix} \mathbf{Q}_{\mathrm{x}}\cdot F_m\cdot(\mathbf{B}+\Delta\mathbf{A}\cdot\mathbf{x}) \\ \mathbf{Q}_{\mathrm{x_f}}\cdot\mathbf{B}_{\mathrm{f}} \end{bmatrix}\cdot r \tag{81}$$

with $\mathbf{a}_{11}$ given by

$$\begin{aligned}\mathbf{a}_{11} &= \left(\mathbf{Q}_{\mathbf{x}}\cdot\mathbf{A}+\mathbf{A}^{T}\cdot\mathbf{Q}_{\mathbf{x}}\right)-F_{m}\cdot\left(\mathbf{Q}_{\mathbf{x}}\cdot\mathbf{B}\cdot\mathbf{C}_{\mathbf{i}}+\mathbf{C}_{\mathbf{i}}^{\mathbf{T}}\cdot\mathbf{B}^{\mathbf{T}}\cdot\mathbf{Q}_{\mathbf{x}}\right)-\\ & F_{m}\cdot\frac{D\cdot T_{s}}{2}\cdot\left(\mathbf{Q}_{\mathbf{x}}\cdot\mathbf{B}\cdot\mathbf{C}_{\mathbf{i}}\cdot\mathbf{A}_{1}+\mathbf{A}_{1}^{\mathbf{T}}\cdot\mathbf{C}_{\mathbf{i}}^{\mathbf{T}}\cdot\mathbf{B}^{\mathbf{T}}\cdot\mathbf{Q}_{\mathbf{x}}\right)-\\ & F_{m}\cdot\left(\mathbf{Q}_{\mathbf{x}}\cdot\Delta\mathbf{A}\cdot\mathbf{x}\cdot\mathbf{C}_{\mathbf{i}}+\mathbf{C}_{\mathbf{i}}^{\mathbf{T}}\cdot\mathbf{x}^{\mathbf{T}}\cdot\Delta\mathbf{A}^{\mathbf{T}}\cdot\mathbf{Q}_{\mathbf{x}}\right)-\\ & F_{m}\cdot\frac{D\cdot T_{s}}{2}\cdot\left(\mathbf{Q}_{\mathbf{x}}\cdot\Delta\mathbf{A}\cdot\mathbf{x}\cdot\mathbf{C}_{\mathbf{i}}\cdot\mathbf{A}_{1}+\mathbf{A}_{1}^{\mathbf{T}}\cdot\mathbf{C}_{\mathbf{i}}^{\mathbf{T}}\cdot\mathbf{x}^{\mathbf{T}}\cdot\Delta\mathbf{A}^{\mathbf{T}}\cdot\mathbf{Q}_{\mathbf{x}}\right)\end{aligned} \tag{82}$$

Except the first two, all the terms in (82) are zero for the elementary, loss-less topologies, i.e., buck, boost, buck-boost since either $\Delta\mathbf{A}$ is zero (buck) and/or $\mathbf{A}_1$ is skew symmetric "in all terms" except the last one of the main diagonal, which is proportional with the load resistance. $\Delta\mathbf{A}$ is also skew symmetric for all basic DC-DC topologies, i.e., buck, boost, buck-boost, Cuk, Sepic and dual Sepic (Zeta), [36], [40]. The derivative of the energy function becomes thus

$$\begin{aligned}\dot{V}_{cl}(\mathbf{x}) &= \frac{1}{2}\cdot\mathbf{x}^{\mathbf{T}}\cdot\left[\left(\mathbf{Q}_{\mathbf{x}}\cdot\mathbf{A}+\mathbf{A}^{T}\cdot\mathbf{Q}_{\mathbf{x}}\right)_{\mathrm{I}}-F_{m}\cdot\left(\mathbf{Q}_{\mathbf{x}}\cdot\mathbf{B}\cdot\mathbf{C}_{\mathbf{i}}+\mathbf{C}_{\mathbf{i}}^{\mathbf{T}}\cdot\mathbf{B}^{\mathbf{T}}\cdot\mathbf{Q}_{\mathbf{x}}\right)_{\mathrm{II}}\right]\cdot\mathbf{x}+\\ & \frac{1}{2}\cdot\mathbf{x}_{\mathbf{f}}^{\mathbf{T}}\cdot\left(\mathbf{Q}_{\mathbf{x}_{\mathbf{f}}}\cdot\mathbf{A}_{\mathbf{f}}+\mathbf{A}_{\mathbf{f}}^{\mathbf{T}}\cdot\mathbf{Q}_{\mathbf{x}_{\mathbf{f}}}\right)_{\mathrm{III}}\cdot\mathbf{x}_{\mathbf{f}}+\\ & \frac{1}{2}\cdot F_{m}\cdot\mathbf{x}^{\mathbf{T}}\cdot\begin{bmatrix}\left(\mathbf{Q}_{\mathbf{x}}\cdot\mathbf{B}\cdot\mathbf{C}_{\mathbf{f}}+\mathbf{C}_{\mathbf{f}}^{\mathbf{T}}\cdot\mathbf{B}^{\mathbf{T}}\cdot\mathbf{Q}_{\mathbf{x}}\right)_{\mathrm{IV}}+\\ +\left(\mathbf{Q}_{\mathbf{x}}\cdot\Delta\mathbf{A}\cdot\mathbf{x}\cdot\mathbf{C}_{\mathbf{f}}+\mathbf{C}_{\mathbf{f}}^{\mathbf{T}}\cdot\mathbf{x}^{\mathbf{T}}\cdot\Delta\mathbf{A}^{\mathbf{T}}\cdot\mathbf{Q}_{\mathbf{x}}\right)_{\mathrm{V}}\end{bmatrix}\cdot\mathbf{x}_{\mathbf{f}}+\\ & \frac{1}{2}\cdot\mathbf{x}_{\mathbf{f}}^{\mathbf{T}}\cdot\left(\mathbf{Q}_{\mathbf{x}_{\mathbf{f}}}\cdot[\mathbf{0}\quad-\mathbf{B}_{\mathbf{f}}]+[\mathbf{0}\quad-\mathbf{B}_{\mathbf{f}}]^{\mathbf{T}}\cdot\mathbf{Q}_{\mathbf{x}_{\mathbf{f}}}\right)_{\mathrm{VI}}\cdot\mathbf{x}+\\ & \left[\mathbf{x}^{\mathbf{T}}\cdot\mathbf{Q}_{\mathbf{x}}\cdot F_{m}\cdot(\mathbf{B}+\Delta\mathbf{A}\cdot\mathbf{x})_{\mathrm{VII}}+\left(\mathbf{x}_{\mathbf{f}}^{\mathbf{T}}\cdot\mathbf{Q}_{\mathbf{x}_{\mathbf{f}}}\cdot\mathbf{B}_{\mathbf{f}}\right)_{\mathrm{VIII}}\right]\cdot r\end{aligned} \tag{83}$$

where the subscripts in roman numerals were added for subsequent cross references. Detailing the terms for the boost converter gives (almost identical for a buck-boost and similar for the buck converter)

$$\dot{V}_{cl}(\mathbf{x}) = -\frac{1}{R_{ld}} \cdot x_2^2 \,\Big|_{\mathrm{I}} - F_m \cdot V_{in} \cdot M \cdot R_i \cdot \left(x_1^2 - \frac{1}{D' \cdot R_{ld}} \cdot x_1 \cdot x_2 \right)_{\mathrm{II}} +$$
$$\left[-\left(1 + \frac{C_1}{C_3}\right) \cdot \frac{1}{R_2} \cdot x_{f1}^2 - \frac{1}{2} \cdot \frac{1}{R_2} \cdot x_{f1}^2 + \frac{1}{2} \cdot \frac{1}{R_2} \cdot x_{f1} \cdot x_{f2} \right]_{\mathrm{III}} +$$
$$\frac{1}{2} \cdot F_m \cdot V_{in} \cdot M \cdot \omega_i \cdot \left(x_1 \cdot x_{f2} + x_2 \cdot x_{f1} - \frac{2}{D' \cdot R_{ld}} \cdot x_2 \cdot x_{f2} \right)_{\mathrm{IV}} +$$
$$\frac{1}{2} \cdot \omega_i \cdot F_m \cdot x_2 \cdot \left(x_2 \cdot x_{f1} - x_1 \cdot x_{f2} \right)_{\mathrm{V}} -$$
$$\left[\frac{1}{2} \cdot C_1 \cdot \left(x_2 \cdot x_{f1} + x_1 \cdot x_{f2} \right) + C_3 \cdot x_2 \cdot x_{f2} \right]_{\mathrm{VI}} +$$
$$\left[F_m \cdot V_{in} \cdot M \cdot \left(x_1 - \frac{x_2}{D'} \cdot \frac{1}{R_{ld}} \right)_{\mathrm{VII}} + \left(C_1 \cdot x_{f1} + C_3 \cdot x_{f2} \right)_{\mathrm{VIII}} \right] \cdot r \qquad (84)$$

where one can now reference which term in (84) corresponds to each of the terms in (83), through the roman numerals indices. For a regulator, $r = 0$, and the last terms disappears; it was kept in because it is used as the control input for the slave converter in the master-slave parallel control scheme, which is analyzed below.

Similarly with the open loop conclusions, relation (84) shows that the nature of the load resistance plays the most important role in determining the sign of the Lyapunov function derivative. It can actually be seen that the first term in the expression of $\dot{V}_{cl}(\mathbf{x})$ is the instantaneous output power with a minus sign. For the resistive load case, obviously, the output power is positive and hence the first term is negative. So is term "II", since x_1 and x_2, the inductor current and capacitor (output) voltage, vary in opposite directions from nominal. The terms multiplied by the compensator capacitors are very small and most of them negative. Each term can be investigated and plotted in greater detail. But, for an individual module that is a passive network with all the poles in the LHP, the converter will be Lyapunov stable, even though it may contain a bounded oscillation due to the PCMC sample and hold effect in the case of an ill design. For the paralleled system, however, even for resistive load, the two converters – each of them being still a constant power load - interact and a more detailed analysis of the energy function derivative terms is needed; this is the subject of the next section.

6.4 Lyapunov Stability for Two Identical Paralleled Converters

Considering two ideal, identical, converters and using the state error Eq. (63), the derivative of a Lyapunov function can be written with the state, **x** replaced with the error, **e**. With this substitution, the expression would have the same terms as in (84) for what pertains to the first row, linear part, of (63) and only the nonlinear terms in the last row of (63) would have to be calculated separately. We obtain

$$
\begin{aligned}
&\dot{V}_{cl}(\mathbf{e}_{\mathbf{x}}) = \frac{1}{2}\cdot\mathbf{e}_{\mathbf{x}}^{\mathbf{T}}\cdot\begin{bmatrix}(\mathbf{Q}_{\mathbf{x}}\cdot\mathbf{A}+\mathbf{A}^{T}\cdot\mathbf{Q}_{\mathbf{x}})_{\mathrm{I}} - \\ F_{m}\cdot(\mathbf{Q}_{\mathbf{x}}\cdot\mathbf{B}\cdot\mathbf{C}_{\mathbf{i}}+\mathbf{C}_{\mathbf{i}}^{\mathbf{T}}\cdot\mathbf{B}^{\mathbf{T}}\cdot\mathbf{Q}_{\mathbf{x}})_{\mathrm{II}}\end{bmatrix}\cdot\mathbf{e}_{\mathbf{x}} + \\
&\frac{1}{2}\cdot\mathbf{e}_{\mathbf{x_f}}^{\mathbf{T}}\cdot(\mathbf{Q}_{\mathbf{x_f}}\cdot\mathbf{A}_{\mathbf{f}}+\mathbf{A}_{\mathbf{f}}^{\mathbf{T}}\cdot\mathbf{Q}_{\mathbf{x_f}})_{\mathrm{III}}\cdot\mathbf{e}_{\mathbf{x_f}} + \\
&\frac{1}{2}\cdot F_{m}\cdot\mathbf{e}_{\mathbf{x}}^{\mathbf{T}}\cdot(\mathbf{Q}_{\mathbf{x}}\cdot\mathbf{B}\cdot\mathbf{C}_{\mathbf{f}}+\mathbf{C}_{\mathbf{f}}^{\mathbf{T}}\cdot\mathbf{B}^{\mathbf{T}}\cdot\mathbf{Q}_{\mathbf{x}})_{\mathrm{IV}}\cdot\mathbf{e}_{\mathbf{x_f}} + \\
&\frac{1}{2}\cdot\mathbf{e}_{\mathbf{x_f}}^{\mathbf{T}}\cdot(\mathbf{Q}_{\mathbf{x_f}}\cdot[\mathbf{0}\quad -\mathbf{B}_{\mathbf{f}}]+[\mathbf{0}\quad -\mathbf{B}_{\mathbf{f}}]^{\mathbf{T}}\cdot\mathbf{Q}_{\mathbf{x_f}})_{\mathrm{VI}}\cdot\mathbf{e}_{\mathbf{x}} + \\
&F_{m}\cdot\mathbf{e}_{\mathbf{x}}^{\mathbf{T}}\cdot\mathbf{Q}_{\mathbf{x}}\cdot\Delta\mathbf{A}\cdot\begin{pmatrix}\mathbf{x}_{2}\cdot\mathbf{C}_{\mathbf{i}}\cdot\left(\mathbf{I}+\frac{D\cdot T_{s}}{2}\cdot\mathbf{A}_{1}\right)\cdot\mathbf{x}_{2} - \\ \mathbf{x}_{1}\cdot\mathbf{C}_{\mathbf{i}}\cdot\left(\mathbf{I}+\frac{D\cdot T_{s}}{2}\cdot\mathbf{A}_{1}\right)\cdot\mathbf{x}_{1}\end{pmatrix}_{\mathrm{IX}} + \\
&F_{m}\cdot\mathbf{e}_{\mathbf{x}}^{\mathbf{T}}\cdot\mathbf{Q}_{\mathbf{x}}\cdot\Delta\mathbf{A}\cdot(\mathbf{x}_{1}\cdot\mathbf{C}_{\mathbf{f}}\cdot\mathbf{x}_{\mathbf{f1}}-\mathbf{x}_{2}\cdot\mathbf{C}_{\mathbf{f}}\cdot\mathbf{x}_{\mathbf{f2}})_{\mathrm{X}} - \\
&\left[\mathbf{e}_{\mathbf{x}}^{\mathbf{T}}\cdot\mathbf{Q}_{\mathbf{x}}\cdot F_{m}\cdot(\mathbf{B}+\Delta\mathbf{A}\cdot\mathbf{x})_{\mathrm{VII}}+(\mathbf{e}_{\mathbf{x_f}}^{\mathbf{T}}\cdot\mathbf{Q}_{\mathbf{x_f}}\cdot\mathbf{B}_{\mathbf{f}})_{\mathrm{VIII}}\right]\cdot r
\end{aligned}
\qquad (85)
$$

or after expliciting the terms,

$$\dot{V}_{cl}(\mathbf{e}_\mathbf{x}) = -\frac{1}{R_{ld}} \cdot e_{x_2}^2 \,|_{\mathrm{I}} - F_m \cdot V_{in} \cdot M \cdot R_i \cdot \left(e_{x_1}^2 - \frac{1}{D' \cdot R_{ld}} \cdot e_{x_1} \cdot e_{x_2} \right)_{\mathrm{II}} +$$
$$\left[-\left(1 + \frac{C_1}{C_3}\right) \cdot \frac{1}{R_2} \cdot e_{x_{f1}}^2 - \frac{1}{2} \cdot \frac{1}{R_2} \cdot e_{x_{f1}}^2 + \frac{1}{2} \cdot \frac{1}{R_2} \cdot e_{x_{f1}} \cdot e_{x_{f2}} \right]_{\mathrm{III}} +$$
$$F_m \cdot V_{in} \cdot M \cdot \omega_i \cdot \left[\frac{1}{2} \cdot \left(e_{x_1} \cdot e_{x_{f2}} + e_{x_2} \cdot e_{x_{f1}} \right) - \frac{e_{x_2}}{D' \cdot R_{ld}} \cdot e_{x_{f2}} \right]_{\mathrm{IV}} -$$
$$\left[\frac{1}{2} \cdot C_1 \cdot \left(e_{x_2} \cdot e_{x_{f1}} + e_{x_1} \cdot e_{x_{f2}} \right) + C_3 \cdot e_{x_2} \cdot e_{x_{f2}} \right]_{\mathrm{VI}} -$$
$$F_m \cdot R_i \cdot (x_{11} \cdot x_{22} - x_{12} \cdot x_{21}) \cdot e_{x_1} \,|_{\mathrm{IX}} + F_m \cdot \omega_i \cdot (x_{11} \cdot x_{22} - x_{12} \cdot x_{21}) \cdot e_{x_{f2}} \,|_{\mathrm{X}} -$$
$$\left[F_m \cdot V_{in} \cdot M \cdot \left(e_{x_1} - \frac{e_{x_2}}{D'} \cdot \frac{1}{R_{ld}} \right)_{\mathrm{VII}} + F_m \cdot \left(e_{x_1} \cdot x_{22} - e_{x_2} \cdot x_{12} \right)_{\mathrm{VII}} + \left(C_1 \cdot e_{x_{f1}} + C_3 \cdot e_{x_{f2}} \right)_{\mathrm{VIII}} \right] \cdot r \tag{86}$$

If we neglect the small terms - also negative - containing the two OpAmp compensator capacitors and resistor, we obtain

$$\dot{V}_{cl}(\mathbf{e}_\mathbf{x}) \cong -\frac{1}{R_{ld}} \cdot e_{x_2}^2 \,|_{\mathrm{I}} - F_m \cdot V_{in} \cdot M \cdot R_i \cdot \left(e_{x_1}^2 - \frac{1}{D' \cdot R_{ld}} \cdot e_{x_1} \cdot e_{x_2} \right)_{\mathrm{II}} -$$
$$F_m \cdot \left[V_{in} \cdot M \cdot \left(e_{x_1} - \frac{e_{x_2}}{D'} \cdot \frac{1}{R_{ld}} \right)_{\mathrm{VII}} + \left(e_{x_1} \cdot x_{22} - e_{x_2} \cdot x_{12} \right)_{\mathrm{VII}} \right] \cdot r + \tag{87}$$
$$F_m \cdot V_{in} \cdot M \cdot \omega_i \cdot \left[\frac{1}{2} \cdot \left(e_{x_1} \cdot e_{x_{f2}} + e_{x_2} \cdot e_{x_{f1}} \right) - \frac{e_{x_2}}{D' \cdot R_{ld}} \cdot e_{x_{f2}} \right]_{\mathrm{IV}} -$$
$$F_m \cdot R_i \cdot x_{22} \cdot e_{x_1}^2 \,|_{\mathrm{IX}} + F_m \cdot \omega_i \cdot x_{22} \cdot e_{x_1} \cdot e_{x_{f2}} \,|_{\mathrm{X}}$$

where the terms "IX" and "X" have been approximated with the new expressions since, in practice, $x_{21} \cong x_{22}$, being the internal variables for each individual module voltage perturbations, Fig. 4.2 and Fig. 4.3, as given by the new modeling technique. These quantities are not accessible for physical measurements, but they greatly alleviate the modeling task. The following calculations will show that a practical control law can be inferred from an engineering perspective, based only on the variables that are available in the real system.

It can be readily seen that the first two terms in (87) are negative. The rest of the terms vary their signs, as the signs of the different error quantities vary. As expected and can be checked on practical examples simulations, [37], the dominant factors within these terms are the ones containing the inductor currents error, e_{x1}, the other ones can be omitted in a first approximation. This is convenient from an engineering standpoint, because each current is accessible for measurement being actually used in PCMC. The product $\omega_i \cdot e_{xf2}$ is nothing but the error in the outputs of the internal OpAmp compensators for each module, $\omega_i \cdot e_{xf2} = e_{vea}$, also accessible for sensing.

If a control law would be derived for the "r" input so the Lyapunov function derivative stays negative, it would need to cancel all the terms except the first two, for example. Based on the above remarks and using (87), we obtain.

$$r \cong \frac{\frac{1}{2}\cdot\omega_i \cdot V_{in} \cdot M \cdot e_{x_{f2}} \cdot e_{x_1} - R_i \cdot x_{22} \cdot e_{x_1}^2 + \omega_i \cdot x_{22} \cdot e_{x_{f2}} \cdot e_{x_1}}{V_{in} \cdot M \cdot e_{x_1} + x_{22} \cdot e_{x_1}}, \tag{88}$$

which can be simplified to

$$r = \frac{\left(\frac{1}{2}\cdot V_{in} \cdot M + x_{22}\right)\cdot e_{v_{ea}} - R_i \cdot x_{22} \cdot e_{x_1}}{V_{in} \cdot M + x_{22}}. \tag{89}$$

Observing also that x_{21} and x_{22} being the internal output voltage variation for each converter they are about half of the system's output voltage perturbation, $x_{21} \cong x_{22} \cong v_o/2$ and using the fact that $V_{od} = V_{in} \cdot M$ is the ideal, desired, DC output voltage, we have that

$$r = \frac{\frac{1}{2}\cdot\left(V_{od} + v_o\right)}{V_{od} + v_o/2}\cdot e_{v_{ea}} - R_i \cdot \frac{\frac{1}{2}\cdot v_o}{V_{od} + \frac{1}{2}\cdot v_o}\cdot e_{x_1}. \tag{90}$$

Adding an adjustable gain factor for each term and since in the real system it is easier to sense the total, DC plus AC, output voltage, $v_{o\_tot} = V_{od} + v_o$ we can further approximate (90) as

$$r = K_1 \cdot \frac{v_{o\_tot}}{v_{o\_tot} + V_{od}}\cdot e_{v_{ea}} - K_2 \cdot R_i \cdot \frac{v_{o\_tot} - V_{od}}{v_{o\_tot} + V_{od}}\cdot e_{x_1}. \tag{91}$$

Even expression (91) can be further simplified in terms of its computation since the main requirement is to feed back the two error quantities, mainly the one of the two currents, with their respective adjustable gains. We obtain, acquainting for the last term sign change,

$$r = K_1 \cdot e_{v_{ea}} + K_2 \cdot R_i \cdot \frac{v_{o\_tot}}{V_{od}} \cdot e_{x_1} \cdot \tag{92}$$

Eqs. (91) and (92) have been obtained after many simplifications and term omissions, due to the complicated, implicit form, of the error equation in the case of non-identical converters. Even in this case, implementing the exact control law that would cancel all the variable sign terms is not feasible in practice and the simulations run, in several instances, also have shown computational issues. However, with a few gain adjustments, (91) and (92) can be used as supervisory controllers for the fuzzy logic controller, i.e. if the latter has a high gain that drives the system unstable. One implementation is to sum the two controllers action, either continuously or at a specified instant, if the current sharing error does not go within the desired boundaries. The commutation indicator function, *I*, can be done through a direct switching action, or by employing a soft switch-in step, [25],

$$u_{tot} = u_{fuzzy} + I \cdot u_{Lyap}, \tag{93}$$

with the indicator function, I=1 if $|\mathbf{x}|$ or $|e_{io}|$ is greater than the desired boundary M and zero otherwise.

If used as a "stand alone", the last two control laws would not offer the performance of the fuzzy controller, being inferred after a lot of simplifications and dynamic-term cancellation. For boost-derived converters, as in this example, the lag effect of the RHPZ in the power stage also has to be considered when choosing the two gains, K_1 and K_2. For example, when a fast fuzzy controller is used, e_{x1} will be different instantaneously than e_{io}, having opposite signs since the inductor current will briefly lag the output one. For a stand alone, slow, Lyapunov controller, the two would have the same sign. In addition and linked to the inductor current error, e_{x1}, the error in the outputs of the internal voltage compensators, e_{vea}, may be slightly shifted, with opposite signs both because the RHPZ lagging effect, but mainly due to the differences between the values of the converter's parasitics and components, e.g. tolerances, offsets, etc. For substantially different converters the sign of K_I may also need to be adjusted. The two current's error will always need to be fed back, even though only a voltage mode controller is employed; this is fairly obvious from an engineering viewpoint even though the simplified relations (91) and (92) may suggest that if PCMC is not used, the term in e_{xI} will not be present. This is only because of the simplifying assumptions and term reductions in the two expressions, and hence all the terms required in practice need to be kept in. Also, one has to recall that we had to write the equations and come up with the models and control laws considering the entire state error, which is more stringent and complicated than considering only the current sharing error. This was dictated by the complicated nature of the system, as the only access path to the current sharing error quantity in a modern control systems framework.

To recapitulate, even after the simplifications required by the complicated form of this type of systems, we were able to come up with a control strategy that offers good bounds for the converter variables, if not also asymptotic stability. The control rule is intuitive from an engineering point of view, since it practically feeds back opposite sign information proportional to the currents error and the difference of the two DC-DC individual modules inner control quantities.

The present research is still work-in-progress for more verifications, [37], which will also highlight possible future research paths for these type of applications. One of the features that complicates the system structure is that the presently established control schemes for the DC-DC converters in the industry are not a linear state feedback, per se. They are a summation of different loops for each state variable, with different gains and/or OpAmp compensators in each loop. This makes the models have a cumbersome form for direct modern control implementations, especially in parallel, distributed power systems. One direction would be to have a linear, stable, full-state feedback for each module, which would then make the plants more amenable to be integrated in a paralleled system, where the same control strategy can be continued. Another approach is to recognize the variable structure features of the DC-DC topologies and to employ a sliding-mode controller, [13], [38], in each module. Associated (or independent) control laws through input-output linearization and coordinate transformations, [34], [35], [38], [39], can also be used.

As a last, but very important note, an additional stabilizing controller will only be needed if a very high gain fuzzy controller is desired. This is because, due to the internal structure of the fuzzy logic compensator described earlier, when the error is at maximum, the control output will saturate Negative Big or Positive Big; and, if we do not use a very high gain, we would have the system's response to a constant command on the DC-DC module reference input. This response may be slower, but it is stable in almost all cases. The fuzzy logic interface is described in the next section.

7 Simulation Results and Comments

Small and large signal simulations have been made on a three PCMC boost converter system supplied from 28 V input voltage that delivers a 44 V output voltage at 22 A load for each converter, thus 66 A nominal load for the system. The converters are chosen with relevant parasitics and tolerances as listed below, so the effect of the unmatched parameters is present in the paralleled system. Their nominal values are f_s=200 kHz, L=10 μH, R_L= 20 mΩ, C=500 μF, R_C=10 mΩ. The internal loop compensator has the same architecture as the CSC in Fig. 2.2 and has the following parameters: $\omega_i=6.5876\cdot10^3$ rad/s, $\omega_z=1.8519\cdot10^3$ rad/s, $\omega_p=85.185\cdot10^3$ rad/s. In the large signal simulations, the OpAmp compensators are assumed supplied from 12 V and operate around a reference voltage of 5.1 V. The master converter module has the nominal parameters - reference voltage, output filter values, internal compensator R and C values - and the parameters of the other two deviate from nominal within the usual 1% to 5% fabrication tolerances.

The system small signal loop gains employing the MSC topology in Fig. 2.1 are plotted in Fig. 7.1 together with the PID improved loop gain.

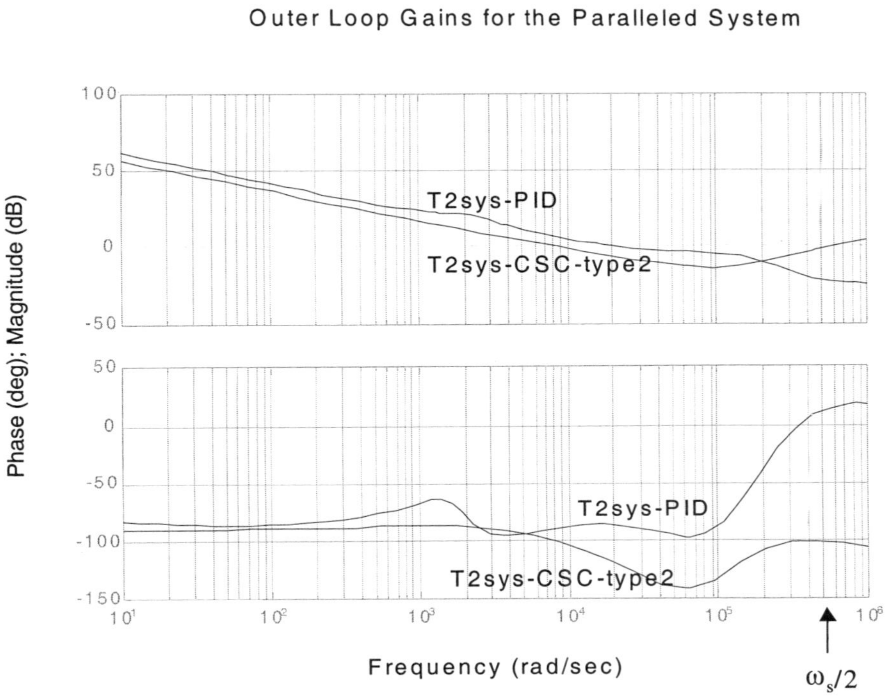

Fig. 7.1. Small signal loop gains for a classical design and an improved PID design.

The BW for the system using the classical CSC is approximately 1.6 kHz, with a PM of 80° and infinite gain margin. The PID design has a BW of 4.7 kHz, with a PM of 100° and also infinite GM. A better dynamic performance for the PID controller can be inferred, as it will be better shown in the large signal, time domain simulated results.

Comparative 25% and 75% nominal load step responses are given in Fig. 7.2-(a) for the classical OpAmp CSC, Fig.7.2(b) for the PID design and Fig. 7.2-(c) for the fuzzy controllers with relatively low gains, 40 and respectively 20, so they do not need a Lyapunov stability adaptation.

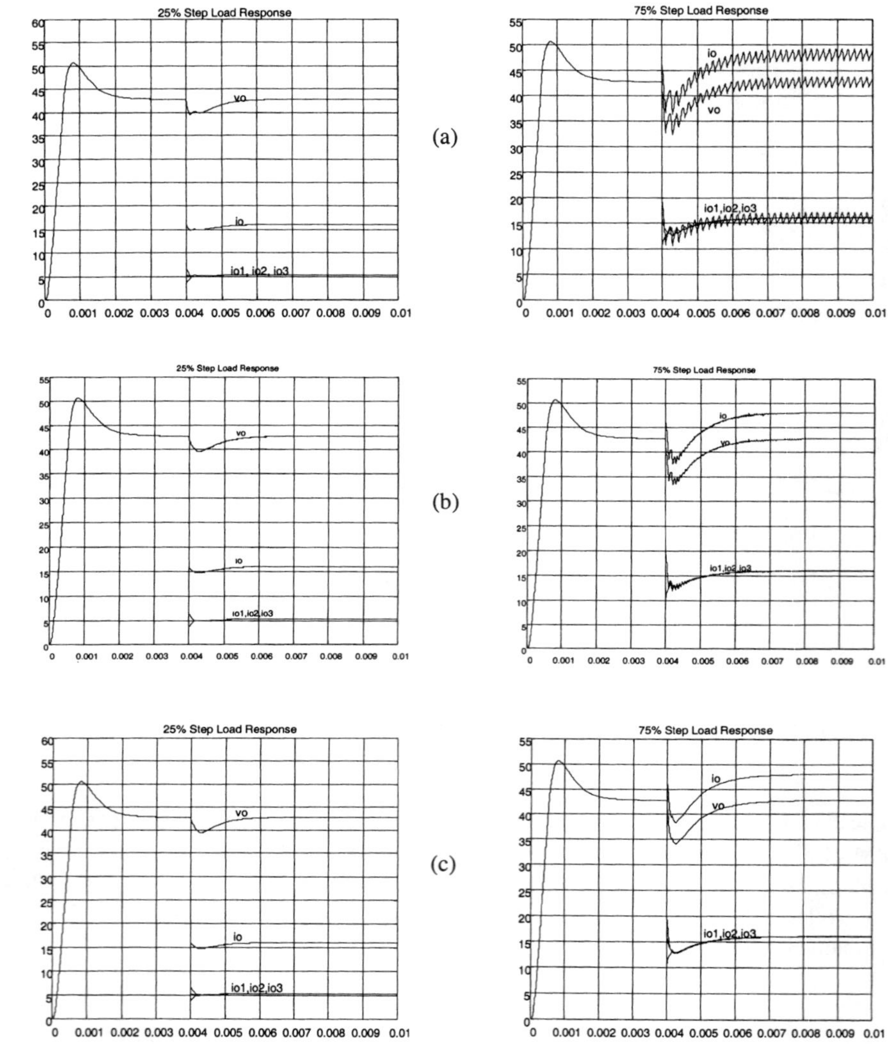

Fig. 7.2. Step load responses for 50% and 75% of the nominal load:

(a)-System with the classical CSC in Fig. 2.1;

(b)-System employing a PID CSC; (c)-System using a Fuzzy Logic CSC.

A substantial improvement in the transient performance is noted in the case of the PID and especially of the fuzzy logic controller. The identical responses for the three step load cases when using the fuzzy control approach shows the validity of fuzzy control in dealing naturally with the complicated and nonlinear character of a system. The low performance index in the traditional cases is due to the fact that a conventional linear controller, designed only on a small signal model, is valid only near the operating point and not for the entire plant trajectory. The fuzzy approach gives also a simple practical way of overcoming modeling complexity by using heuristic judgment. It can also be noticed that for each step load, the voltage waveforms for the three controllers are close, since in general, the output voltage dynamics are influenced only by the internal loops of the master converter, the slave converters exhibiting practically no voltage gain.

7.1 Control Effort

Fig. 7.3 depicts the CSC compensator outputs plots for the same cases shown in Fig. 7.2. One can again see that the fuzzy controller gives the minimal needed effort to control the system current sharing. The control voltage barely saturates in this case, Fig. 7.3-(c). In comparison, the use of the compensator in Fig. 2.2, is the slowest and requires the most control effort, the PID controller gives an intermediary performance. The noise-like areas of the waveform are given by numerical errors in Matlab. Cleaner waveforms can be obtained by increasing the precision of the numerical methods and the resolution with which the graphs are printed, however the simulation time becomes very long and impractical.

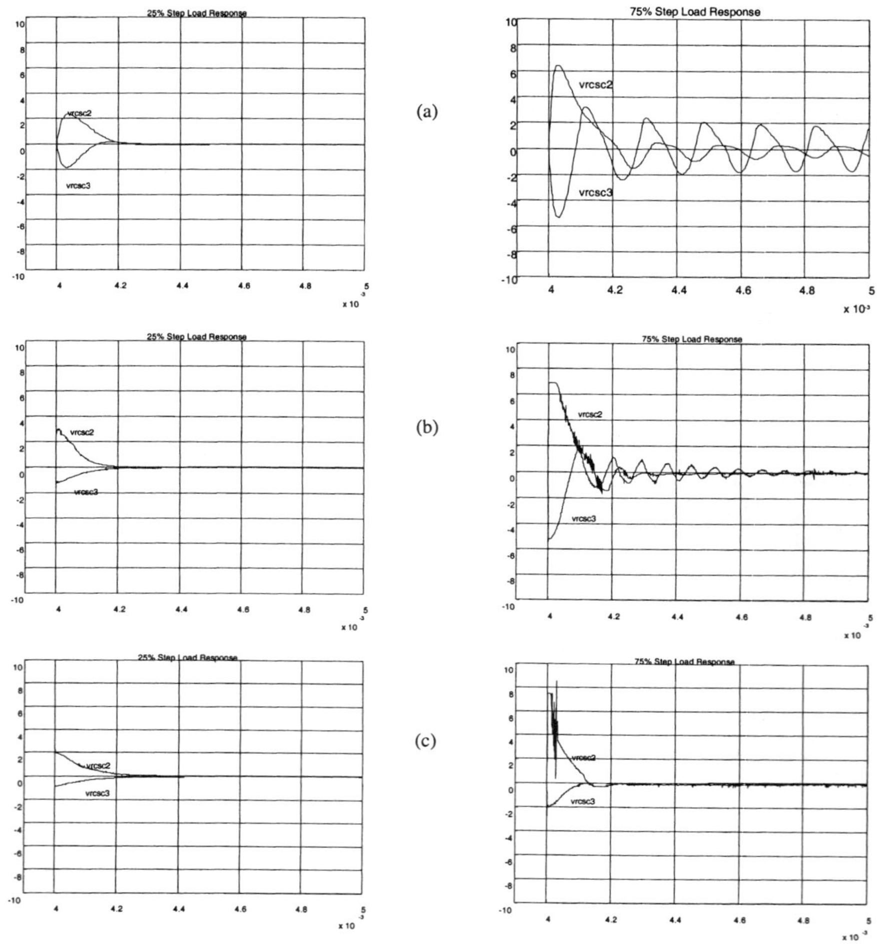

Fig. 7.3. CSC Control Effort: (a)-System with the classical CSC in Fig. 2.1;
(b)-System employing a PID CSC;
(c)-System using a Fuzzy Logic CSC.

To ensure stability, the Lyapunov descent path supervisory controller highlighted above was used in an oscillating system with higher gains for the fuzzy compensator, which is illustrated in Fig. 7.4.

(a)

(b)

(c)

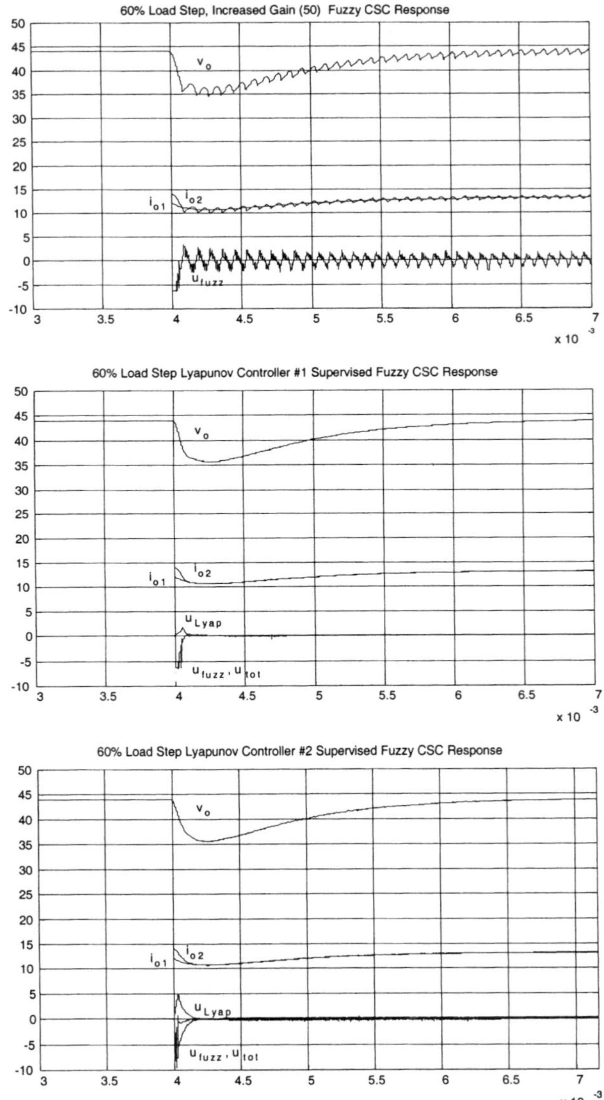

Fig. 7.4. Lyapunov stability adaptation waveforms: (a)-Stand alone Fuzzy Logic Controller;

(b)-Combined FLC and Lyapunov supervisory controller (91), K_1=0.05, K_2=2;

(c)-Combined FLC and Lyapunov supervisory controller (92), K_1=0.05, K_2=0.5.

Two converters sharing a 44 A load, hence a higher load than before for each module, were used in a 60% step load case simulation with an oscillatory behavior given by a high gain fuzzy controller, i.e., 50, Fig. 7.4 (a). The control laws (91) – Fig. 7.4 (b) and (92) – Fig. 7.4 (c), were appended continuously in a summed, supervisory controller. It can be seen that the Lyapunov supervisory controller has an immediate regulatory action and brings the error to zero. This is done by keeping the total gain from exceeding the value that would bring the system to oscillate, e.g., 50, through the algebraic sum of the two controllers. On an off-line gain adjustments are possible, based on each practical application.

7.2 Conclusions

A Lyapunov-supervised stable fuzzy logic controller for the master-slave current sharing loop of a paralleled DC-DC system has been discussed with considerably improved large signal performance over the presently employed, small signal, compensators, both in system response and control effort. The design uses a PID expert to derive the fuzzy inference rules, and simulation results show a good parameter insensitive transient response in load-step responses from 25% to 75% of the nominal load. The technique benefits also from the heuristic approach to the problem that overcomes the complexity in modeling such systems and, hence, offers a practical engineering tool, amenable to both analog and digital implementations.

The drawbacks of the fuzzy control could be a lower noise immunity together with different errors, e.g. quantization errors associated with any digital controller, for example. This can be addressed on an application specific basis either through conventional, analog or digital, filtering or by building filtering capabilities in the fuzzy controller itself, e.g. by using triangular or Gaussian fuzzifiers instead of the singleton fuzzifiers, [25]. Quantization and other types of errors are present in any practical implementation, but more and more precise and high speed processor are becoming available to the practicing engineer. The main objective of this study is to introduce the new fuzzy logic control application, develop the associated mathematical theory and prove the concept and its advantages through comparative simulation verification with the existing, classical, methods. Only a full practical implementation study would fully clarify if different errors and noise raise any substantial issues; this is left as a subject for future research work.

References

1. van Wyk J.D and Lee F.C. (1999), Power electronics technology at the dawn of the new millennium - status and future, *Proceedings of IEEE Power Electronics Specialist Conference*, pp. I-3-I-13, Charleston, SC.
2. Tomescu B. and VanLandingham H. (1999), Improved large signal performance of paralleled DC-DC converters current sharing using fuzzy logic Control, *IEEE Trans. Power Electronics*, **14**, no. 3, pp. 573 -577.
3. Middelbrook R.D. and Cuk S. (1983), A general unified approach to modeling switching converter power stages, *Advances in Switched Mode Power Conversion*, Tesla Co. **1**, pp. 73-89.
4. Vorperian V. (1989), Simplified analysis of PWM converters using the model of the PWM switch, Part I: continuous conduction mode, Part II: discontinuous conduction mode, VPEC Seminars, September.
5. Sun J. (2000), Unified averaged switch models for stability analysis of large distributed power systems, *Proceedings of IEEE Applied Power Electronics Conference*, pp. 249-255, New Orleans, LA.
6. Panov Y., Rajagopolan J., and Lee F.C. (1997), Analysis and design of *N* paralleled DC-DC converters with master-slave current sharing control, *Proceedings of IEEE Applied Power Electronics Conference*, pp. 436-442, Atlanta, GA.
7. Erickson R.W., Cuk S. and Middlebrook R.D. (1982), Large-signal modeling and analysis of switching regulators, *Proceedings of IEEE Power Electronics Specialist Conference*, pp. 240-250, Cambridge, MA.
8. Harada K. and Nabeshima T. (1981), Large signal transient response of a switching regulator, *Proceedings of IEEE Power Electronics Specialist Conference*, pp. 388-394, Boulder, CO.
9. Salut G., Marpinard J.C., and Valentin M. (1985), Large signal feedback control for power switching conversion, *Proceedings of IEEE Power Electronics Specialist Conference*, pp. 741-750, Toulouse, France.
10. Sanders S.R. and Verghese G.C. (1992), Lyapunov-based control for switched power converters, *IEEE Trans. Power Electronics,* **7,** no. 1, pp. 17-24.
11. Chen F. and Cai X.S. (1989), Design of feedback control laws for switching regulators based on the bilinear large-scale model, *Proceedings of IEEE Power Electronics Specialist Conference*, pp. 468-475, Milwaukee, WI.
12. Derese I. and Noldus E., Design of linear feedback laws for bilinear systems, *International Journal of Control*, **31**, no. 2, pp. 219-237.
13. Malesani L., Rossetto L., Spiazzi G., and Tenti P. (1992), Performance optimization of Cuk converters by sliding-mode control, *Proceedings of IEEE Applied Power Electronics Conference*, pp. 395-402, Boston, MA.
14. Hiti S. and Borojevic D. (1995), Robust nonlinear control for boost converters, *IEEE Trans. Power Electronics*, **10**, no. 6, pp. 651-658.
15. Mattavelli P., Rossetto L., Spiazzi G., and Tenti P. (1995), General purpose fuzzy controller for DC-DC Converters, *Proceedings of IEEE Power Electronics Specialist Conference*, pp. 723-730, Atlanta, GA.
16. Ueno F., Inoue T., Oota I., and Sasaki M. (1991), Regulation of Cuk converters using fuzzy controllers, *Proceedings of International Telecommunications Energy Conference*, pp. 261-267, Kyoto, Japan.

17. So W.C., Tse C.K., and Lee Y.S. (1996), Development of a fuzzy logic controller for DC-DC converters: design, computer simulation and experimental evaluation, *IEEE Trans. Power Electronics*, **11**, no.1, pp. 24 -32.
18. Schefler M., Bellini A., Rovatti R., Zafarana A., and Diazzi C. (1996), A fuzzy analog controller for high- performance microprocessor power supply, *Proceedings of 4th European Congress on Intelligent Techniques and Soft Computing*, pp. 1000-1004, Aachen, Germany.
19. Rajagopolan J., Xing K., Guo Y., and Lee F.C. (1996), Modeling and dynamic analysis of paralleled DC-DC converters with master/slave current sharing control, *Proceedings of IEEE Applied Power Electronics Conference*, pp. 678-684, San Jose, CA.
20. Rajagopolan J., Panov Y., Tomescu B., and Guo Y. (1995), Stability assessment of distributed power systems used in ISSA, *NASA LeRC Quarterly Report*, Virginia Tech - VPEC, December.
21. Thottuvelil V.J. and Verghese G. (1997), Analysis and control design of paralleled DC-DC converters with current sharing, *Proceedings of IEEE Applied Power Electronics Specialist Conference*, pp. 638-646, Atlanta, GA.
22. Siri K., Lee C.Q., and Wu T.F. (1992), Current distribution control for parallel-connected converters, *IEEE Trans. Aerospace and Electronic Systems*, **AES-28**, no. 3, pp. 829-850.
23. Luo S., Ye Z., Lin R.L., and Lee F.C. (1999), A classification and evaluation of paralleling methods for power supply modules, *Proceedings of IEEE Power Electronics Specialist Conference*, Charleston, SC, pp. 901-908.
24. Tomescu B. and VanLandingham H. (1997), Neuro-fuzzy multi-model control using Sugeno inference and Kohonen tuning in parameter space, *Proceedings of IEEE International Conference on Systems, Man and Cybernetics*, pp. 1028-1032, Orlando, FL.
25. Wang L.X. (1997), *A Course in Fuzzy Systems and Control*, Prentice Hall, Upper Saddle River, NJ.
26. Lewis L.R., Cho B.H., Lee F.C., and Carpenter B.A., Modeling and analysis of distributed power systems: modeling, analysis and design of PWM converters, **2**, *VPEC Publication Series*, pp. 505-514.
27. Brown A.R. and Middlebrook R.D. (1981), Sampled-data modeling of switching regulators, *Proceedings of IEEE Power Electronics Specialist Conference*, pp. 349-369, Boulder, CO.
28. Feng X., Ye Z., Xing K., Lee F.C., and Borojevic D. (1999), Impedance specification and impedance improvement for DC distributed power systems, *Proceedings of the IEEE Power Electronics Specialist Conference*, pp. I-889-I-894, Charleston, SC.
29. Ridley R., Cho B. H., and Lee F.C., (1988), *Analysis and interpretation of loop gains of multiloop-controlled switching regulators, IEEE Trans. Power Electronics,* **3**, no. 4, pp. 489-498.
30. Ridley R. (1991), A new, continuous-time model for current-mode control, *IEEE Trans. Power Electronics*, **6**, no. 2, pp. 271-280.
31. Tymerski R. and Li D. (1993), State-space models for current programmed pulsewidth-modulated converters, *IEEE Trans. Power Electronics*, **8**, no. 3, pp. 271-278.
32. Tomescu B. and VanLandingham H. (1999), Disturbance rejection and robustness considerations in DC-DC converters, *Proceedings of the IEEE Power Electronics Specialist Conference*, pp. 1204-1210, Charleston, SC.

33. Jordan M. (1991), Load share IC simplifies power supply design, *Proceedings of High-Frequency Power Conversion Conference*, pp. 65-76, Toronto, Canada.
34. Isidori A. (1992), *Nonlinear Control*, Springer Verlag, Berlin, Germany.
35. Khalil H. (1992), *Nonlinear Control*, Macmillan Publishing Company, New York, NY.
36. Mitchell D. (1988), *DC-DC Switching Regulator Analysis*, McGraw-Hill Book Company, New York, NY.
37. Tomescu B, (2001), *On the use of fuzzy logic to control paralleled DC-DC converters*, Ph.D. Dissertation, Virginia Polytechnic Institute and State University, Blacksburg, VA.
38. Sira-Ramirez H. (1987), Sliding motions in bilinear switched networks, *IEEE Trans. Circuits and Systems*, **34**, no. 8, pp. 919-933.
39. Sira-Ramirez H. (1989), Exact linearization in switched mode DC to DC power converters, *International Journal of Control*, **50**, no. 2, pp. 511-524.
40. Severns R.P. and Bloom G.E. (1985), *Modern DC-to-DC Switchmode Power Converter Circuits*, Van Nostrand-Reinhold Publishing Company, New York, NY.

Soft Computing-Based Optimal Operation in Power Energy System

Akimoto Kamiya[*], Masakazu Kato[*], Kazue Shimada[*], and Shigenobu Kobayashi[†]

[*] Toshiba Corporation

[†] Tokyo Institute of Technology

Keywords: *power system, thermal power plant, short-term load forecast, economic load dispatch, unit commitment, start-up scheduling, artificial neural networks, genetic algorithms, optimization, active constraints.*

1 Introduction

Fossil fuels, chiefly coal, oil and natural gas, currently account for more than 60 % of the primary energy used for electricity generation worldwide. This share will continue to increase steadily along with the growing global electricity demand [48]. There is therefore great demand for optimal operation of power energy systems aimed at reducing fossil fuel consumption. Such optimal operation could not only save fuel cost but also reduce CO_2 emission, which is considered the main contributor to global warming.

A power energy system normally consists of a number of power plants distributed over a wide area, connected by an electric power grid. As shown in Fig. 1, optimal operation of such a system can generally be divided into two levels [37]. The higher level is carried out at the load dispatch center, while the lower level is carried out at individual power plants. Note that, in this chapter, a power plant is referred as a power generating unit. In the load dispatch center, weather data and historical load data are used for short-term load forecasting to predict the load demand of the power system with a lead time of a few hours to weeks. Subject to the short-term load demand, the economic load dispatching system optimizes power plants' base load set points as well as start-up/shutdown times, aimed at minimizing the overall quantity of fossil fuels consumed by the power plants. Based on the base load set points and start-up times given by the economic load dispatch center, optimal start-up scheduling is performed at the power plant level in order to minimize the start-up time by generating an optimal start-up schedule for the power plant control systems.

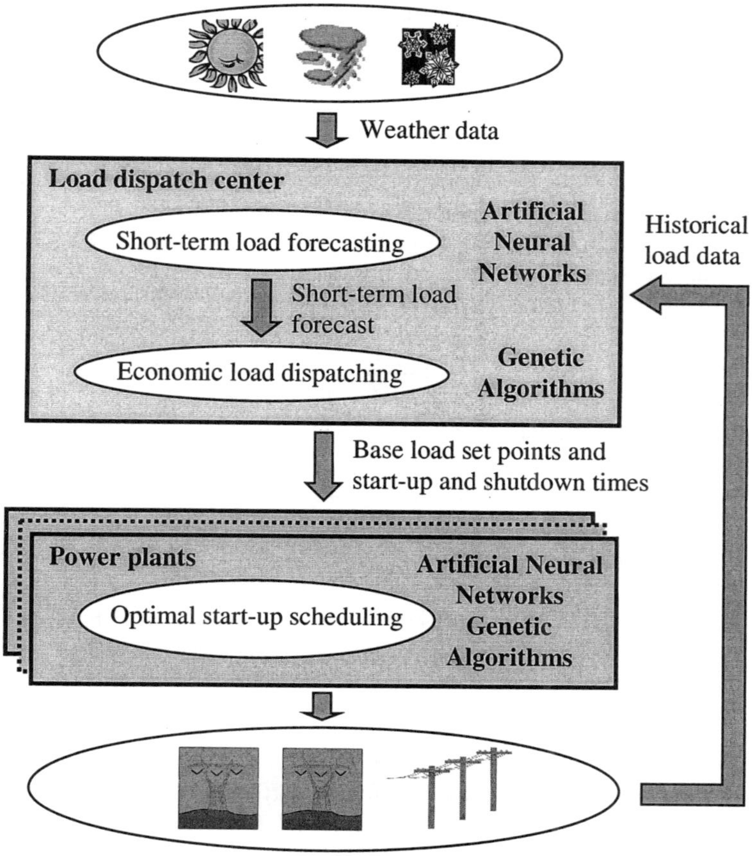

Fig. 1. Soft Computing techniques used for power energy operation system.

To achieve such an optimal system, however, involves the short-term load forecasting problem, the unit commitment problem and the optimal start-up scheduling problem. All these problems are highly nonlinear and the optimization problems contain a number of local optima, making them difficult to solve by the conventional "hard computing" techniques alone. Applications of soft computing—artificial neural networks and genetic algorithms, aimed at achieving an optimal load dispatch and automatic power plant start-up control system, are described in this chapter.

2 Short-Term Load Forecasting

2.1 General

STLF (Short-Term Load Forecasting) has always been an important task at the load dispatch center. STLF results are used for controlling and scheduling power systems for economic operation, aimed at minimizing fossil fuel consumption. They are also used as inputs to load flow studies and contingency analyses to achieve stable and reliable power system operation. Optimal power system operation depends completely on the accuracy of the STLF. Due to the nonlinearity and randomness of the power system load, accurate STLF is difficult problem to achieve. Several STLF models have been proposed [7,20,21,22,39], including traditional models such as the regression method and modern models such as ANN (Artificial Neural Networks). The regression model approach has been used at load dispatch centers for a long time. Implementation of such a model, however, requires complicated modeling techniques and enormous computation effort to produce acceptable results. ANN is capable of generalization and learning without pre-assuming any specific model. Interests in applying ANN to STLF began in late 1980. Recently, STLF based on ANN [23,25] or ANN with a fuzzy heuristic approach [38] has been introduced successfully into load dispatch centers for actual power system operation with good and acceptable forecasting accuracy.

In load forecasting, ANN is usually used for extracting the relationship between weather variables and load demand for a particular day type and season. However, the load demand is also affected by other variables such as the annual load demand trend, which is independent of weather variables but instead is dependent on the economic situation and other factors of that year. Depending on the annual load demand trend, even if the weather of a particular period in the past year is the same as that of the current year, the load demand between them may be different. This chapter describes an ANN trained by a two-stage training method [26] for compensation of the annual load demand trend. The trained ANN is used for daily peak load forecasting, which serves as basis for the daily load demand curve forecasting and more importantly for assuring an appropriate reserve supply margin.

Note that the load demand is also influenced by special events including holiday, sports events, major television programs and others. Since the occurrence time of such special events is irregular, they are manually adjusted by power system operators.

2.2 Outline of an Artificial Neural Network

A multi-layer perception model with a backpropagation learning algorithm is widely applied to load forecasting. The ANN used for STLF is outlined in Fig. 2. As shown, the ANN consists of three layers. Each layer contains units as follows.

a) The input layer contains three units, each having an identity function. This approach considers the six candidate weather variables listed below, based on which three is chosen for the input units by preliminary experiments.

   (1) Temperature at 10 AM of the forecasted day,
   (2) Temperature at 3 PM of the forecasted day,
   (3) Maximum temperature of the forecasted day,
   (4) Average temperature of 3 AM, 10 AM and the maximum temperature of the forecasted day,
   (5) Humidity of the forecasted day, and
   (6) Maximum temperature of the previous day.

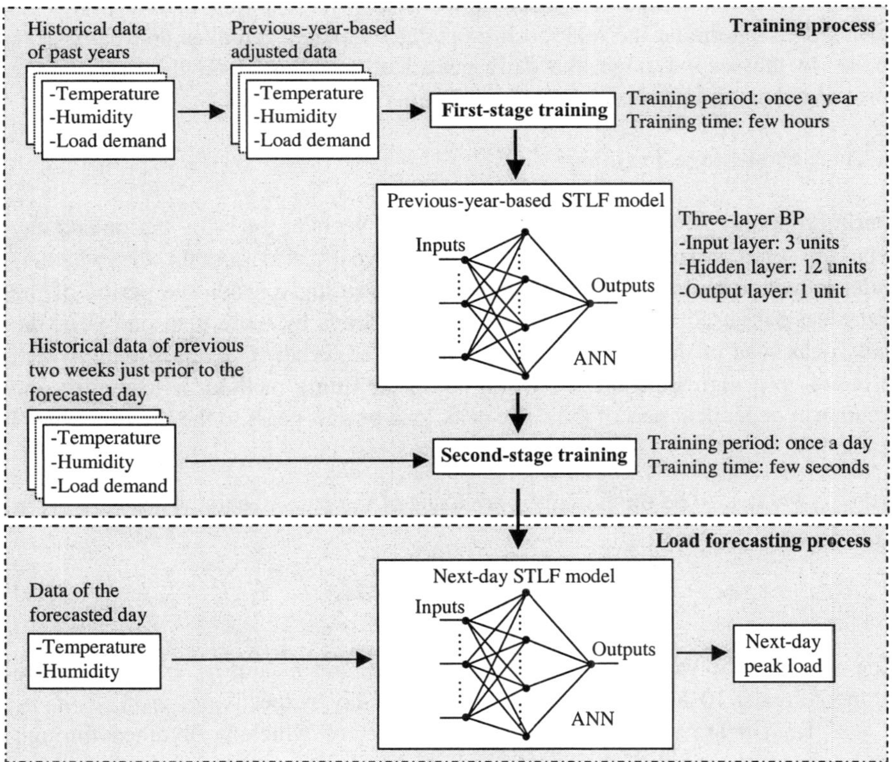

Fig. 2. An STLF model based on an ANN with a two-stage training method.

b) The hidden layer contains twelve units, each having a sigmoidal function. The number of hidden units is decided by preliminary experiments.

c) The output layer has one unit with a linear function, which represents the peak load of the day.

As shown in Fig. 2, the operations of ANN for STLF consist of two processes: training and forecasting. During the training process, a two-stage training method is used for training the ANN. The ANN training is based on the historical data of past years and the data of the two weeks just prior to the forecasted day. During the load forecasting process, given a set of weather variables of the forecasted day, the peak load of the forecasted day is predicted by using the trained ANN.

2.3 Two-Stage Training Method

The power system daily peak load is assumed to be a linear combination of two parts: a weather-dependent part and a weather-independent part. Based on this assumption, the two-stage training method is applied to determine the contribution of these two parts in separate stages. In the first stage, historical data of past years are used to determine the relationship between weather variables and daily peak load. In the second stage, the daily peak load is adjusted on the basis of the historical data of the two weeks just prior to the forecast day.

2.3.1 First-Stage Training

During first-stage training, historical data of past years of the same season and day type are used. The historical data of past years also contain the weather-independent part of the peak load demand. Accordingly, when the period of the historical data used for training in this stage expands by more than one year, the daily peak load of the historical data have to be adjusted. The adjustment, which is based on a regression model with a nonlinear fitting method, is to modify the weather-independent part of the daily peak load of past years to the previous year's level.

Let $\hat{Y}'_i$, $i = 1,\ldots, n$ be the fit daily peak loads of the past n years. These are given by:

$$\hat{Y}'_i = a_i T_i^3 + b_i T_i^2 + c_i T_i + d_i, \tag{1a}$$
$$\text{where } T_i = e_i t_{i1} + f_i t_{i2},\ i = 1,\ldots, n, \tag{1b}$$

T_i, $i = 1,\ldots, n$ is weighted average temperature, and t_{i1} and t_{i2}, $i = 1,\ldots, n$ are temperatures at 10 AM and 3 PM of the relevant day, respectively. a_i, b_i, c_i, d_i, e_i, f_i, $i = 1,\ldots, n$ are sets of coefficients, each set of which is obtained through nonlinear fitting based on the corresponding year's temperature and the daily peak load data.

Let $\hat{Y}_i$, $i = 2,\ldots, n$ be the adjusted daily peak load of the past $n - 1$ years. These are adjusted to the previous year $i = 1$ as follows:

$$\hat{Y}_i = Y_i - D_i, \tag{2a}$$

$$\text{where } D_i = average(\hat{Y}'_i) - average(\hat{Y}'_1),\ i = 2,\ldots, n, \tag{2b}$$

Y_i, $i = 1,\ldots, n$ are the actual daily peak loads of the past n years and D_i, $i = 2,\ldots, n$ are the differences between the weather-dependent daily peak loads of the past $n - 1$ years ($i = 2,\ldots, n$) and the previous year ($i = 1$). $average(\hat{Y}'_i)$ means the average of the fit daily peak loads of the corresponding year ($i = 1,\ldots, n$).

The actual daily peak loads of the previous year Y_1, the adjusted daily peak loads of past $n - 1$ years $\hat{Y}_i$, $i = 2,\ldots, n$, and the corresponding weather variables are used to train the ANN at the first stage.

2.3.2 Second-Stage Training

During second-stage training, historical data within the two weeks just prior to the forecasted day are used. Second-stage training is to train the ANN, which has been trained according to first-stage training, to learn the weather-independent part of the daily peak load of the current year. Although the nonlinear sigmoid function is widely used as the output function for the ANN, a linear function is instead used for the STLF. A linear function is used to control the learning rate within a preset limit. This is to keep the ANN from overfitting to the small set of second-stage training data, so as to preserve the weather-dependent part of the daily peak load, which the ANN learned during first-stage training. Furthermore, for the same reason (to avoid overfitting), during second-stage training, the weights between the input layer and the hidden layer are fixed.

Let W_k be the preset limit of the change in weights between the hidden layer and the output layer, and Δ be the limit of the change in gradient of the input-output function. W_k and Δ are defined by:

$$W_k = \max_{j=1,\ldots,N_2} \left| w_{jk}^{2'} - w_{jk}^{2} \right|, \tag{3}$$

$$\Delta = \max_{i=1,\ldots,N_1} \left\{ \max_{I_l \le I_i \le I_h} \left| \frac{\partial o'_k}{\partial I_i} - \frac{\partial o_k}{\partial I_i} \right| \right\}, \tag{4}$$

where N_2 is the number of units of the hidden layer, w_{jk} is the pre-second-stage-training value of the weights connecting the j-th unit of the hidden layer and the k-th unit of the output layer, w'_{jk} is the post-second-stage-training value of the weights, $I_i, i = 1,\ldots, N_1$ is the i-th unit of the input layer, I_l and I_h are the lower and upper limits of those inputs, respectively, N_1 is the number of units of the input layer, $o_k, k = 1,\ldots, N_3$ is the pre-second-stage-training output of the k-th unit of the output layer, and $o'_k, k = 1,..., N_3$ is the post-second-stage-training output of the same unit.

Through a mathematical proof, Eq. (5) is obtained [27].

$$\Delta = W_k \times \max_{i=1,...,N_1} \sum_j \frac{|w_{ij}|}{4}, \tag{5}$$

where w_{ij} is the weight connecting the i-th unit of the input layer and j-th unit of the hidden layer. Therefore, by selecting a proper value W_k, the change in gradient in the input-output function can be controlled.

2.4 Simulation Results

Three consecutive years (1988–1990) of historical data from Hokuriku Electric Power Company, Japan, are used for training and testing the ANN. Daily peak load forecasting is performed for weekdays in summer. Historical data of the years 1988 and 1989 are used for first-stage training, while historical data of the year 1990 are used for second-stage training and testing. Two weeks of historical data prior to the forecasted day are used for the second-stage training.

The choice of weather variables would considerably affect the STLF performance. In order to select a proper set of weather variables, a number of simulation tests has been performed with different combinations of weather variables, as shown in Section 2.2.a). As an example, this chapter presents simulation results with the two sets of weather variables shown below.

Set 1:
- a) Temperature at 10 AM of the forecasted day,
- b) Temperature at 3 PM of the forecasted day, and
- c) Maximum temperature of the previous day.

Set 2:
- a) Average temperature of 3 AM, 10 AM and the maximum temperature of the forecasted day,
- b) Maximum temperature of the previous day, and
- c) Humidity of the forecasted day.

To evaluate the performance of STLF, the popularly used *MAPE* (Mean Absolute Percentage Error) is applied as defined below.

$$MAPE = \frac{1}{N} \sum_{i=1}^{N} \frac{|actual\ load(i) - forecast\ load(i)|}{actual\ load(i)} \times 100\ \%, \tag{6}$$

where N is the number of observations. The ANN's test results in terms of *MAPE* with weather variable set 1 over different weight change limits W_k (Eq. (3)) for the second-stage training are as shown in Table 1. Note that the weights of ANN are normalized to 0 to 1. $W_k = 0$ means that the change in the input-output gradients Δ (Eq. (4)) is forbidden during the second-stage training except the bias unit. Meanwhile, $W_k = 1$ is equivalent to changes in weight without limit.

Table 1. STLF test results of ANN with weather variable set 1 over different weight change limits W_k.

Year	Limit (W_k)	*MAPE*
1990	0.0	2.2
	0.1	2.1
	1.0	2.4

This test result suggests that the weight change limit should be set to a small value for the second-stage training. This means the weather-independent part of the daily peak load should be learned mainly with the bias unit, which is also independent of the weather variables.

The *MAPE* values for ANN compared with the existing regression model are given in Table 2. In this simulation test, ANN with weather variable set 2 and weight change limit $W_k = 0.1$ is used.

Table 2. STLF test results of ANN with weather variable set 2 and weight change limit $W_k = 0.1$, and the existing regression method.

Year	Method	*MAPE*
1990	ANN	1.9
	Regression	2.2

The simulation results shown in Table 1 and Table 2 indicate that using the average temperature and humidity of the forecasted day effectively improves forecasting accuracy. The average temperature has the advantage that it does not change abruptly as the temperatures at a particular time do. Although the most important parameter in predicting load is temperature [38], the simulation results above indicate that it is effective to use the humidity of the forecasted day. Thus, along with temperature, humidity should also be applied to the STLF task. Albeit slightly, the test results shown in Table 2 indicate that the forecasting error of the ANN is smaller than that of the existing regression model. It verified that the ANN can produce good accuracy and is a promising technique for the STLF problem.

As shown in Table 3, there are other ANN schemes used to overcome the overfitting problem in learning the most recent available information for the STLF task. For instance, Khotanzad, et al. applied two forecasters, one forecasts the base load and the other forecasts the change in load. The final forecast is computed by adaptive and linear combination of these two forecasts. The average *MAPE* of daily peak load over ten utilities across North America is 2.12 % [23]. Wang Feng, et al. developed an adaptive schema that adjusts the weights during on-line forecasting based on the most recent training data. The *MAPE* value with this approach for North China Electric Power System is about 2.5 % [25]. Hwang, et al. proposed to keep some appropriate weights of the ANN at some constant values, while the rest are updated, based on the most recent available information. The *MAPE* value with this method is 3.6 % [46], which has been verified being better than the traditional cross-validation method [24]. Note that even with the same STLF algorithm, the performance would vary depending on different types of power system load profiles. Nevertheless, as shown in Table 2 and Table 3, the two-stage training method as proposed in this chapter is competitive compared with other ANN-based STLF algorithms.

Table 3. Other ANN-based STLF.

Reference	[23]	[25]	[46]
MAPE	2.12	2.5	3.6

3 Unit Commitment

3.1 General

The unit commitment problem in a power system involves determining the start-up and shutdown schedule of a group of thermal units to meet short-term load forecast. The objective is to minimize operation costs, including fossil fuel costs and start-up costs, while observing a large set of operation constraints. The unit commitment problem is a complex mathematical optimization problem with both integer and continuous variables.

Dynamic Programming (DP), Mixed Integer Programming (MIP) and Lagrangian Relaxation (LR) method have been studied extensively as possible solutions to this problem [8,9,10,11]. Although DP and MIP can produce an optimal schedule, this approach suffers from the "curse of dimensionality" because the problem grows exponentially with the number of generating units to be committed [11]. LR with DP is a realistic and efficient method for solving a large-scale and highly nonlinear unit problem [12]. This is because the execution time with this approach varies linearly with both units and number of stages (or time periods) [10]. However, it cannot handle "coupling constraints" explicitly. "Coupling constraints" mean constrained operations of a set of specified units. These include BTG (Boiler, Turbine and Generator) constraints, power system flow constraints, fuel consumption constraints from the same fuel stockpile base and so on. BTG constraints are applied to prohibit simultaneous start-up/shutdown of two or more units at the same power station. Currently, it takes about two days for a human domain expert to tackle these coupling constraints in exploring a feasible schedule with low operation costs. An approach based on expert systems has been proposed for dealing with coupling constraints [28]. Although they can take practical coupling constraints into account, they often have difficulty in dealing with large-scale problems within a realistic computation time.

Recently, Genetic Algorithms (GA) have been applied successfully to the combinatorial optimization problem, including the unit commitment problem, and encouraging results have been reported [1,2,29]. This chapter describes a practical and advanced method of solving the unit commitment problem by integrating soft-computing GA with hard-computing LR and DP [3,30]. Fusion of soft computing and hard computing techniques has been verified to provide innovative solutions for a wide range of demanding real-world problems [40]. In this unit commitment problem, LR with DP performs local search, efficiently finding a schedule that yields a minimum operation cost. This schedule, however, may not satisfy the coupling constraints. In order to satisfy the coupling constraints, GA, acting as a global search, is applied to generate a set of local constraints based on the schedules given by LR with DP. The local constraints are to assign some units to a specified operation state for a particular time period. In turn, LR with DP is used to produce a schedule based on a set of local constraints given by GA. The new schedule given by LR with DP also yields a minimum operation cost, but it still may not satisfy the coupling constraints in another way. Thus, GA is again applied. This process repeats until a termination condition is detected. This approach, GA combined with LR and DP, is aimed at efficiently finding a near-optimal or better-than-human-expert schedule within an acceptable computation time or a time range of about one hour.

3.2 Lagrangian Relaxation Method with Dynamic Programming

3.2.1 Problem Formulation with LR

In the practical unit commitment problem, the following constraints must be observed.

a) Demand and supply balance,

b) Maximum and minimum unit output,
c) Minimum down time (power unit stand-still time),
d) Minimum up time (power unit running time),
e) BTG constraints (prohibit simultaneous unit start-up and shutdown at the same station),
f) Power system flow constraints, and
g) Fuel consumption constraints from the same fuel stockpile base.

The unit commitment problem with the first four constraints (demand and supply balance, maximum and minimum unit output, minimum down time, and minimum up time constraints) can be formulated as follows.

Minimize (fuel cost + start-up cost)

$$= F\left(P_j^t, u_j^t\right) = \sum_{t=1}^{T} \sum_{j=1}^{N} \left(u_j^t f_j\left(P_j^t\right) + s_j u_j^t \left(1 - u_j^{t-1}\right)\right). \tag{7a}$$

$$\text{Subject to } \sum_{j=1}^{N} P_j^t = D^t, t = 1, \ldots, T, \tag{7b}$$

$$u_j^t \underline{P_j} \leq P_j \leq u_j^t \overline{P_j}, t = 1, \ldots, T, j = 1, \ldots, N, \tag{7c}$$

minimum down- and up-time constraints, (7d)

where T: number of time periods over the planning horizon,
N: number of thermal units,
D^t : load demand over time period t,
P_j^t : output of unit j over time period t,
$\underline{P_j}, \overline{P_j}$: minimum and maximum output of unit j,
f_j, s_j : fuel cost function and start-up cost function of unit j, and
u_j^t : 1 if unit j is in-service over time period t, 0 otherwise.

By applying a set of Lagrangian multipliers λ^t, t = 1,..., T to the demand and supply balance constraint (Eq. (7b)), a Lagrangian function can be formulized as follows:

$$\text{Minimize } L\left(P_j^t, u_j^t\right) = F\left(P_j^t, u_j^t\right) - \sum_t^T \left(\lambda^t \left(\sum_{j=1}^{N} P_j^t - D^t\right)\right). \tag{8}$$

By fixing each multiplier λ^t, t = 1,..., T, the problem can be decomposed into N sub-problems, each corresponding to each unit. For each unit, the sub-problem is given by:

$$\text{Minimize } F\left(P_j^t, u_j^t\right) - \sum_{t=1}^{T} \lambda^t P_j^t = \sum_t^T \left(u_j^t f_j\left(P_j^t\right) - \lambda^t P_j^t\right) + \sum_{t=1}^{T} \left(s_j u_j^t \left(1 - u_j^{t-1}\right)\right) \tag{9a}$$

$$\text{Subject to } u_j^t \underline{P_j} \leq P_j^t \leq u_j^t \overline{P_j}, \tag{9b}$$

minimum-down and up-time constraints. (9c)

3.2.2 Dynamic Programming

Each sub-problem as shown by Eq. (9) can be solved individually by DP. Assume that, for a particular unit, the minimum up time is two intervals and the minimum down time is four intervals. For this unit, the state transition graph used for DP can be constructed as shown by Fig. 3.

Since the minimum up time constraint is two intervals, Node A, which represents a unit that has been in-service for one interval, can transit only to Node B. This indicates that it has been in-service for more than one interval. From Node B, possible transitions are to Node C, which represents a unit being kept in-service for more than one interval, or to Node D, which is for a unit that is out-of-service for one interval. Since the minimum down time constraint is four intervals, Node E can transit only to Node F, which represents the unit being kept out-of-service for two intervals. Node G can transit to Node H or Node I. Node H indicates that the unit has been in-service for one interval, while Node I means that the unit remains out-of-service for more than four intervals.

For each node, a fuel cost with a Lagrangian term is given by:

$$\text{Minimize } u_j^t f_j\left(P_j^t\right) - \lambda^t P_j^t, \tag{10a}$$

$$\text{Subject to } u_j^t \underline{P_j} \leq P_j^t \leq u_j^t \overline{P_j}\,. \tag{10b}$$

For each transition, a start-up cost is given by:

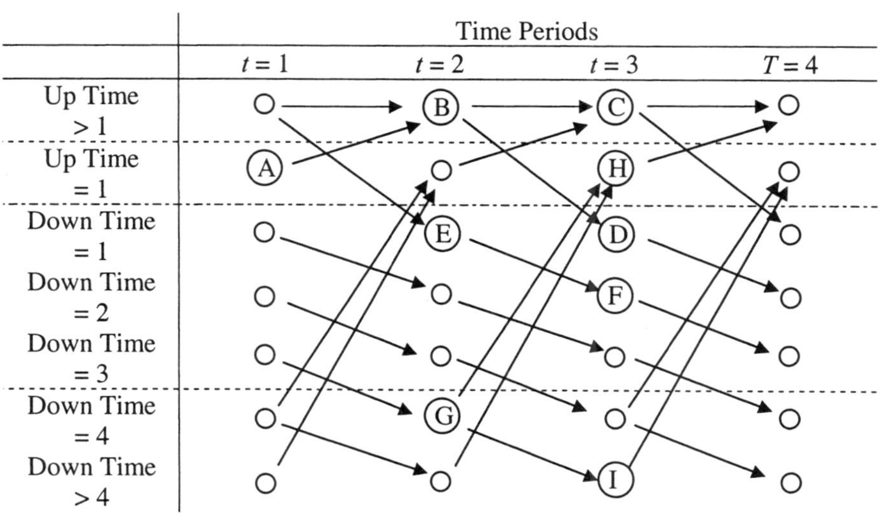

Fig. 3. State transition graph used for DP for solving the sub-problem, each corresponding a generating unit.

$$\text{Start-up cost} = s_j u_j^t \left(1 - u_j^{t-1}\right). \tag{11}$$

DP is applied to find a path from an initial node (t = 1) to the final node (t = T), which will consume a minimum fuel cost and start-up cost.

3.2.3 Modification of Lagrangian Multipliers

Since the Lagrangian multipliers are fixed only approximately, the demand and supply balance constraint (Eq. (7b)) may not be satisfied. In order to satisfy the constraint, the Lagrangian multipliers are modified by an iteration method as shown below.

$$\lambda^t{}_{k+1} = \lambda^t{}_k - \frac{\Delta^t{}_k}{a + bk}, \tag{11a}$$

$$\text{where } \Delta_k^t = \sum_{j=1}^{N} P_j^t\left(\lambda_k^t\right) - D^t, \tag{11b}$$

a, $b > 0$ is a constant. Each time the Lagrangian multipliers $\lambda^t{}_k$ are modified, P_j^t $(\lambda^t{}_k)$ is updated by DP based on Eq. (9). This process repeats, until $\Delta^t{}_k$ approaches to an acceptably small value for all time periods t = 1,…, T.

3.3 Genetic Algorithms with LR and DP

3.3.1 Algorithm

Fig. 4 shows the algorithm with respect to GA combined with LR and DP. The algorithm starts by generating an initial schedule by using LR with DP. The initial schedule satisfies the constraints of demand and supply balance, maximum and minimum unit output, minimum down time, and minimum up time. However, it may not satisfy other operation constraints, namely, the coupling constraints of BTG, power system flow, fuel consumption from the same fuel stockpile base, and so on. As shown in Fig. 4, the initial schedule is tested to determine whether it satisfies the coupling constraints. If it does, it is an optimal schedule. Otherwise, an initial population is formed by randomly generating a set of strings, whose number equals the population size. Each string represents a set of local constraints and is configured as shown below.

$$\left(t_1 s_1 \ldots t_i s_i \ldots t_n s_n\right), \tag{12}$$

where t_i is a binary number representing the time period during which the operation state of a power station is specified, s_i is a binary number standing for those power stations assigned to a specified operation state, and n is the number of time periods during which coupling constraints are violated. The reason why s_i is used to represent a power station instead of a power unit is to make the string simple, that is, to have a short string for an efficient string processing. The choice of which unit should start up or shutdown within a power station s_i is based on economic efficiency.

Based on each string given as a set of local constraints, a modified schedule is generated by LR with DP. Each modified schedule is evaluated by a fitness function as shown below:

$$fitness = \frac{1}{cost \times 10^{-6} + penalty}, \tag{13}$$

where *cost* means operation cost including fuel cost and start-up cost (Eq. (7a)), and *penalty* expresses the number of time periods during which local constraints are violated. Through reproduction, crossover and mutation operations based on the current population of strings and its fitness (*fitness*), a new population of strings is generated. This process repeats until a termination condition is detected. The termination condition is defined by: a) the number of generations that exceeds a preset number, or b) the population that has converged to an identical string.

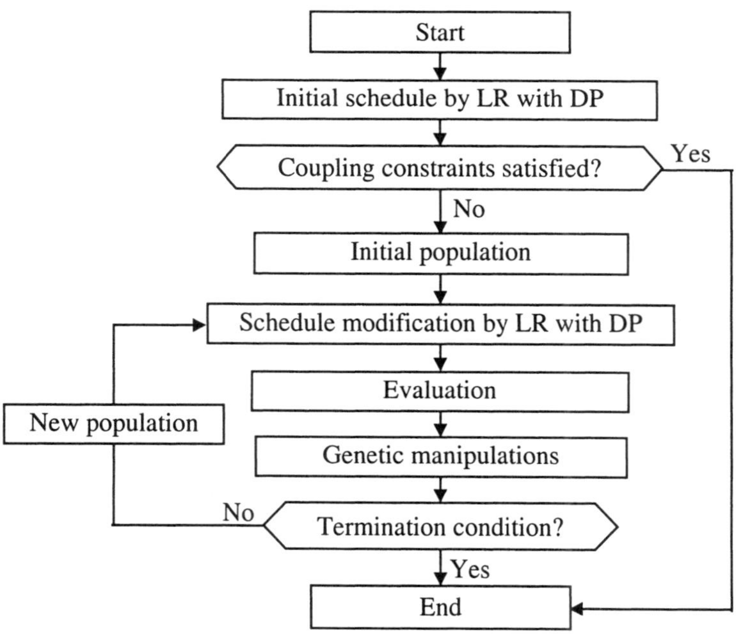

Fig. 4. GA combined with LR and DP for solving unit commitment problem.

3.3.2 Other Considerations for Efficient Search

In order to accelerate the search process, the algorithm incorporates the following approaches.

1) Use of Heuristics

Modifications of a schedule generated by LR with DP could be performed for any time period over the planning horizon. However, the chance of improving a schedule with this time period, which is far away from that for which coupling constraints are violated, would be very slim. As such, the modifications are limited to those time periods within two time periods before or after that for which the coupling constraints are violated. Use of such heuristics has the advantage of achieving a shorter string due to a smaller figure of t_i (Eq. (12)), resulting in an efficient string manipulation so as to accelerate the search process.

2) Modification of Fitness Function

As generations evolve, the individuals (which violate coupling constraints) with lower fitness value would gradually die out, and only those (which satisfy coupling constraints) with higher fitness value would survive. During such later search processes, the difference among fitness values of individuals becomes smaller, the competition is less strong, and thus the search tends to be a random process. In order to efficiently find a near-optimal schedule, the fitness function given by Eq. (13) is modified as:

$$fitness' = fitness - \alpha \tag{14}$$

where *fitness'* is the modified fitness, *fitness* is the raw fitness, and α is the average raw fitness over all generations so far. If *fitness'* is negative, it is set to a small positive value.

3) Limitations of Crossover Points

In general, in order to maintain the diversity of a population, it is desirable to generate as many different types of individual as possible through crossover or mutation operations. In this regard, the crossover points should not be limited. However, in this case, it would result in generation of numerous fatal individuals. For example, an individual is fatal if it includes a power station s_i=111 for a planning problem that only has five power stations. For this reason, the crossover points are limited to the positions between time period t_i and power station s_i. Thus, generation of fatal individuals can be minimized that would increase search efficiency.

4) Fatal Individuals

Even though the crossover points are limited to particular positions as mentioned above, there are still chances of generating fatal individuals through crossover or mutation operations. The fatal one will be discarded and the crossover or mutation operations will continue until a normal individual is generated.

3.4 A Numerical Example

3.4.1 Power System

Table 4 shows an example of a power system with ten units located in five power stations. The next-day load demand forecast for the power system is shown in Fig. 5.

Table 4. Example of a power system with ten units located in five power stations.

Stations	Units	Max MW	Min MW	Min Up Time (hr)	Min Down Time (hr)	Fuel Cost			Start-up Cost	
						a	*b*	*c*	*P*	*q*
A	A1	375	100	1	1	0.65	17 800	102 000	8.5	375
	A2	375	100	1	1	0.56	18 500	94 000	8.5	375
	A3	500	50	1	3	0.65	16 000	127 000	20.0	25
B	B1	220	50	1	6	2.70	17 400	58 000	6.0	90
	B2	220	50	1	6	0.05	21 900	38 000	4.5	90
	B3	220	50	1	6	3.12	17 300	52 000	8.0	40
C	C1	500	160	1	1	0.05	19 600	103 000	75.0	880
D	D1	220	60	1	6	0.05	20 700	59 000	11.5	40
	D2	220	60	1	6	0.63	19 300	63 000	11.5	40
E	E1	700	175	1	1	0.63	16 300	163 000	5.0	550

Note: Fuel Cost = $aP^2 + bP + c$.
Start-up Cost = $p + q\tau$, where τ is down time.

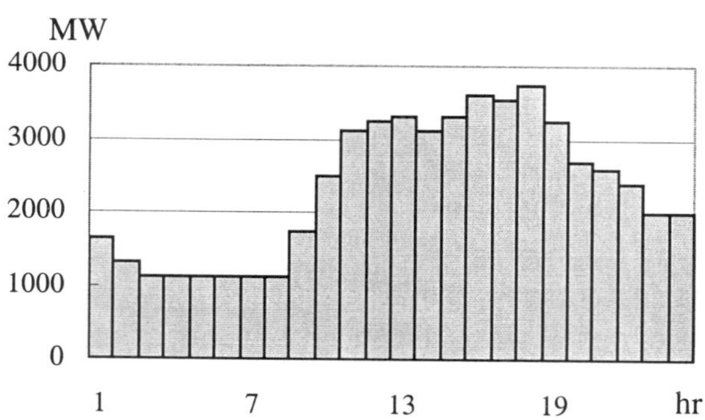

Fig. 5. Next-day load demand forecast with a power system as shown in Table 4.

The operation constraints given in Table 4 are maximum and minimum unit output (Max MW, Min MW) and minimum down time and minimum up time (Min Up Time, Min Down Time). The two coupling constraints below are also applied to this numerical example.

1) BTG Constraints

These constraints are applied to prohibit two or more units of the same station to start up or shutdown during the same time period.

2) Power System Flow Constraints

Assume that the power stations as shown in Table 4 are divided into two power systems: West System and East System. The power system flow constraints are given as:

$$\left|\sum_i P_{Wi}^t - \sum_j P_{Ej}^t\right| \leq \beta \, , \tag{15a}$$

$$\text{where } Wi \in \{\text{A, B, C}\}\text{: West System,} \tag{15b}$$

$$Ej \in \{\text{D, E}\}\text{: East System,} \tag{15c}$$

P_{Wi}^t is output of West System's unit Wi, P_{Ei}^t is output of East System's unit Ej, and β (= 1 500 MW) is the power system flow limit between West and East Systems. The constraints given by Eq. (15) are used to limit the maximum power system flow for all time periods t.

As for the BTG constraints, if power system flow constraints are violated, GA is applied to assign power stations to a specified operation state in order to get rid of the violated constraints. The violation of power system flow constraints, however, can also be resolved by adjusting the unit output. Accordingly, if the modified schedule (given by LR with DP) violates the power system flow constraints, the output at time period t when the constraints are violated is adjusted by solving the problem below.

$$\text{Minimize (fuel cost)} = \sum_{j=1}^{N} u_j^t f_j\left(P_j^t\right), \tag{16a}$$

$$\text{Subject to } \sum_{j=1}^{N} P_j^t = D^t \, , \tag{16b}$$

$$\left|\sum_i P_{Wi}^t - \sum_j P_{Ej}^t\right| \leq \beta \tag{16c}$$

$$u_j^t \underline{P_j} \leq P_j \leq u_j^t \overline{P_j}, j = 1,\ldots,N \, . \tag{16d}$$

The evaluation for the schedule, which violates the power system flow constraints, is based on the adjusted output. However, if a feasible solution to the problem (Eq. (16)) cannot be obtained, a penalty is added to the fitness function as mentioned above (Eq. (13)).

3.4.2 Schedules

Fig. 6 shows the initial schedule generated by LR with DP based on the power system and the load demand forecast given above. As shown in the figure, BTG constraints are violated during time period 9 and power system flow constraints during time periods 10, 18, 19 and 20. In order to get rid of the violated constraints, strings are constructed with respect to these time periods. In this case, each string consists of two parts. The generation of the first part of a string is based on time periods 9 and 10, and the second part is based on time periods 18, 19 and 20. Note that, according to Eq. (12), each string should have consisted of five parts, corresponding to the number of time periods during which the coupling constraints are violated. However, time periods 9 and 10, and 18, 19 and 20, respectively, are consecutive, and modification of a time period would affect the operation state of neighboring time periods. Accordingly, the modification is carried out in such a way that consecutive time periods are combined to construct one part of a string.

Fig. 7 shows the candidate blocks, which are obtained from the initial schedule. Each part (first and second) of a string is constructed by randomly selecting two blocks: one from the time period blocks and the other from the station blocks. These two parts (the first and second part) are combined to form a string. An example of a constructed string is shown as follows.

<u>An example of constructed string: 001001|010011</u>

- First part (001001): modification of unit start-up and shutdown at station A with time period 7 and adjustment of unit output.

Stations	Units	Time Periods 1	2	3	4	5	6	7	8	9	10	11	12	13	14	15	16	17	18	19	20	21	22	23	24
	A1	0	0	0	0	0	0	0	0	1	1	1	1	1	1	1	1	1	1	1	1	1	1	1	0
A	A2	0	0	0	0	0	0	0	0	1	1	1	1	1	1	1	1	1	1	1	1	1	0	0	0
	A3	0	0	0	0	0	0	0	0	0	1	1	1	1	1	1	1	1	1	1	1	0	0	0	0
	B1	1	1	1	1	1	1	1	1	1	1	1	1	1	1	1	1	1	1	1	1	1	1	1	1
B	B2	1	1	1	1	1	1	1	1	1	1	1	1	1	1	1	1	1	1	1	1	1	1	1	1
	B3	1	1	1	1	1	1	1	1	1	1	1	1	1	1	1	1	1	1	1	1	1	1	1	1
C	C1	1	1	1	1	1	1	1	1	1	1	1	1	1	1	1	1	1	1	1	1	1	1	1	1
D	D1	0	0	0	0	0	0	0	0	0	1	1	1	1	1	1	1	1	1	0	0	0	0	0	0
	D2	0	0	0	0	0	0	0	0	0	0	1	1	1	1	1	1	1	0	0	0	0	0	0	0
E	E1	1	1	1	1	1	1	1	1	1	1	1	1	1	1	1	1	1	1	1	1	1	1	1	1
Amount of power flow (MW) which violates power system flow constraints.											69.6								70.9	226.6	184.1				

Note: 1 means in-service, and 0 out-of-service.

Fig. 6. An initial schedule generated by LR with DP.

<u>Candidate blocks for the first part of a string relative to time periods 9 and 10</u>

Time period 7: 001	Power station A: 001
Time period 8: 010	Power station B: 010
Time period 9: 011	Power station C: 011
Time period 10: 100	Power station D: 100
Time period 11: 101	Power station E: 101
Time period 12: 110	

<u>Candidate blocks for the second part of a string relative to time periods 18, 19 and 20</u>

Time period 16: 001	Power station A: 001
Time period 17: 010	Power station B: 010
Time period 18: 011	Power station C: 011
Time period 19: 100	Power station D: 100
Time period 20: 101	Power station E: 101
Time period 21: 110	
Time period 22: 111	

Fig. 7. Candidate blocks obtained from the initial schedule for the construction of strings.

Stations	Units	Time Periods																							
		1	2	3	4	5	6	7	8	9	10	11	12	13	14	15	16	17	18	19	20	21	22	23	24
	A1	0	0	0	0	0	0	0	1	1	1	1	1	1	1	1	1	1	1	1	1	1	1	1	0
A	A2	0	0	0	0	0	0	0	0	1	1	1	1	1	1	1	1	1	1	1	1	1	0	0	0
	A3	0	0	0	0	0	0	0	0	0	1	1	1	1	1	1	1	1	1	0	0	0	0	0	0
	B1	1	1	0	0	0	0	0	0	1	1	1	1	1	1	1	1	1	1	1	1	1	1	1	1
B	B2	1	1	1	1	1	1	1	1	1	1	1	1	1	1	1	1	1	1	1	1	1	1	1	1
	B3	1	1	1	1	1	1	1	1	1	1	1	1	1	1	1	1	1	1	1	1	1	1	1	1
C	C1	1	1	1	1	1	1	1	1	1	1	1	1	1	1	1	1	1	1	1	1	1	1	1	1
D	D1	0	0	0	0	0	0	0	0	0	1	1	1	1	1	1	1	1	1	1	1	0	0	0	0
	D2	0	0	0	0	0	0	0	0	0	0	1	1	1	1	1	1	1	0	0	0	0	0	0	0
E	E1	1	1	1	1	1	1	1	1	1	1	1	1	1	1	1	1	1	1	1	1	1	1	1	1

Note: 1 means in-service, and 0 out-of-service.

Fig. 8. A near-optimal schedule found by GA with LR and DP.

- Second part (010011): modification of unit start-up and shutdown at station C with time period 17 and adjustment of unit output.

Fig. 8 shows a near-optimal schedule, which is found by GA with LR and DP, subject to all operation constraints, including coupling constraints. As a result of preliminary experiments, parameter settings of GA are: population size 200, crossover rate 0.5, mutation rate 0.1, and maximum generations 100.

This example, as shown above, illustrates simulation results of a one-day planning horizon with only two coupling constraints, which are BTG constraints and power system flow constraints, being included. Reference [3] verified that GA with LR and DP could solve the unit commitment problem with all coupling constraints over a seven-day planning horizon in 38 minutes and 17 seconds of computing time with an engineering workstation TOSHIBA AS7000. This computing time should be acceptable to power system operators.

4 Power Plant Start-Up Scheduling

4.1 General

Computerized power plant automation in Japan began around 1968, when the first DDC (Direct Digital Control) system for turbine run-up control was designed and successfully put into operation [31]. Since then, power plant automation has expanded and developed into fully computerized automation systems where only one operator is required to operate a large-scale thermal power plant.

A large-scale thermal power plant is a complex system where the control system consists of tens of thousands of input/output points and hundreds of control loops. For such a large-scale automation system, COPOS (Computerized Optimum Plant Operation System) [32], a crisp and event-oriented ES (Expert System) has been developed in order to overcome the limitations of traditional automation-software production techniques that use the flow-chart method. A large-scale thermal power plant control system is based on a hierarchy structure. The ES technique is applied at the top level to provide the human-machine interface for the power plant operation, to monitor the whole power plant system and to carry out the coordination tasks by generating set-point and kick-action commands for lower-level controllers. The lower-level controllers are used to control each power plant subsystem. At this level, there are two types of control: regulation control, e.g., EHC (Electro-Hydraulic Control) for turbine run-up DDC [33] and sequence control, e.g., boiler clean-up sequence used prior to initial firing of the boiler.

Although these conventional HC (Hard Computing) techniques have so far played a key role in computerized power plant automation systems, integration of these HC techniques with the emerging SC (Soft Computing) [40] has been proposed for on-line scheduling in order to achieve the next-generation optimal automation system [34]. As shown in this chapter, SC for the power plant start-up scheduling problem is implemented through the integration of reinforcement-learning-based ANN [35] and GA [47] capable of stochastic searching, learning and generalization. Prior to power plant start-up, the SC searches and learns the optimal schedules for different start-up conditions. During the start-up process, SC generates and searches the optimal or near-optimal schedule for the ES, which controls the power plant start-up process based on the given schedule. It has been verified that the optimal or near-optimal schedule can be obtained within tens of seconds, a time range which is acceptable for power plant operation. The optimal schedule can reduce the start-up time by approximately 10% with respect to warm start-up mode.

4.2 Concept

Fig. 9 shows a conceptual diagram of the next-generation automation system, which is aimed at achieving an optimal full-scale automation system for thermal power plant operation. An SC-based scheduler, located at the top level of the hierarchical control system, is applied to generate an on-line optimal control plan for the ES-based supervisor. The ES-based supervisor monitors the whole power plant and carries out coordination tasks by generating set-point and kick-action commands for the controllers. The controllers located at the low level are used for regulation control and sequence control of each individual power plant subsystem. The role of these controllers is to respond to dynamic change of the process, based on set-point and kick-action commands given by the ES-based supervisor.

The control system based on the DCS (Distributed Control System) architecture for a large-scale thermal power plant is configured as shown in Fig. 10.

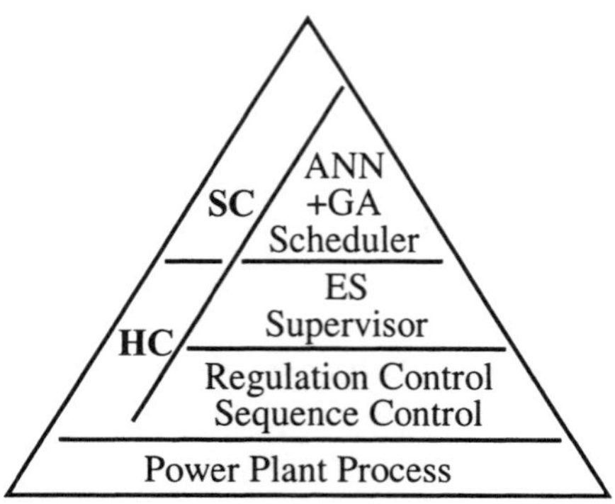

Fig. 9. Conceptual diagram of the integration of SC and HC for the next-generation power plant automation system.

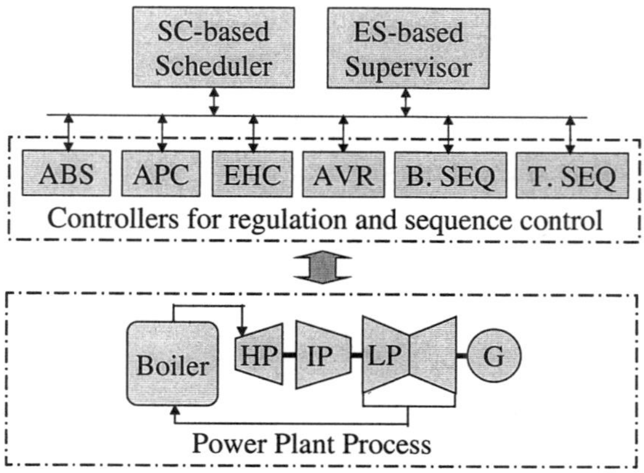

ABS: Automatic Burner System, APC: Automatic Plant Control,
EHC: Electro-Hydraulic Control, AVR: Automatic Voltage Regulator,
B. SEC: Boiler Sequencer, T. SEC: Turbine Sequencer.

Fig. 10. Configuration of next-generation power plant automation system based on DCS (Distributed Control System) architecture.

4.3 Start-Up Scheduling Problem

Power plant start-up scheduling is aimed at minimizing start-up time while limiting maximum turbine-rotor stresses to an acceptable level. A shorter start-up time not only reduces fuel and electricity consumption during the start-up process, but also increases its capability to adapt to changes in power system load demand. As shown in Fig. 10, a large-scale thermal power plant consists of a boiler, a set of high-pressure (HP), intermediate-pressure (IP) and low-pressure (LP) turbines, and a generator. During the start-up process, as shown in Fig. 11, the steam temperature rises after initial firing in the boiler. When the steam temperature reaches the rolling-off temperature X_{10}, the turbine starts rolling and then the turbine speed is accelerated to the rated speed with low speed hold time X_2, rated speed hold time X_3 and at an accelerated rate X_1. After synchronizing, the generator load is increased to an initial load at a loading rate X_4 and, after holding for an initial load hold time X_5, is increased again to the rated load at different loading rates X_6, X_7, X_8, and X_9. The scheduling problem is to decide the schedule parameters X_i, i = 1,..., 10 in order to minimize the start-up time T from boiler initial firing to the generator's rated load, while restricting the maximum stresses σ_{m_j}, j = 1,..., 4 within the preset limits σ_{l_j}, j = 1,..., 4. The subscript j stands for specific locations where thermal stresses are developed most critically. These are the HP turbine rotor surface (j = 1), the IP turbine rotor surface (j = 2), the HP

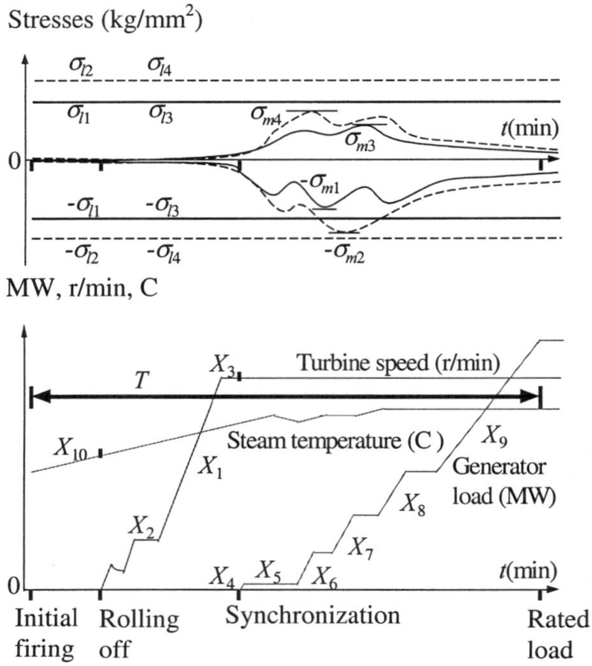

Fig. 11. Power plant start-up schedule.

turbine rotor bore (j = 3) and the IP turbine rotor bore (j = 4). The start-up scheduling problem can therefore be formulated as a function optimization problem with constraints below.

$$\min_{X_i, i=1,\ldots,10} T(X_1,\ldots, X_i,\ldots, X_{10}), \tag{17a}$$

$$\text{subject to } X_i \in \{X_i^1,\ldots, X_i^{k_i},\ldots, X_i^{s_i}\},\ i=1,\ldots, 10, \tag{17b}$$

$$\sigma_{m_j}(X_1,\ldots, X_i,\ldots, X_{10}) \le \sigma_{lj}, j=1,\ldots, 4, \tag{17c}$$

$$\text{where } \sigma_{mj}(X_1,\ldots, X_i,\ldots, X_{10}) = \max_{0\le t\le T} \left|\sigma_j(X_1,\ldots, X_i,\ldots, X_{10}, t)\right|, j = 1,\ldots, 4, \tag{17d}$$

$$\sigma_j (X_1,\ldots, X_i,\ldots, X_{10},t)=g_j(X_1,\ldots, X_i,\ldots, X_{10}), j = 1,\ldots, 4, \tag{17e}$$

$T(X_1,\ldots, X_i,\ldots, X_{10})$ is power plant start-up time; X_i^1 and $X_i^{s_i}$ are the lower and upper bounds of X_i, respectively; s_i is the size of a set of possible values $X_i^{k_i}$ of X_i (the size for each schedule parameter is shown in Table 5 of Section 4.6); g_j, j = 1,..., 4 expresses a dynamic simulation model used to simulate the turbine start-up process with the schedule parameters X_i, i = 1,..., 10 serving as inputs and the thermal stresses σ_j, j = 1,..., 4 over time t serving as outputs. The details of the dynamic simulation model can be found in [5]. Based on thermal expansion or compression, thermal stress values can be positive or negative, and the maximum of their absolute values is to be limited as shown by Eq. (17d).

4.4 SC-Based Scheduler

At present, power plant start-up scheduling is peformed by turbine design engineers during the turbine design stage [16] and installed into the power plant computer for start-up control [42]. This approach, however, involves design costs and it is difficult to obtain optimized start-up schedules. An optimal schedule for the start-up process has been formulated [43] by assuming that the relationship between thermal stress and control input is linear and time invariant. This formulation cannot be applied in real operation because the actual process is highly nonlinear and time variant. An on-line fuzzy expert system has been proposed that codes the expertise of turbine design engineers into fuzzy rules for power-plant operation support [17], but building such an expert system for practical use would involve knowledge acquisition costs.

It has been proved that power plant start-up scheduling contains active constraints [49], that is, the optimal schedule lies on the edge of the feasible space. The edge where the optimal schedule is assumed to stay is defined by these feasible schedules ($X_1, \ldots, X_i, \ldots, X_{10}$) which satisfy:

$$(\exists(j)\ \sigma_{mj}(X_1, \ldots, X_i, \ldots, X_{10}) = \sigma_{lj}) \text{ or } (\forall(i) X_i = X_i^1 \text{ or } X_i = X_i^{s_i}), \tag{18}$$

where $i = 1, \ldots, 10$ and $j = 1, \ldots, 4$. Therefore, instead of the entire space, searching along the edge would increase the search efficiency significantly without missing the optimum. If the problem is unimodal and differentiable, then the conventional active set strategy based on the gradient projection method can be applied to effectively find the optimal solution by following exactly the edge of the feasible space [49]. This scheduling problem is, however, neither unimodal nor differentiable, and therefore the traditional gradient-based optimization method is difficult to apply. Based on the observations that very often the global solution lies on the edge of the feasible region, closed-boundary approach in evolutionary systems has been proposed to precisely search the boundary area between feasible and infeasible regions of the search space for those multimodal or non-differentiable problems [44,45]. In [44], evolution operators—crossover and mutation—are closed. That is, they transform boundary point(s) to new boundary point(s). In [45], a repair operator is added to move an offspring generated by crossover and mutation operations to the boundary, thus allowing the use of any crossover and mutation. However, since the constraint calculation for this scheduling problem needs dynamic simulation which is highly nonlinear, it is difficult to formulate such closed-boundary evolution operators or repair operator in order to generate child schedules precisely on the edge or to move a schedule back precisely to the edge.

Fig. 12 outlines the search model, which is based on the integration of **RE**inforcement-learning-based A**N**N and **G**A with **E**nforcement operation (RENGE). RENGE is proposed as an on-line search model to address those nonlinear optimization problems with local optima and active constraints for which searching precisely on the edge is difficult.

Because a schedule cannot be calculated immediately from maximum stresses, the enforcement operation uses a dynamic simulation and a linear inverse calculation to update a given schedule by an iteration method. The dynamic simulation uses the turbine dynamic model (Eq. (17d)) based on a given schedule ($X_1,\ldots, X_i,\ldots, X_{10}$) to calculate the nonlinear constraint functions, i.e., the maximum stresses $\sigma_{m_j}(X_1,\ldots, X_i,\ldots, X_{10})$, $j = 1,\ldots, 4$. For the linear inverse calculation, inverse enforcement gains A_{ij}, $i = 1,\ldots, 10, j = 1,\ldots, 4$ as defined below are used.

$$A_{ij} = \frac{X_i'' - X_i}{\sigma_{m_j}''\left(X_1,\ldots,X_i'',\ldots,X_{10}\right) - \sigma_{m_j}\left(X_1,\ldots,X_i,\ldots,X_{10}\right)}, \tag{19}$$

where X_i'', $i = 1,\ldots, 10$ are the candidate schedule parameters to replace the parameter X_i, and $\sigma_{m_j}''(X_1,\ldots, X_i'',\ldots, X_{10})$, $j = 1,\ldots, 4$ are the linear maximum stresses.

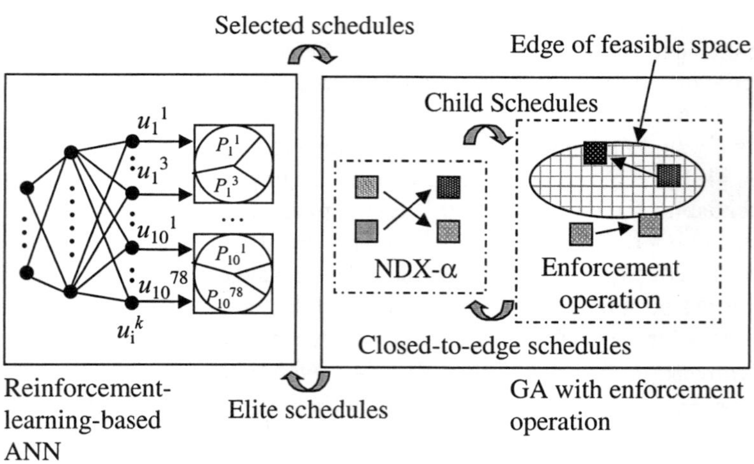

Fig. 12. Integration of reinforcement-learning-based NN and GA with enforcement operation for on-line scheduling. NDX-α means Normal Distribution Crossover-α used for the GA's crossover operation.

Although maximum stresses σ_{m_j} (X_1,..., X_i,..., X_{10}) are highly nonlinear, based on a theoretical proof [5], the schedule can be moved closer to the edge at each iteration of the enforcement operation, provided that the linear inverse enforcement gains A_{ij} are adapted by:

$$A_{ij}=\frac{\phi}{\mu_{\lambda_{ij}}\left(1+\varepsilon_{0\lambda_{ij}}^2\right)},\ 0<\phi<2 \qquad (20a)$$

$$\text{where } \varepsilon_{0\lambda_{ij}}=\frac{\varepsilon_{\lambda_{ij}}}{\mu_{\lambda_{ij}}}. \qquad (20b)$$

$\mu_{\lambda_{ij}}$ and $\varepsilon_{\lambda_{ij}}$ are an unbiased estimated average and standard deviation of maximum stress coefficients λ_{ij}, i=1,...,10, j = 1,..., 4, respectively. The unbiased estimation is based on M latest sets of λ_{ij}, i=1,...,10, j = 1,..., 4, which are defined by:

$$\lambda_{ij}=\frac{\sigma_{m_j}\left(X_1,\ldots,X_i+\Delta X_i,\ldots,X_{10}\right)-\sigma_{m_j}\left(X_1,\ldots,X_i,\ldots,X_{10}\right)}{\left(X_i+\Delta X_i\right)-\left(X_i\right)}. \qquad (21)$$

The search performance of GA can be improved significantly with the enforcement operation to limit the search along the edge of the feasible space.

In Fig. 12, NDX-α is applied for the GA's crossover operation, generating a pair of child schedules based on a normal distribution, whose variance is defined by the product of the distance between their parents and the step size α [36]. NDX-α is used with a small step size, in order to focus the search along the edge by generating schedules not far away from the edge. In local search, the search process is prone to repeat searching the same solutions that had been searched recently. In order to increase the search efficiency further, in addition to enforcement operation, we have introduced a tabu search strategy [18]. The objective of this strategy is to forbid certain iteration moves in order to avoid cycling and to make early escape from a local optimal point possible, thus reducing the iteration cost and increasing the capability of searching the global optimal. It has been shown that local-search-based GA with tabu search strategy performs better than stand-alone GA, stand-alone tabu search, and simulated annealing [19] with or without tabu search strategy [15]. In the following descriptions, GA means GA with tabu search startegy.

In order to find an on-line optimal schedule in response to various start-up conditions within an acceptable time, GA with enforcement operation is further integrated with a reinforcement-learning-based ANN. As mentioned above, a reinforcement-learning-based ANN is capable of searching, learning and generalization. The input layer units of the ANN represent the start-up conditions. The output layer units are stochastic parameters $u_i^{k_i}$, i = 1,..., 10; k_i = 1,..., s_i from which the schedule parameters X_i, $i = 1,\dots, 10$ are generated in such a way that the greater the stochastic parameter's value $u_i^{k_i}$ is, the higher probability the corresponding output parameter value $X_i^{k_i}$ is selected for the output parameter X_i (Eq. (22)).

$$p_i^{k_i} = \frac{\exp\left(u_i^{k_i}\right)}{\sum_{n=1}^{s_i} \exp\left(u_i^{n}\right)}, \tag{22}$$

where $p_i^{k_i}$, $i = 1,\dots, 10$; $k_i = 1,\dots, s_i$ are the probability of schedule parameter values $X_i^{k_i}$ being selected for the schedule parameter X_i. Each stochastic parameter u_i^k corresponds to each schedule parameter value $X_i^{k_i}$. During the learning process, a reinforcement signal is given to ANN when an elite schedule, which is a feasible schedule with the shortest start-up time of those schedules obtained so far, is generated. The updating rules of the connection weights with the given reinforcement signal are based on the stochastic gradient method [35] with a backpropagation approach [6].

It has been pointed out that reinforcement-learning-based ANN with the stochastic gradient method would converge to a local optimum [35]. In order to escape from local optima and to accelerate the learning and searching process, reinforcement-learning-based ANN is combined with GA [6], which is capable of searching the global optimum. Prior to power plant start-up, by providing the elite, GA guides the reinforcement-learning-based ANN to search and learn optimal schedules with respect to a number of representative start-up conditions. During start-up, given a set of start-up conditions, the trained reinforcement-learning-based ANN generates promising schedules that serve as the GA's initial population. With the promising initial population, which has a good chance to contain building blocks [47], the search for the optimal or near-optimal schedules can be accelerated substantially.

4.5 ES-Based Supervisor and Lower-Level Controllers

An ES-based supervisor located at the middle level of the hierarchical control system is used to monitor the whole power plant and to carry out the tasks of coordination based on the optimal schedule generated by the SC-based scheduler. The control rules of the ES-based supervisor are expressed in an if-then format. An example of control rules embedded in the ES-based supervisor to control the lower-level controllers based on the optimal schedule parameters X_i, i=1,..,10 is shown in Fig. 13. Lower-level controllers based on traditional DDC or sequence control techniques are applied to respond to the dynamic change in the process. A control block diagram of EHC is shown in Fig. 14 as an example of a lower-level controller with the set point based on the optimal schedule.

R1: If boiler firing = 1 & steam temperature ≥ X_{10},
then output low-speed set point and acceleration-rate set point (X_1) to EHC (Electro-Hydraulic Control).

R2: If turbine speed ≥ rated speed & hold time X_3 elapsed,
then start synchronizing sequencer.

R3: If number of boiler feed-water pumps in service ≥ 2 & generator load ≥ 50%,
then output rated-load set point and loading-rate set point (X_9) to APC (Automatic Plant Control).

Fig. 13. Example of control rules embedded in the ES-based supervisor to control the lower-level controllers based on the optimal schedule parameters.

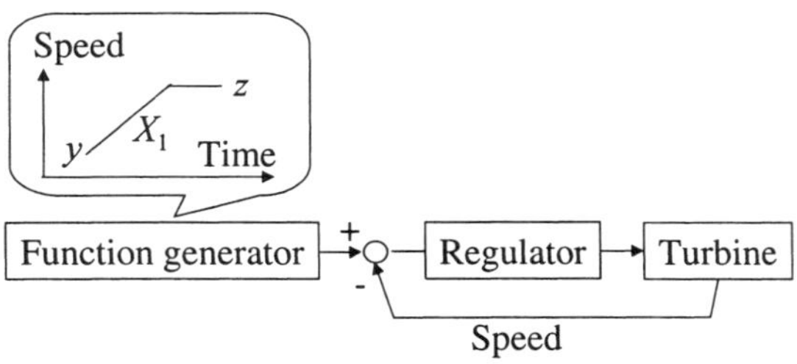

Fig. 14. Control block diagram of lower-level controllers (EHC) to control the turbine speed from initial speed y to low speed z with an optimal acceleration rate X_1 given by the ES-based supervisor.

4.6 Simulation Results

The start-up schedule parameters and their optimized values over different stress limit sets are shown in Table 5 [6]. The schedules shown in Table 5 are considered optimal based on consistent results of a number of experiments with different search models (GA, tabu search, simulated annealing, and the combinations) and parameter settings conducted so far [4,5,6,15]. Stress limit set 2 includes optimized and designed values. The start-up time of the optimized schedule is about 32 minutes shorter than the designed schedule performed by a human expert. For a 350 MW LNG (Liquefied Natural Gas) thermal power plant, this optimal schedule can reduce the energy loss during warm start-up by 30 Gcal.

Warm start-up means that the turbine rotor is warm when the turbine starts rolling. For DSS (Daily Start-up Stop), this reduction in energy loss would result in the reduction of annual start-up costs by an estimated US$232 000 and the annual CO_2 emission by 294 tons.

Table 6 [4,6] shows the test results of the search performance with respect to the degree of integration: stand-alone GA (Model 1), GA with enforcement operation (Model 2) and the trained RENGE (Model 3) based on the integration of reinforcement-learning-based ANN and GA with enforcement operation. Prior to the test, the RENGE is trained by stress limit set 2, set 6 and set 10.

Since the computing cost is spent mostly on the dynamic simulation calculations (Eq. (17e)), about one CPU second per simulation, as shown in Table 6, the number of simulations is used to measure the search performance of each search model. The number of trials is eight for Model 1 and 100 for both Model 2 and Model 3. Each trial is performed until an optimal or near-optimal schedule is found or until the maximum generations (30 000) have passed. Near-optimal schedules are defined by those feasible schedules whose start-up time is no longer than 1% of the optimal schedule. These near-optimal schedules are acceptable for the actual power plant operation.

As shown in Table 6, a GA with an enforcement operation (Model 2 and Model 3) succeeds in finding the optimal or near-optimal schedules for all trials and all stress limit sets and requires a small number of simulations to perform such searching. On the other hand, a GA without an enforcement operation (Model 1)

Table 5. Test results of optimal schedules over stress limit sets with respect to turbine warm start-up mode [6]. Opt: Optimized schedule; Des: Human Designed schedule.

X_i	Possible values $X_i^{k_i}$	Size s_i	Larger←Stress limit sets→Smaller						
			set 1	**set 2**		set 3...8	set 9	set 10	set 11
			Opt	**Des**	**Opt**	Opt	Opt	Opt	Opt
X_1	120,180,360	3	360	**180**	**360**	360	360	360	360
X_2	0,1,...,28	29	0	**0**	**0**	0	0	0	0
X_3	5,6,...,37	33	5	**9**	**5**	5	5	5	5
X_4	0.5,1.0	2	1.0	**1.0**	**1.0**	1.0	1.0	1.0	1.0
X_5	0,1,...,40	41	0	**12**	**0**	0	23	25	28
X_6	0.5,1.0	2	0.5	**1.0**	**1.0**	1.0	1.0	1.0	1.0
X_7	0.5,1.0	2	1.0	**1.0**	**0.5**	0.5	0.5	0.5	0.5
X_8	1.0,1.5	2	1.0	**1.0**	**1.0**	1.0	1.0	1.0	1.0
X_9	1.5,2.0,2.5,3.0	4	3.0	**2.0**	**3.0**	3.0	2.0	2.0	1.5
X_{10}	349,350,...,426	78	349	**376**	**349**	349	374	380	382
T	---	---	210.6	**243**	**211.6**	211.6	252.8	257.1	269.3

Table 6. Test results of different search models with respect to turbine warm start-up mode [4,6]. Model 1: Stand-alone GA; Model 2: GA with enforcement operation; Model 3: Trained RENGE; Number of trials: 8 (Model 1) or 100 (Model 2 and 3); Opt No.: Number of optimal or near-optimal schedules has been found; Ave: Average; Std: Standard deviation; Max: Maximum; Min: Minimum.

Stress limits	Model 1					Model 2					Model 3				
	No. of simulations				Opt No.	No. of simulations				Opt No.	No. of simulations				Opt No.
	Ave	Std	Max	Min		Ave	Std	Max	Min		Ave	Std	Max	Min	
Set 1	12 789	9 373	27 078	468	7	28	18	87	4	100	1	0	1	1	100
Set 2	4 495	7 937	16 394	161	4	17	10	49	5	100	1	0	1	1	100
Set 3	1 391	1 106	3 330	251	8	28	18	85	4	100	1	0	1	1	100
Set 4	1 394	869	2 488	284	8	33	28	159	5	100	1	0	1	1	100
Set 5	1 515	822	2 922	653	8	29	20	99	6	100	1	0	1	1	100
Set 6	1 056	502	2 017	525	8	40	18	101	7	100	1	0	1	1	100
Set 7	2 493	1 599	6 150	1 431	8	53	46	223	15	100	2	1	8	2	100
Set 8	---	---	---	---	0	64	80	417	13	100	6	3	13	2	100
Set 9	6 949	5 382	17 447	2 310	8	1 237	1 213	6 167	44	100	5	12	58	2	100
Set 10	8 433	6 968	17 725	2 442	8	1 006	754	3 286	16	100	1	0	1	1	100
Set 11	2 213	1 473	4 760	543	8	396	340	1 589	12	100	8	5	49	3	100
Overall	5 719	18 680	27 448	161	6.8	266	1 472	6 167	4	100	3	13	58	1	100

fails to find the optimal or near-optimal schedules for all trials within maximum generations with stress limit set 8, and requires a large number of simulations for such search with the remaining sets. Note that in order to reduce experimental time, only eight trials are applied to each stress limit set for the GA without enforcement operation (Model 1). As verified by this test result, it is evident that, by focusing the search along the edge, an enforcement operation can significantly improve the search performance. With the trained RENGE (Model 3), the number of simulations decreases further, to only one to eight on average. This result shows that trained RENGE can quickly find the optimal or near-optimal schedules with the generalization capability provided by ANN. This number of simulations is equivalent to one to eight CPU seconds (Sun SPARC station 20). This time range satisfies the on-line scheduling requirement for power plant operation.

5 Conclusions

This chapter has described the use of soft computing techniques for optimal operation of power energy system, solving the Short-Term Load Forecasting (STLF) problem, unit commitment problem, and optimal start-up scheduling problem.

An ANN with a two-stage training method has been used for the STLF problem. This method is based on the assumption that power system load demand consists of two parts: a weather-dependent part and a weather-independent part. The first-stage training is for determining the weather-dependent part, while the second-stage training is for determining the weather-independent part. An ANN based on two-stage training can achieve a result with 1.9 % *MAPE* (Mean Absolute Percentage Error), which is slightly smaller than those obtained with the existing regression model. The two-stage training method has been successfully applied to Hokuriku Electric Power Company, Japan. Other successful employments of ANN-based STLF for actual power system operation can also be found in North America and China [23,25]. Neural networks are such flexible models that the task of designing an ANN-based forecasting system for a particular application is, however, far from easy. Despite the successful employments above, the ANN scheme has drawbacks compared to the traditional regression method. One is the trial-and-error-based parameter tuning tasks accompanying a time-consuming training processes based on a great amount of historical weather and load data. Applications of ANN to the STLF problem are still in the initial stage. Further research and experience gathered from practical applications are required to reduce parameter-tuning time and to automate the tuning processes while maintaining a high level of forecasting accuracy.

A practical and advanced search model, GA with LR and DP has been applied to efficiently solve the unit commitment problem with coupling constraints. The coupling constraints include BTG constraints (prohibition of simultaneous unit start-up/shutdown at the same station), power system flow constraints, fuel consumption constraints from the same fuel stockpile base and so on. To the chapter authors' knowledge, this is the first ever approach—GA with LR and DP—to successfully solving the unit commitment problem with BTG constraints included. However, in order to further improve the search performance, instead of LR, the authors are currently employing an Augmented Lagrangian (AL) relaxation method for the unit commitment problem. With this method, quadratic penalty terms associated with the local constraints such as load demand are added to the objective function to improve the algorithm's convergence. Recent studies have confirmed that AL is faster and more efficient than LR in dealing with unit commitment problem [13,14,41].

For power plant start-up scheduling problem, integration of reinforcement-learning-based ANN and GA with enforcement operation (RENGE) has been discussed. In order to accelerate the GA search process, the enforcement operation, which is based on the iteration of constraints and inverse constraints calculation, is applied to move the solutions generated by GA toward the edge where the optimal solution is assumed to stay. A GA with an enforcement operation has been verified to improve the search performance for the power plant scheduling problem far better than a stand-alone GA. Test results have verified that the trained RENGE can improve the search performance to find the optimal or near-optimal schedule in only one to eight CPU seconds on average with Sun SPARC station 20. With higher-performance computers being used in actual power plants, even less CPU time can be expected. The accuracy of the dynamic simulation model used for calculating the maximum stresses would, however, affect the performance of RENGE. The development of model parameter tuning tool, for which GA is also a promising technique, is required before the proposed approach can be effectively used for the power plant start-up operations.

References

1. Chen C.P., Liu C.W., and Liu C.C. (2000), Unit commitment by Lagrangian relaxation and genetic algorithms, *IEEE Transactions on Power Systems*, **15**, no. 2, pp. 707-714.
2. Richter, Jr.,C.W. and Sheble G.B. (2000), A profit-based unit commitment GA for the competitive environment, *IEEE Transactions on Power Systems*, **15**, no. 2, pp. 715-721.
3. Takata T., Takahashi J., Yokoi H., Nakano H., Kato M., Aoyagi M., Shimada K., and Arai J. (1999), Advanced method for unit commitment problem using genetic algorithm and mathematical programming, *Transactions of IEE Japan*, **B-119**, no. 3, pp. 333-343 (in Japanese).
4. Kamiya A., Kawai K., Ono I., and Kobayashi S. (1999), Adaptive edge search for power plant start-up scheduling, *IEEE Transactions on Systems, Man, and Cybernetics-Part C: Applications and Reviews*, **29**, no. 4, pp. 518-530.
5. Kamiya A., Kawai K., Ono I., and Kobayashi S., Theoretical proof of edge search strategy applied to power plant start-up scheduling, *IEEE Transactions on Systems, Man, and Cybernetics-Part B* (Under Review).
6. Kamiya A., Kimura H., Yamamura M., and Kobayashi S. (1998), Power plant start-up scheduling: A reinforcement learning approach combined with evolutionary computation, *Journal of Intelligent and Fuzzy Systems* **6**, pp. 99-115.
7. Hippert H.S., Pedreira C.E., and Souza R.C. (2001), Neural networks for short-term load forecasting: A review and evaluation, *IEEE Transactions on Power Systems*, **16**, no. 1, pp. 44-45.
8. Dillion T.S., Edwin K.W., Kochs H.D., and Taud R.J. (1978), Integer programming approach to the problem of optimal unit commitment with probabilistic reserve determination, *IEEE Transactions on Power Apparatus and Systems*, **PAS-97**, no. 6, pp. 2154-2166.
9. Peng C.K., Sheble G.B., and Albuyeh F. (1981), Evaluation of dynamic programming based methods and multiple area representation for thermal unit commitment, *IEEE Transactions on Power Apparatus and Systems*, **PAS-100**, no. 3, pp. 1212-1218.
10. Virmani S., Adrian E.C., Imhof K., and Mukherjee S. (1989), Implementation of a Lagrangian relaxation based unit commitment problem, *IEEE Transactions on Power Systems*, **4**, no. 4, pp. 1373-1380.
11. Sheble G.B. and Fahd G.N. (1994), Unit commitment literature synopsis, *IEEE Transactions on Power Systems*, **9**, no. 1, pp. 128-135.
12. Merlin A. and Sandrin P. (1983), A new method for unit commitment at Electricite de France, *IEEE Transactions on Power Apparatus and Systems*, **PAS-102**, no. 5, pp. 1218-1225.
13. Wang S.J., Shahidehpour S.M., Kirschen D.S., Mokhtari S., and Irisarri G.D. (1995), Short-term generation scheduling with transmission and environmental constraints

using an augmented Lagrangian relaxation, *IEEE Transactions on Power Systems*, **10**, no. 3, pp. 1294-1301.

14. Abdul-Rahman K.H., Shahidehpour S.M., Aganagic M., and Mokhtari S. (1996), A practical resource scheduling with OPF constraints, *IEEE Transactions on Power Systems*, **11**, no. 1, pp. 254–259.
15. Kamiya A., Ono I., Yamamura M., and Kobayashi S. (1997), Thermal power plant start-up scheduling with evolutionary computation by using an enforcement operator and tabu strategy, *Journal of Japanese Society for Artificial Intelligence*, **12**, pp. 100-110 (in Japanese).
16. Hanzalek F.J. and Ipsen P.G. (1966), Thermal stresses influence starting, loading of bigger boilers, turbines, *Electrical World*, **165**, pp. 58-62.
17. Matsumoto H., Eki Y., Kaji A., Nigawara S., Tokuhira M., and Suzuki Y. (1993), An operation support expert system based on on-line dynamics simulation and fuzzy reasoning for startup schedule optimization in fossil power plants, *IEEE Transaction. on Energy Conversion*, **8**, no. 4, pp. 674-680.
18. Glover F., Taillard E., and De Werra D. (1993), A user's guide to tabu search, *Annals of Operations Research*, **41**, pp. 3-28.
19. Kirtpatrick, S., Gelatt, Jr.,C.D., and Vecchi M.P. (1983), Optimization by simulated annealing, *Science*, **220**, pp. 671-680.
20. Amjady, N. (2001), Short-term hourly load forecasting using time-series modeling with peak load estimation capability, *IEEE Transactions on Power Systems*, **16**, no. 3, pp. 498-505.
21. Papalexopoulos A.D. and Hesterberg T.C. (1990), A regression-based approach to short-term system load forecasting, *IEEE Transactions on Power Systems*, **5**, no. 4, pp. 1535-1547.
22. Mastorocostas P.A., Theocharis J.B., and Bakirtzis A.G. (1999), Fuzzy modeling for short term load forecasting using the orthogonal least squares method, *IEEE Transactions on Power Systems*, **14**, no. 1, pp.29-36.
23. Khotanzad A., Afkhami-Rohani R., and Maratukulam D. (1998), ANNSTLF-artificial neural network short-term load forecaster-generation three, *IEEE Transactions on Power Systems*, **13**, no. 4, pp.1413-1422.
24. Prechelt L. (1998), Automatic early stopping using cross validation: Quantifying the criteria, *Neural Networks* **11**, no. 4, pp. 761-767.
25. Wang F., Yu E. K., Liu Y. Q., and Yan C. S. (1998), Short-term load forecasting based on weather information, *Proceedings of IEEE International Conference on Power System Technology*, pp. 572-575, Beijing, China.
26. Shimakura Y., Fujisawa Y., Maeda Y., Ono M., Fann J.Y., and Fukusima N. (1993), Short-term load forecasting using an artificial neural network, *Proceedings of the Second International Forum on Applications of Neural Networks to Power Systems*, pp. 233-238, Yokohama, Japan.
27. Shimakura Y., Fujisawa Y., Kishi Y., Kunugi M., Shimada K., Kawahara Y., Fann J.Y., and Ono M. (1992), Short-term load forecasting using an artificial neural network, *IEE Japan Power System Technology Meeting*, PE-92-11 (in Japanese).
28. Osaka S., Amano M., and Kawakami J. (1988), An expert system for power generation scheduling, *Proceedings of International Workshop on Artificial Intelligence for Industrial Applications*, pp. 557-562, Hitachi, Japan.

29. Rudolf A., and Bayrleithner R. (1999), A genetic algorithm solution to the unit commitment problem of a hydro-thermal power system, *Proceedings of the 13th Power Systems Computations Conference*, pp. 745-752, Trondheim, Norway.
30. Ohta T., Matsui T., Takata T., Kato M., Aoyagi M., Kunugi M., Shimada K., and Nagata J. (1996), Practical approach to unit commitment problem using genetic algorithm and Lagrangian relaxation method, *Proceedings of IEEE International Conference on Intelligent Systems Applications to Power Systems*, pp. 434-440, Orlando, FL.
31. Kawai K., Takizawa Y., and Watanabe S. (1998), Advanced automation for power-generation plants—past, present and future, *Proceedings of the 7th IFAC/IFIP/IFORS/IEA Symposium in Analysis, Design and Evaluation of Man-Machine Systems*, pp. 239-244, Kyoto, Japan.
32. Tanaka S., Ohta K., Minoura T., and Kogure Y. (1975), New concept software system for power generation plant computer control—COPOS, *Proceedings of International Conference on Power Industry Computer Applications*, pp. 267-275.
33. Kakehi A., and Hatsumi H. (1996), Development of new digital electro-hydraulic control system for steam turbine, *Proceedings of the 4th International Conference on Control, Automation, Robotics and Vision*, pp. 2106-2110, Singapore.
34. Kamiya, A., Kakei A., Kawai K., and Kobayashi S. (1999), Advanced power plant start-up automation based on the integration of soft computing and hard computing Techniques, *Proceedings of IEEE International Conference on Systems, Man, and Cybernetics*, pp. I-380-I-385, Tokyo, Japan.
35. Kimura H., Yamamura M., and Kobayashi S (1995), Reinforcement learning by stochastic hill climbing on discounted reward, *Proceedings of the 12th International Conference on Machine Learning*, pp. 295-303, Lake Tahoe, CA.
36. Ono I., Yamamura M., and Kobayashi S. (1996), A genetic algorithm with characteristic-preserving for function optimization, *Proceedings of the 4th International Conference on Soft Computing*, pp. 511-514, Iizuka, Japan.
37. Schellstede G., and Wagner H. (1986), A software package for unit commitment and economic dispatch, *Proceedings of IFAC Symposium on Power Systems and Power Plant Control*, pp. 264-271, Beijing, China.
38. Lewis, III, H.W. (2001), Intelligent hybrid load forecasting system for an electric power company, *Proceedings of the IEEE Mountain Workshop on Soft Computing in Industrial Applications*, pp. 23-27, Blacksburg, VA.
39. H. Chen and J. Liu (1998), A weighted multi-model short-term load forecasting, *Proceedings of IEEE International Conference on Power System Technology*, pp. 557-560, Beijing, China.
40. Ovaska S.J., Dote Y., Furuhashi T., Kamiya A., and VanLandingham H.F. (1999), Fusion of soft computing and hard computing techniques: A review of applications, *Proceedings of IEEE International Conference on Systems, Man, and Cybernetics*, pp. I-370-I-375, Tokyo, Japan.
41. Murillo-Sanchez C., and Thomas R.J. (1998), Thermal unit commitment including optimal AC power flow constraints, *Proceedings of the Thirty-First Hawaii International Conference on System Sciences*, **3**, pp. 81-88, Koahala Coast, HI.
42. Karashima N., Suzuki M., Abe M., and Kogure Y. (1981), Newly developed comprehensive automation technique applied to Hirono thermal power station, *Proceedings of American Power Conference 43rd Annual Meeting*, pp. 153-168, Chicago, IL.

43. Bednarski S. and Shen C.N. (1973), Analysis and algorithm for a minimax problem with thermal stress applications, *14th Joint Automatic Control Conference*, **20**, no. 22, pp. 765-775.
44. Schoenauer M. and Michalewicz Z. (1996), Evolutionary computation at the edge of feasibility, *Voigt H.-M., Ebeling W., Rechenberg I., and Schwefel H.-P. (Eds.), Proceedings of the 4th Int. Conf. on Parallel Problem Solving from Nature, Springer*, LNCS 1141, pp. 245-254.
45. Schoenauer M. and Michalewicz Z. (1997), Boundary operators for constrained parameter optimization problems, *Proceedings of the 7th International Conference on Genetic Algorithms*, pp. 322-329, East Lansing, MI.
46. Hwang R.C., Huang H.C., Chen Y.J., and Hsieh J.G. (1998), Power load forecasting by neural network with a new learning process for considering overtraining problem, *Proceedings of International Conference on Energy Management and Power Delivery*, **1** , pp. 317-322, Singapore.
47. Goldberg D.E. (1989), *Genetic Algorithms in Search, Optimization & Machine Learning*, Addison Wesley, Reading, MA.
48. (1999), *International Energy Annual 1999*, U.S. Department of Energy, Energy Information Administration (EIA), DOE/EIA-0219(99), Washington, DC.
49. Gill P.E., and Murray W. (Eds.) (1974), *Numerical Methods for Constrained Optimization*, Academic Press, Burlington, MA.

Neural Network-Based Adaptive Protection System

Adrian Halinka[*], Pawel Sowa[*], Michal Szewczyk[*], and Les M. Sztandera[†]

* Technical University of Silesia

† Philadelphia University

Keywords: *electrical power systems, power generating units, power system protection, adaptive systems, neural networks, multi-layer perceptron, supervised learning.*

1 Introduction

Practical successes have been achieved with neural network models in a variety of domains, including energy-related industry. The large, complex design space of electrical power systems (EPS) is only minimally explored in current practice. The satisfactory results that nevertheless have been obtained testify that neural networks are a robust modeling technology; at the same time, however, the lack of a systematic design approach implies that the best neural network models generally remain undiscovered for most applications. This chapter presents an approach to an adaptive protective systems problem in complex power generating units. First, we demonstrate the complex interdependencies between various parameters of EPS protection systems. Then, we present an approach, based on protection and adaptation criteria, for designing a neural network based adaptive protection system.

Electrical Power Systems (EPS) comprise units characterized by a complex topology, both in respect of the quantity and the variety of electrical appliances, which constitute the unit, as well as the number of operating modes, which differ as to their configurations and functions.

From the point of view of the EPS protection systems, sweeping protection of such complex units necessitates the application of advanced adaptive protection systems.

The reliability of protection systems is determined mostly by the accuracy of decisions classifying the current operating mode of the protected unit as one of the two categories of events: normal or fault. This reliability is dependent on meeting several conditions, first of all on the possibility of acquirement and acquisition of a large number of data (binary and analog) concerning the protected unit, the speed at which data are processed and decisions are made, the ability to adapt their innate properties to the current operating mode of the unit, and the immunity to any disturbances and faults.

Digital technology makes it possible to adopt a new approach to the issue of identification of operating modes in a protected unit as well as the issue of correct adaptations of protection functions. The benefits of employing digital technology in complex protection systems are connected with the number and the variety of the acquired data, the possibility to process these data by using heuristic techniques, the speed at which the data are processed, as well as the digital pre-processing of the measurement signals, which, for example, eliminates disturbance factors.

Several basic protection functions, which are blocked in analog relays, are here activated, improving the overall quality and reliability with which the unit is protected from the consequences of disturbances. This happens because both measurement algorithms as well as the algorithms realizing particular protection functions are capable of adapting to the current frequency (in the operating modes of the unit which are characterized by the frequency of measurement signals, changing within a wide range, e.g. frequency start-up) and to the operating mode.

2 Protection Systems in Complex EPS Units

Taking advantage of the benefits offered by digital technology, one can create a concept of digital managing systems which are dedicated to complex EPS units and which are characterized by the following features:

a) They are able to identify correctly the current operating mode of the unit, i.e. the realization of the *structure criterion.*

b) They provide high accuracy and speed of the protection algorithms, both in all possible operating modes of the protected unit and in the frequency of the input measurement signals which changes within a wide range, i.e. the realization of the *protection criterion.*

c) The structure of protection system can be adapted quickly to the changing conditions and operating modes of the protected unit i.e. the realization of the *adaptation criterion.*

d) They are capable of a fast analysis of the in-flowing alarm signals and the data concerning responses of protection functions, which enables one to predict the place of disturbance, and to follow any necessary procedure aiming at the minimization of the consequences of such disturbances, i.e. realization of the *prevention-restitution criterion.*

Reversible hydrogenerators, which operate in conjunction with a unit transformer, have been here adopted as an example of a complex power unit. An adaptive managing system, that employs the above-mentioned criteria, has been proposed.

An example of an adaptive managing system shown against the background of the protected unit is presented in Figure 1. Each of the modules realizes the *protection* and the *adaptation criteria* and the digital protection systems with the open configuration, regularly equipped with nine input measurement channels, are used as the systems realizing these criteria. Additionally, Figure 1 shows places where input signals for particular systems are measured. The system is equipped with three basic modules (ID1, ID2, ID3) and one master system (GID), which realize the *structure criterion.* Four systems analyzing alarm signals and predicting the place of disturbance (FID) have been employed to realize the *prevention-restitution criterion.*

Because of the complexity of the unit, the adaptive managing system has been split into modules dedicated to particular components of the unit. Decentralizing operation of the system into particular modules makes it possible to separate functions responsible for particular criteria, which increases the number of indispensable processors, but which also significantly decreases the load and the time in which the whole system responds to changes in the protected unit.

Placing the dedicated modules nearer the chosen components of the protected unit increases the accuracy and reliability of the data acquisition. Particular modules exchange information that has already been processed (for example, identification of the operating mode of the protected unit's component). That makes possible a prediction-based initial analysis of the protected unit's operating mode. The decrease in the number of operations realized by processors is also achieved through the division of the identified operating modes into groups and subgroups of active protection functions. Taking into consideration the level of complexity of a given operating mode, there is also a possibility of combining a particular group of protection functions into sets.

3 Digital System Responsible for Identification of Current Operating Mode

An accurate identification of operating modes in the protected unit on the basis of the data entering the system (the *structure criterion*) is the basic condition for correct functioning of the system.

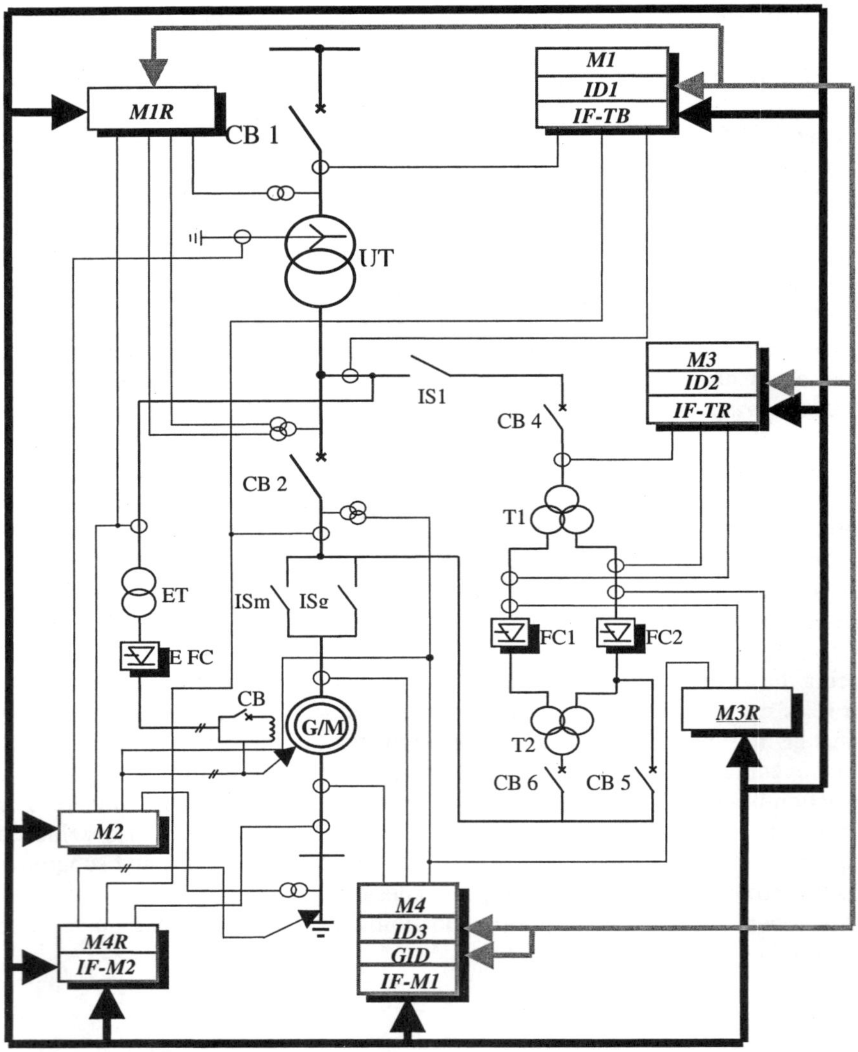

Fig. 1. Reverse hydro-generator adaptive protective system where:

- ID1, ID2, ID3, GID - digital current operation mode identification system;
- M1, M1R, M2, M3, M3R, M4, M4R - modules that are responsible for protection and adaptation criteria;
- IF-TB, IF-TR, IF-M1, IF-M2 - modules that are responsible for prevention and restitution criteria.

Three types of data acquirement are proposed [1] for identification purposes:

- Using binary inputs representing the position of connectors in the protected unit (their quantity should be sufficient for an unequivocal identification of the current configuration of the protected unit).

- Utilizing the output of measurement algorithms, for example the voltage measurement algorithm, frequency measurement algorithm (in the chosen places within the unit).

- Exchanging of binary data between the modules that constitute the protection system.

Each module identifies (by binary or binary-analog data concerning the position of the switches) the current configuration of the nearest component of the protected unit. That makes it possible to subdivide the identification of the operating mode of the whole unit into modules dedicated to a given function, which, in turn, depending on the situation exchange data among themselves.

Next, a sample module responsible for the identification of the operating mode of the protected unit will be presented. This module employs Artificial Neural Network (ANN) structures.

The reversible hydrogenerator with the configuration presented in Figure 1 is proposed as the protected unit.

The protected unit has been divided into three parts for which ANN structures responsible for the initial identification of the global operating mode have been determined. This was done on the basis of the operating modes of its constituent parts. Input signals comprise of: binary and the binary-analog signals, exchange of data among given networks, as well as the information from the measurement algorithms.

Figure 2 presents a fragment of the scheme illustrating the structure of the system responsible for the identification of the operating mode, which is based on ANNs. This is the structure of a feed-forward network of the multilayer perceptron (MLP) type, with three layers. The main advantage of such a solution is that in this case there is no explicit need to define relations between particular pieces of information (signals).

During the teaching process, the network optimizes its structure by choosing the number of neurons and appropriate weights and biases between them [2]. The process of adding neurons is explained in Section 6. This is done in such a way as to be able to identify and classify situations by which the network was taught, as well as situations that are their generalizations. They are different in respect of certain input signals, which, however, does not result in classifying them as a different category of events.

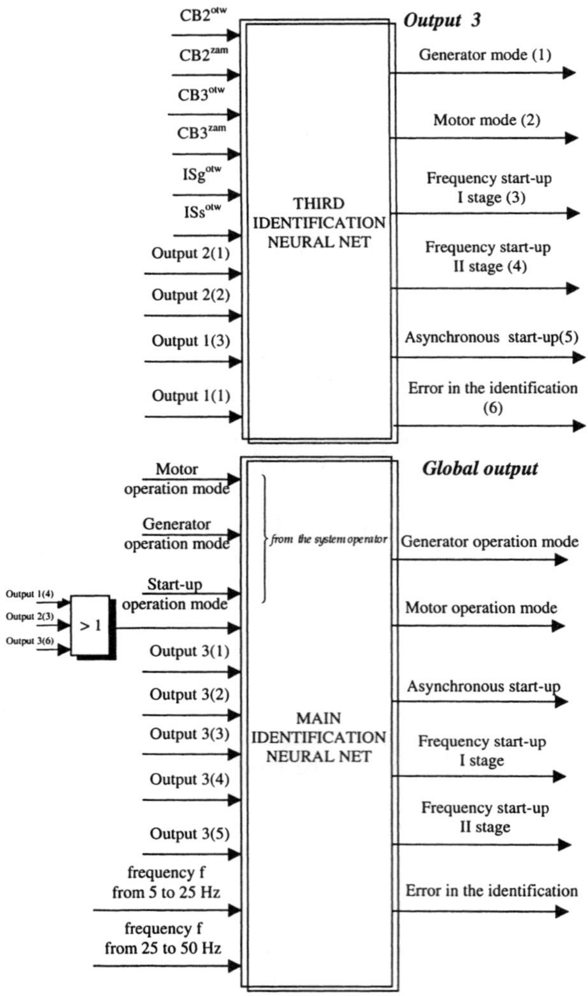

Fig. 2. Part of current operation mode identification unit based on the artificial neural networks of the MLP type.

The main advantage of such a structure is the fact that it eliminates certain situations, such as incompatibility of binary signals or the case of missing signals, which are the primary sources of the "error" signal generated by the identification system in binary logic. When we adopt the above mentioned structure of the identification system, the proper determination of the teaching base becomes the issue of paramount importance. The network must be taught by the greatest possible number of events that belong to all possible categories of events, i.e., operating modes of the unit.

In the case of the unit under consideration, any subsequent identification of the protected unit's operating mode is performed every second or every third period of the basic component of the current frequency. This occurs in the main identification unit in the case of all types, or stages, of start-up. It is necessitated by the need for the activation of the correct sets of protection functions, depending on the current frequency and the unit's operating mode. The protection system must be equipped with separate identification components that contain their own fast processors and memory buffers because of a high repeatability and complexity of the identification process.

In modes such as the generator mode, motor mode and stand-by (out-of-service) unit, identification can be repeated in much longer intervals, for example every five minutes. Identification is repeated in all components of the system, and it is synchronized both by the master identification output module and the communication system of the main machine protection.

The identification system is restarted immediately in the following situations:

- Activation or operation of the active protection function.
- Recognition of error by the main identification unit (GID).
- There is information about opening of the main circuit breaker of the system CB2, if the circuit breaker was previously switched on.

4 Adaptive Protection Systems

Adaptive protection systems that realize the *protection* and *adaptation criteria*, dedicated to complex EPS units, usually have a diffused structure. Such a structure is achieved through the subdivision of the unit into smaller protected fragments – similarly as in the case of the system responsible for the identification of operating modes in the protected unit. Making such an assumption makes it possible to establish protection systems that perform protection functions (with adaptive properties) and measurement functions. They also perform functions responsible for communication with adjacent systems, that is, with systems responsible for the identification of the operating mode (ID, GID), with systems responsible for the prediction of the place and type of disturbance (FID), or with the master system. In this system, a very important role is played by the module responsible for the adaptation of the digital protection system to the current operating mode of the protected unit. The protected mode is placed in the fragment of the protection unit dedicated to the synchronous machine (Figure 3). This module is the central managing system that controls protection systems dedicated to particular fragments of the unit. It also activates and oversees functioning of the measurement and protection algorithms established for the synchronous machine unit. The accuracy and reliability of the functioning of the module responsible for the adaptation of the digital protection system, that guarantees that protection functions are appropriately realized, are determined by the quality of the signals generated in the system responsible for the identification of the operating mode of the protected unit.

The adaptive criterion subdivides the set of active protective algorithms into four main groups:

- A set of algorithms that are active in the generator-operating mode.
- A set of algorithms that are active in the motor operating mode.
- A set of algorithms that are active in the frequency start-up.
- A set of algorithms that are active in the asynchronous start-up.

Due to the above setup, it is easy to choose a set of active algorithms dedicated to a given operating mode of the unit. It is achieved by using one control binary signal that activates an appropriately programmed unit of the protection system.

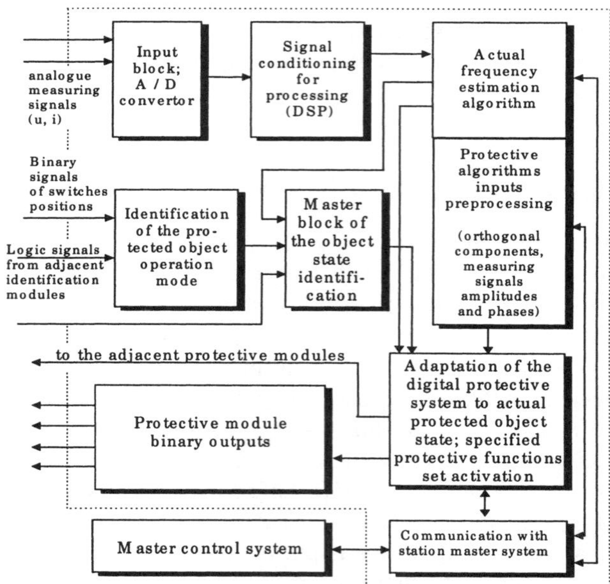

Fig. 3. Scheme of the protective system dedicated to the synchronous machine.

5 An Analysis of Alarm Signals and the Localization of the Place of Disturbance

The managing system's ability to undertake preventive actions (*prevention criterion*) is of paramount importance from the point of view of the correct functioning of the protected unit. It is especially useful in situations when alarm signals are received. This would indicate the occurrence of disturbances in the protected unit or in the systems co-operating with it, for example, a loss of synchronicity, a fluctuation of frequency, a current overloading, an asymmetrical load, or a damage of the cooling system of transformers. Procedures of this kind attempt to maintain functioning of the unit in disturbance conditions by influencing the control systems. Procedures of this kind comprise, among other:

- The reduction of the generator's load.
- Shifting from the automatic to manual voltage control in case when there are alarm signals about the loss of excitation (e.g., when the excitation circuit-breaker is closed). This would indicate a significant danger of a damage in the Automatic Voltage Regulation (AVR) unit.
- Shifting to manual voltage control when there are data indicating current overload of rotor, e.g., overexcitation. This could indicate a possible breakdown of the voltage limiters in the AVR unit.

In the situation when functioning of a given unit or one of its components cannot be maintained, and consequently, it must be switched off, it is possible to create an algorithm of a fast restitution of the unit after the disturbance has been dealt with (the *restitution criterion*). The following applies especially to generating sets such as plants with gas-driven turbines [3, 4] and pump-storage plants with reversible hydrogenerators, or to the plants where there is a high degree of redundancy of components (e.g., so it is possible to use start-up units from another machine, or having two hydrogenerators operating with only one unit transformer).

6 Teaching and Testing of the ANN-Based System

Artificial neural network models offer an attractive paradigm: they learn to solve problems from examples. These models achieve good performance via massively parallel nets composed of non-linear computational elements, sometimes referred to as units or neurons. With each neuron is associated a number, referred to as the neuron's activation. Similarly, with each connection between neurons, a number is also associated, called its weight. These resemble the firing rate of a biological neuron and the strength of a synapse (connection between two neurons) in the brain. A neuron's activation depends on the activations of the neurons connected to it and the interconnection weights. Neurons are often arranged into layers. Input layer neurons have their activations externally set.

Neural network learning algorithms can be divided into supervised and unsupervised:

- Supervised neural networks need and external “teacher” during the learning phase, which goes before the recalling (utilization) phase.
- Unsupervised neural networks “learn” from correlations of the input.

In this research one of the most popular and successful neural network architectures, a Multi-Layer Perceptron (MLP) was used.

An MLP network comprises a number of identical units organized in layers, with those on one layer connected to those on the next layer so that the outputs of one layer are fed-forward as inputs to the next layer. MLP neural networks are typically trained using a supervised training algorithm known as “back propagation”.

Typically, an MLP is used to classify patterns based on input vectors (a set of features taken from an example of the problem) presented to the network. An MLP network designed to discriminate between n classes would have n outputs. The network is trained with training cases such that a “one” on a particular output unit corresponds to a particular input class, while all other output units are “zeros”. When used in recall mode, features are presented to the neural network and the output unit with the most significant output indicates the class of the input pattern.

The MLP architecture can be extended to produce a continuous-valued output that is a function of its inputs, resulting in a trainable function generator. If this configuration were used to model a process that is known to evolve with time, time would be one of the neural network inputs. However, a limitation of an MLP in recall mode is that it has no “memory” of previous inputs. Hence, to model a system that depends upon the time history of input variables, it is necessary to include the time history in the neural network input vector.

For example, an MLP network can be used as a predictor for time series data. For many time series, the value of the next sample in the series can be predicted as a function of a number of previous samples. To use an MLP to predict a value xt in a time series, the values xt-1, xt-2 ... xt-n form the input to the neural network. Existing time series data is used to train the neural network. In some cases, the network predictions are improved by including long-term running averages in the neural network input vector. An MLP neural network may be trained on data originating either from a real world system or from a sophisticated model of the process. In the latter case, the neural network learns to “model” the model. While a neural network is unlikely to provide a sophisticated model, it can often provide outputs of adequate accuracy over a limited range of input conditions, with the advantage of requiring far less computation than other modeling methods. It is also capable of modeling processes where the underlying principles are not fully understood.

It is generally accepted that the performance of a well-designed MLP neural network is comparable with, but generally no better than can be obtained using good classical statistical techniques. MLP networks score over classical techniques in their much reduced development time, their ability to adapt to changing situations, and their ability to make use of related information.

The process of teaching neural networks aims at finding a proper correlation between input signals and the target response of the network. The teaching process results in the correction of weights and biases of particular neurons of the network in relation to their initial values.

In our simulation tests, which have been carried out, the method of teaching with a teacher has been used. Systems responsible for the identification of the operating mode were made on the basis of the Multi Layer Perceptron (MLP) structures, and the algorithm of error back propagation was used as the algorithm responsible for updating weights.

The most important questions in determining a feedforward neural network architecture are how to calculate the number of nodes in hidden layers, and how to calculate the number of hidden layers. Lippmann [5] argued that a network with two hidden layers could solve arbitrary classification problems. Irie and Miyake [6] showed that a one hidden layer back propagation network with an infinite number of nodes in the hidden layer could also solve arbitrary mapping problems. However, these results have little practical value. Lippmann [5] also argued that the nodes in a hidden layer corresponded to separate decision regions into which training examples were mapped. Kung and Hwang [7] used the algebraic projection approach to specify how each node should be created. In those approaches, however, we have to know the properties of the training data, such as decision regions or pattern properties, thus their applications are limited. The determination of feedforward neural network architectures has been an important area in neuroengineering research.

Practical approaches for dynamic neural network architecture generation have been sought by Sirat and Nadal [8], and Bichsel and Seitz [9]. In those architectures, at the end of a training process, all training examples are recognized and a neural network architecture is generated. In particular, the "tiling" algorithm of Sirat and Nadal [8] generates a feedforward network architecture by adding nodes and layers in a sequential manner. However, the algorithm does not give us the exact sequence in which a node is added to achieve the optimal classification of training examples. A similar algorithm of Bichsel and Seitz [9] uses information entropy to determine generation of nodes and hidden layers. Other algorithm of a special interest, the ID3 algorithm of Quinlan [10] dynamically generates a decision tree using information entropy functions. Studies by

Dietterich et al. [11] and Fisher and Mckusick [12] revealed strong evidence that information entropy could be used as a criterion for determining the number of hidden layers in feedforward neural network architectures. Sztandera and Cios [13] also addressed the problem of a dynamic generation of neural network architectures. The results presented here are based on outcomes obtained by Sztandera [2], Sirat and Nadal [8], Bichsel and Seitz [9], and Sztandera and Cios [13]. The algorithm used generates nodes and hidden layers until a learning task is accomplished. The algorithm operates on continuous data and equates a decision tree with a hidden layer of a neural network. A learning strategy used in this approach is based on minimization of the entropy function. This minimization of entropy translates into adding new nodes to the network until the entropy is reduced to zero. When the entropy is zero then all training examples are regarded as correctly recognized.

After applying the above described approach, it was established that the first layer would contain 15 neurons, the second one 10, and the number of neurons in the third layer would be equal to the number of outputs of a given module in the identification system. Hyperbolic tangent was used as the function activating the first and the second layer. The linear function, on the other hand, is the activating function for the third layer.

In the process of teaching neural networks, the order in which teaching patterns are presented is the issue of great importance. Situations wherein all patterns belonging to one class, and then patterns belonging to the next class, etc. are presented, are to be avoided because the network forgets the patterns it has previously learnt.

Generally, patterns belonging to the teaching base have been divided into two groups:

- The group that contains signals carrying information about the operating mode of the unit.

- The group that carries erroneous signals.

In the teaching process, the principle of turn-taking teaching was adopted (first the pattern carrying information about a given operating mode, then the pattern carrying error information). A variety of operating modes have been taken into consideration, i.e. generator, motor, and asynchronous start-up. However, it was assumed that some of the patterns would not be provided at the input of the neural network in order to check the neural network's ability to generalize events.

Establishing initial weights is also a very important issue in the teaching process. In the case of MATLAB program with the Neural Network toolbox, the *initff* command is responsible for establishing initial weights. This command generates initial weights and biases of the network, making use of the first pattern in the teaching base. Taking into consideration the fact that the algorithm of back propagation does not guarantee that a global minimum of a function will be found, it was necessary to provide random input signals, belonging, however, to the range, which they really assume. The user of the program introduces the initial input vector. In the situation when we suspect that the local minimum was achieved, the teaching process must be repeated. Figure 4 presents sample results of tests of the sub identification realized by the third identification neural network (see Figure 2).

Three cases have been presented. At the net's input, a correct sequence of the signals is provided that represents a second stage of the frequency start-up in the protected machine [Fig. 4(a)]. The strongest signal of the network's response is achieved at the fourth output, responsible for the identification of the second stage of the frequency start-up [Fig. 4(b)]. The sequence of input signals contained an error (signals representing the position of the main circuit-breaker of the generator CB2 [Fig. 4(c)] were incompatible). Also in this case the strongest signal is generated at the fourth output of the network [Fig. 4(d)]. Despite the erroneous input information, the identification is still correct.

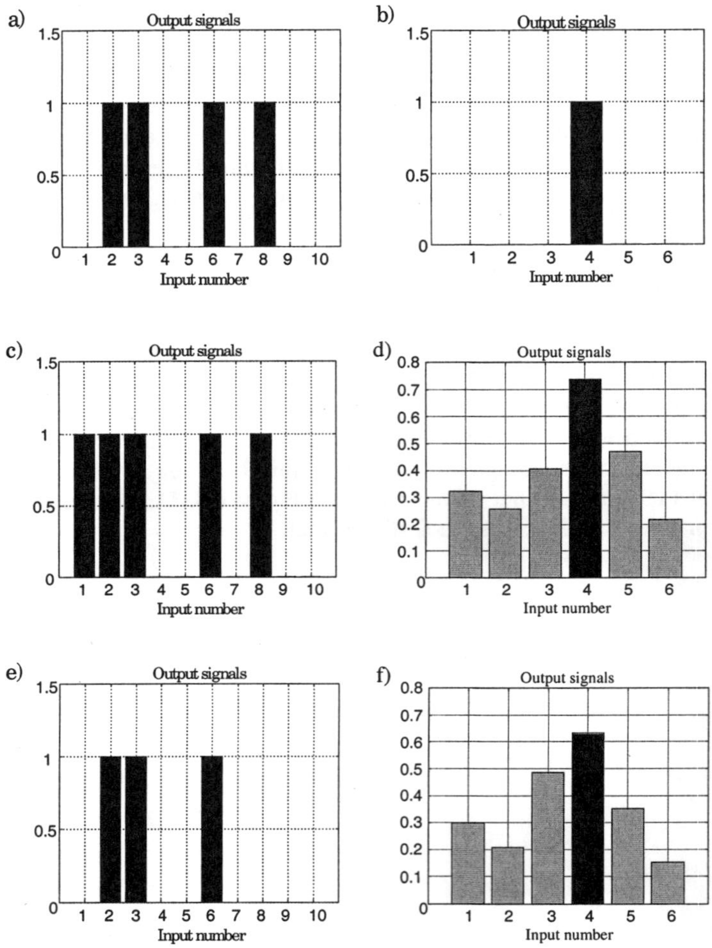

Fig. 4. Responses of the third identification net b), d), f) to the signals with [c), e)] or without [a)] errors.

7 Conclusion

The managing system presented in this chapter performs protection, measurement, adaptation and identifying functions, using the network of relations among the constituent modules. Establishing the four basic criteria for such systems, that is *the protection criterion, the adaptation criterion, the structure criterion, and the prevention-restitution criterion,* significantly increases the range of their possibilities. The systems of this kind, based on the intelligent identification systems, are capable of an automatic adaptation of their measurement and protective functions to the variable working conditions of the protected unit. Those conditions include changes of the connections among particular components of the unit, as well as the changes of the parameters that characterize the input measurements (frequency, voltage, current, etc.). Such systems are dedicated first of all to the generating sets with a high-degree redundancy of components, for example the combined plants with steam- and gas-driven turbines [3, 4], where it is easy to replace one component with another. Systems of this kind are characterized by a complex structure and a short start-up time, thus, becoming fast sources of power control. Due to these properties, it is possible to work out autonomous algorithms of the adaptive prevention-restitution automation.

References

1. Sowa P., Halinka A., Witek B., and Szewczyk M. (1998), Identification methods of complex power system units operation mode, *Proceedings of the International Workshop on Complex Power Systems*, Villeurbanne, France, pp. 21-26.
2. Sztandera L. M. (1994), Dynamically generated neural network architectures, *Journal of Artificial Neural Systems,* **1**, pp. 41-66.
3. Buck D. (1995), Das elektrische system von ABB - kombikraftwerken, *ABB Technik,* **2**, pp. 15-23.
4. Halinka A., Winkler W., and Witek B. (1997), Adaptive relaying for gas-driven turbine and hydroturbine generators, *Proceedings of the 59thAnnual Meeting\of the American Power Association,* Chicago, IL, pp. 21-26.
5. Lippmann, P. (1987), An introduction to computing with neural nets, *IEEE Transactions on Acoustics, Speech, and Signal Processing,* **44**, pp. 4-22.
6. Irie B. and Miyake S. (1988), Capabilities of three-layered perceptrons, *Proceedings of IEEE International Conference on Neural Networks*, pp. 641-648, San Diego, CA.
7. Kung S. Y. and Hwang J. N. (1988), An algebraic projection analysis for optimal hidden units size and learning rates in back-propagation learning, *Proceedings of IEEE International Conference on Neural Networks*, pp. 363-370, San Diego, CA.
8. Sirat J. A. and Nadal J. P. (1990), Neural trees: a new tool for classification, *Network,* **1**, pp. 423-438.
9. Bichsel M. and Seitz P. (1989), Minimum class entropy: a maximum information approach to layered networks, *Neural Networks,* **2**, pp. 133-141.
10. Quinlan J. R. (1986), Induction of decision trees, *Machine Learning*, **1**, pp. 81-106.

11. Dietterich T.G., Hild H., and Bakiri G., (1990), A comparative study of ID3 and back propagation for English text-to-speech mapping, *Proceedings of the 7th International Conference on Machine Learning*, pp. 36-41, Austin, TX.
12. Fisher D. H. and Mckusick K. B. (1989), An empirical comparison of ID3 and back-propagation, *Proceedings of the 11th International Joint Conference on Artificial Intelligence*, pp. 788-793, Detroit, MI.
13. Sztandera L. M. and Cios K. J. (1997), Ontogenic neuro-fuzzy algorithm: F-CID3, *Neurocomputing*, **14**, pp. 383-402.

Part 2: Emerging Applications

Fuzzy Logic-Based Traffic Controller

Jarkko Niittymäki, Riku Nevala, and Marko Mäenpää

Helsinki University of Technology

Keywords: *fuzzy logic, traffic control, traffic simulation, fuzzy traffic control applications, traffic signal controller, signalized pedestrian crossing, phase control, multi-objective control, areal traffic signal control, traffic and transportation engineering.*

1 Introduction

Many subjects in transportation engineering are often characterized as subjective, ambiguous and vague. In traffic signal control, several traffic flows compete for the same time and space, while each is often assigned its own respective priority. Normally, optimization operates upon several criteria simultaneously, like average delays, maximum queue lengths and percentages of stopped vehicles. So it is likely that fuzzy control is very competitive at complicated real intersections where the use of traditional optimization methods is problematic. In practice, for traffic safety reasons, uniformity is the goal signal control strives to achieve. This goal sets limitations on both the cycle time and the phase arrangements. Hence, traffic signal control in practice is based on tailor-made solutions and adjustments made by the traffic planners. The modern programmable signal controllers with a great number of adjustable parameters are well suited to this process. For good results, both an experienced planner, and fine-tuning in the field, are needed. Fuzzy control has proven to be successful in problems where the process can be controlled by an experienced human operator, but exact mathematical modeling of the problem is difficult or impossible. Thus, traffic signal control is a suitable task for fuzzy control. Indeed, one of the oldest examples of the potential of fuzzy control is a simulation of traffic signal control at an intersection of two one-way streets. Even in this very simple case, the fuzzy control was at least as good as the traditional adaptive control [32].

The opportunities offered by fuzzy logic for controlling traffic have been the subject of our FUSICO-research project centered at the Laboratory of Transportation, Helsinki University of Technology. Particular areas of interest include the theoretical analysis of fuzzy traffic signal control, developing generalized fuzzy rules for traffic signal control using linguistic variables, validating fuzzy control principles, and developing a fuzzy adaptive signal controller.

2 Traffic Signal Control and Fuzzy Logic

The impetus for the introduction of the earliest traffic signals was to ensure safety at intersections by keeping conflicting traffic flows apart. Nowadays, the effectiveness of traffic signal design can be assessed by applying performance criteria to the operation of intersections, i.e., by assessing desirable outcomes. Hence, a decrease in delay, in the number of stops, in fuel consumption, pollutant emissions, noise, vehicle operating costs, queue length and personal time are all desirable. An increasing consideration for pedestrians, bicycles and public transport vehicles is also desirable, as is an increased attention to safety in general.

The functioning of traffic signal control has a significant effect on the environment. Traditionally, the main performance measure for judging the effectiveness of traffic signal control systems has been the reduction of vehicle delay and stops [42]. Other considerations such as convenience and safety for pedestrians or bus priority have usually been incorporated into the systems later in their development as pragmatic features required for on-street operation. Traditionally they have not been used as performance measures during the development and optimization phases [3].

The difficulty of the traffic control process lies in the fact that the process must be repeated at very short time intervals. For instance, when tuning the green phase duration, the decision process using new input data from varying traffic situations is activated after each green phase extension, which usually last for only a few seconds. Secondly, because traffic conditions in the immediate future cannot be predicted precisely, the control action is based on optimizing the current state only; as a result, the individual control actions do not, in the long term, necessarily yield the optimal conditions, such as minimum overall delay or stopping percentage. Thirdly, detectors like inductive loops and infrared detectors, cannot capture details of prevailing conditions on the approaches, such as those relating to vehicle types and speed changes, as well as humans. When the intersection is complex in terms of geometric design, channelization, and types of vehicles to be handled, the control process must consider many usually mutually conflicting objectives, of which safety is the most important. How to place weights among these objectives is also an important and subjective issue. Thus, signal control deals with a complex multi-objective and multi-constraint problem, in which optimization is based on recent information only.

The basic idea of fuzzy signal control is that of modeling the control on the basis of human expertise and knowledge rather than on the basis of a precise modeling of the process itself. Fuzzy-logic based controllers are designed to capture the key factors in controlling a process without requiring many mathematical functions. Due to this fact, they have many advantages in real time applications. The controllers have a simple computational structure, since they do not require many numerical calculations. The "if-then-else" logic of their inference rules does not require much computational time. Also the controllers can operate on a large range of inputs, since different sets of control rules can be applied to them. If the system's related knowledge is represented by simple if-then-else fuzzy rules, then a fuzzy-based controller can control the system with efficiency and ease.

The design of fuzzy signal controller requires expert knowledge and experience of traffic control. The knowledge is needed in the formulation of the linguistic protocol that generates the control input to be applied to the traffic control system [5]. So, the fuzzy control rules can be derived from expert experience and control-engineering knowledge.

Fuzzy signal control has been developed in the context of fuzzy inference. Fuzzy inference is the inference process based on the multi-value logic of inference; in other words, the truth-values of input and rules of the inference process are not singular (yes or no), rather, they are multi-valued. As a result, the truth of the conclusion is given by a value between 0 and 1, which more closely resembles the inference performed by a human, and represents the ambivalence of the human decision process. The essence of this inference process is the use of fuzzy sets to represent the input and the rules expressing the relations between inputs and outputs.

3 Literature Review of Fuzzy Traffic Signal Control

3.1 Isolated Traffic Signals

The first known attempt to use fuzzy control in traffic signal control was made by Pappis and Mamdani (1977), who made a theoretical simulation study of a fuzzy logic controller at an isolated signalized intersection (2+2 lanes, one-way intersection) [32]. In their study, Pappis and Mamdani compared their fuzzy method to an ideal adaptive signal control, in which the cycle time was optimized for every traffic situation with Webster's algorithm [42]. According to the results, the fuzzy controller was equal to or slightly better than the method used for comparison. In this study, the heuristic approach to the control problem was employed, which resulted in a set of linguistic control statements. The basic ideas of the theory of fuzzy sets were used for the quantitative interpretation of these instructions as well as for the decision making process. The fuzzy control instructions are of the form:

if	*T*(ime) is long
and	*A*(rrivals) are more than many
and	*Q*(ueue) is less than medium
then	*E*(xtension) is long
else, etc.	

A total of 25 rules were used. Every rule is a fuzzy relation between the inputs *T*, *A*, *Q* and the output *E*. The connectives "and" and "else" are interpreted as the operators "min" and "max", respectively. Since in this application each fuzzy rule is represented by a four-dimensional array, and as five rules operate at each intervention, the fuzzy algorithm employed to decide the control action at each intervention is represented by the union of five such arrays. At each intervention the five rules are invoked ten times (i.e., for each of the next 10 s). It is worth noting that the detectors are located sufficiently far away from the intersection to enable the traffic situation data to be available for each of the next 10 s.

Kim (1994) has studied the fuzzy algorithms of the isolated intersection, discussing also the problems of turning traffic [9]. The fuzzy algorithm adjusts the duration of the green traffic light by evaluating the traffic conditions at the end of each phase. The green time for each phase is divided into a number of interventions. Every fuzzy rule is parameterized by the process state variables in conjunction with the linguistic values devoted to these variables. The output variable corresponds to the "extension" of the green time and each fuzzy control rule has the following typical form:

if	*S.left* is large
and	*S.thru* is less than medium
and	*C.left* is more than small
and	*C.thru* is any
then	*EXTENSION* is long

etc.

S.left and *S.thru* are fuzzy variables related to the number of vehicles present within about 60 meters distance behind the stopping line of the phase having the right of way. *C.left* and *C.thru* are fuzzy variables of the halted traffic. The simulated result shows up to 5–33 % improvement in the efficiency of traffic control using fuzzy logic. See also [39].

Favilla et al. (1993) presented a fuzzy traffic controller with adaptive strategies for fuzzy urban traffic control systems combined with two different defuzzification and decision-making criteria. The simulation results indicated that the fuzzy control was adaptive for different traffic situations [6].

Trabia et al. (1999) presented a fuzzy logic-based adaptive traffic signal controller for an isolated four-approach intersection with through and left-turning movements. The controller had an ability to make adjustments to signal timing in response to observed changes in the approach flow. The information was used in a two-stage fuzzy logic procedure to determine whether to extend or terminate the current signal phase. In the first stage, the controller estimated the traffic intensity on each approach. This intensity information was in turn used during the second stage to determine whether to extend or terminate the current phase. The controller produced shorter vehicle delays than the traffic-actuated controller, while maintaining the percentage of stopped vehicles at the same order [37].

Sayers et al. (1995, 1998) aimed to develop a flexible signal controller that could be configured to incorporate the objectives appropriate to the situation in which it is to be used. They used the multi-objective genetic algorithm (MOGA) optimization technique to derive optimal solutions for fuzzy control [35,36].

3.2 Coordinated Traffic Signals

Nakatsuyama et al. (1984) clarified that the fuzzy logic phase controller is valid in signal control of two successive intersections on an arterial road under certain conditions, such as when a large number of vehicles are using the intersection. The fuzzy logic phase controller is composed of fuzzy control statements, which determine the termination of green or amber periods. Co-operation of a fuzzy logic controller and a fuzzy logic phase controller results in good performance in all situations, especially when the number of cars changes within a large margin, as before or after the rush hour. A fuzzy logic controller or a fuzzy logic phase controller have to be interchanged with each other in evaluating the performance of controllers with respect, for example, to queues or arrived vehicles, using a fuzzy algorithm. The fuzzy phase controller co-operating with the fuzzy logic controller is easily applicable to the control of the traffic signals located successively on the one-way arterial roads. The fuzzy rules, which determine the termination of a green or amber period, resemble the rules of Pappis-Mamdani [32]. The interchange between a fuzzy logic controller and a fuzzy logic phase controller is executed by fuzzy logic statements. An example of these rules is presented below [17]:

if A_1 is not small
and A_2 is medium
then C is coordinated
else if A_1 is not large
and A_2 is any
then C is isolated
else etc.

Kim (1994) has also studied the fuzzy controlled coordination. The fuzzy rule base needs to take into account the possible blockage at the intersection due to potentially heavy traffic conditions. To prevent intersection blockage some modifications can be made to the "isolated" fuzzy control policy. In general, the rules are:

if "blockage-conditions" are about to occur
then to prevent the blockage do not allow traffic into the link from the upstream intersection, but allow traffic only out of the link that has the right of way.
else apply the same fuzzy algorithm used at the isolated intersections.

The results of fuzzy coordination have shown about a 10 % larger capacity than the results of conventional coordination [9].

3.3 Control at Network Level

The signal timing at an intersection in a traffic network is defined by three parameters: cycle time, phase split and offset. Chiu (1992) used fuzzy decision rules on the basis of local information only to adjust these three parameters [4]. The rules for adjusting cycle time, phase split and offset are decoupled so these parameters are adjusted independently.

Cycle time is adjusted to maintain a good degree of saturation on the approach with highest saturation. The saturation degree is defined as vehicles on a particular approach divided by the capacity of that approach. The inputs of the seven cycle rules for adjusting the cycle time are the highest degree of saturation on any approach (*highest_sat*) and the highest degree of saturation on its competing approaches (*cross_sat*). The output of the rules is the amount of adjustment to the current cycle time. For example

if *highest_sat* is high and *cross_sat* is high then *cycle change* is medium.

Phase split is adjusted to maintain equal degrees of saturation on competing approaches. The inputs to these 15 rules are the difference between the highest degree of saturation on the east-west approaches and the highest degree of saturation on the north-south approaches (*sat_diff*), and the highest degree of saturation on any approach (*highest_sat*). The output of the rules is the amount of adjustment to the current east-west green period. For example,

if *sat_diff* is medium and *highest_sat* is high then *green change* is medium.

Offset is adjusted to coordinate adjacent signals in a way that minimizes stops of the dominant traffic flow. The inputs to the rules are the normalized difference between the traffic volume (*vol_diff*) in the dominant direction and the average volume in the remaining directions, and the required time adjustment (*req_adjust*) relative to the adjustable amount of time. For example,

if *vol_diff* is high and *req_adjust* is high then *allow adjustment* is low.

The effectiveness of the fuzzy method has been studied by simulations. The results have showed slight improvements (about five %-units) in the share of waiting time of the total travel time, when only 3 intersections are adaptive, but the improvement becomes significant (10–20 %-units) when all intersections in the network are adaptive.

4 Fuzzy Applications and Control Algorithms of Traffic Signal Control in Finland

4.1 Introduction

Control of traffic signal timing requires a continuous evaluation of whether to (1) terminate the current phase/signal-group and change to the next most appropriate phase/signal-group, or (2) extend the current phase/signal group. The controller gathers information continuously (or at regular, 1 to 5 second intervals), evaluates the status of each approach, and selects the most appropriate option. Like many control problems, this control process involves the following elements: input, processor, output, desired goal, evaluation criteria and feedback loop. In feedback control, input is the desired state of the system and the information about the current state. The processor is the knowledge (or rule) base that provides the appropriate decisions according to the given input, i.e., to continue or terminate the current phase. Output is the predicted consequences of the control prescribed by the processor. The desired goal is the target and it establishes the tolerable conditions before the current phase needs to be changed. Evaluation criteria and the feedback loop represent the process of comparing the output and the target; the output is sent back to become part of the new input in the next time increment [1].

The fuzzy signal control process consists of seven parts: current traffic situation with signal status, detection or measuring (crisp input), traffic situation modeling, fuzzification interface, fuzzy inference (fuzzy decision making), defuzzification, and signal control actions (for example, extension or termination of the signal group) (Fig.1).

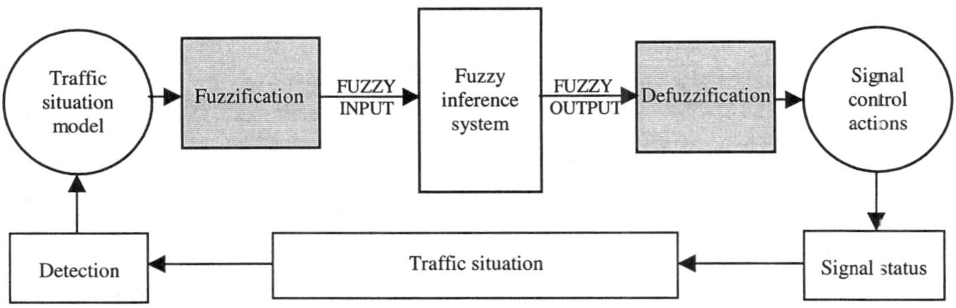

Fig. 1. Structure of fuzzy traffic signal control.

4.2 Signalized Pedestrian Crossing

Perhaps the simplest traffic signal control problem is controlling a pedestrian crossing. The main issue of control here is, given the prevailing conditions of vehicle arrivals on the roadway and of the pedestrians wishing to cross the street, when to terminate the current phase and give green to the other approach. The objectives are maximum safety, minimum waiting time for pedestrians and minimum delay to vehicle movements.

For the pedestrian crossing two types of control are commonly used: manual control by a pedestrian crossing guard, and an automated demand actuated traffic signal. With the manual control, a crossing guard (or policeman) applies a set of "if...then..." rules. The rules are linguistic rather than numerical; they are based on words, for example, if a pedestrian has been waiting for a long time, then the current green to the roadway needs to be terminated and green should be given to the pedestrian. The entire set of control processes are executed on the basis of the exchange of linguistic information, including the evaluation of actual and target conditions [19].

Let us assume that a fuzzy logic controlled pedestrian crossing is to be installed on a two-lane, two-way, roadway. It has two vehicle detectors on each of the approaches and a pedestrian push-button on each side of the crossing. One vehicle detector is placed at the crossing and the other detector is 100 m from it. These two vehicle detectors provide information on the number of approaching vehicles.

The signal operates as follows: when there is no pedestrian wishing to cross the street, the signal phase is green for the vehicular traffic (called the rest phase). When a pedestrian arrives, and presses the button (or pad detector), the controller evaluates the state of vehicle arrivals and the waiting time of the pedestrians. If specified criteria are met, the current phase is terminated and a green phase is provided for the pedestrians; otherwise the current phase continues.

The necessary variables are as follows:

INPUT:	The waiting time of the first pedestrian (*WT*).
	The total number of approaching vehicles on both approaches (*A*).
	The discharge ratio (*S*, time headway between two vehicles).
OUTPUT:	To continue the current phase (*E*) or to terminate the current phase (*T*).

The fuzzy controller is a multi-objective controller with the following objectives:

<u>Minimum pedestrian waiting time</u>: the pedestrian wishing to cross must be accommodated as soon as possible.

<u>Minimum delay to the vehicular movement</u>: the delay to the vehicles should be kept to a minimum. Vehicles should not be delayed for an unreasonable length of time.

<u>Maximum safety for both vehicles and pedestrians</u>: when a group of vehicles is approaching, the group should be preserved and passed. This is important from two aspects: to avoid rear end collisions and to create a gap in the downstream flow to provide better opportunities for the pedestrians at downstream crossings to cross.

The general format of the rules is as follows:

If *WT* is <u>(*X1*)</u> and *A* is <u>(*X2*)</u> and *S* is <u>(*X3*)</u>, then <u>*E* (or *T*)</u>

where *WT* is the waiting time, *A* is the approaching traffic volume, *S* is the discharge gap, and *X1*, *X2* and *X3* are natural language expressions of the condition of respective variables. *E* and *T* mean continue and terminate, respectively. *WT* is divided into three fuzzy sets: "short", "long" and "very long". A is divided into three fuzzy sets: "very few", "some" and "many". *S* is divided into two sets: "small" and "large". The number of rules depends on the number of combinations of fuzzy sets for *X1*, *X2* and *X3*.

In applying these rules, the basic signal system requirements must be considered. They are the minimum green times for pedestrians (10 s) and for the roadway (5 s). Hence, once a command to terminate the current phase is issued then the phase will run for the minimum required time. Thereafter, the controller reviews the situation and executes the above rules every second.

4.3 Two-Phase Control

The general rule at the signalized intersection is "the fewer phases, the better", because in every phase change some effective green time is lost in the safety margins needed between conflicting green phases. Thus, the ideal number of phases is two, which is also the lowest number of phases possible in a meaningful traffic signal installation. It is still true that there are more two-phase intersections than any other type. This is because signals with two phases do the basic job of assigning the right-of-way, and from then on leave the motorist and pedestrian on their own [31]. In our fuzzy application, two detectors are located per approach lane. The location of the first one is 100 m from stopping line (D10), while that of the second is at the stop line (D01). This means that we know how many vehicles are approaching the stop line within next 6–8 s. The main goals of our fuzzy rules are to adjust the cycle time and to divide the cycle into the green times of phases.

There are only two input variables for the fuzzy rule base:

A = the number of approaching vehicles at the moment *t* (veh), or vehicles between detectors D01 and D10 of green phase approaches

Q = the number of queuing vehicles at the moment *t*, or stopped vehicles on red phase approaches between detectors D01 and D10.

The fuzzy rules of our application are of the following form [28]:

After minimum green (for example, 5 seconds)
    if *A* is zero then terminate immediately
or   if *A* is a few and *Q* is a few then *EXT* is short (3 s)
or   if *A* is more than a few and *Q* is any then *EXT* is medium (6 s)
or   if *A* is many and *Q* is any then *EXT* is long (9 s).

After the first extension (first extension, *x* seconds + minimum green, 5 seconds)
    if *A* is zero then terminate immediately
or   if *A* is a few and *Q* is a few then *EXT* is short (3 s)
or   if *A* is medium and *Q* is any then *EXT* is medium (6 s) ...etc.

The number of the successive rule bases of this kind is five. The parallel additional rule for two-phase control is "if Q (length) is too long change immediately". It is possible to use the safety rules together with these rules, but the main goals of these rules are to adjust the cycle time and to divide the cycle into the green times of phases. The preliminary idea of this rule base is that the main traffic can be emphasized by using membership functions that are different for minor and major streets. Genetic algorithms were used also in optimizing the membership functions of fuzzy control. The results show that the optimization process would not necessarily bring extra benefit to the system if the intersection being optimized is simple and contains nothing unusual. On the other hand, this kind of method could be useful in situations were the geometry of the intersection differs from normal. Situations like these could occur for example at complex intersections with many lanes, or at intersections were the detector location is different from the recommended 80-100 meters.

4.4 Multi-Phase Control

The fuzzy rule base of multi-phase control works at three levels (Fig. 2).

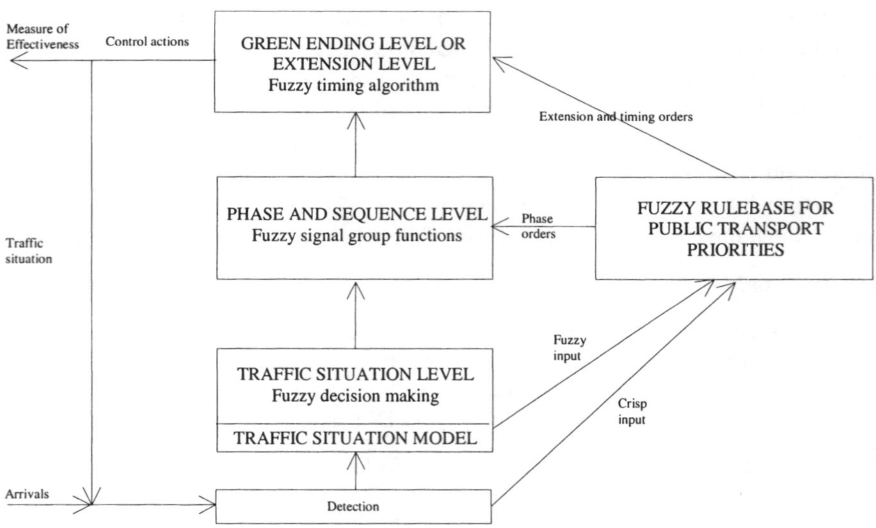

Fig. 2. Three levels of fuzzy control.

The tasks and objectives of each level are:

1. *Traffic situation level*; the traffic situation is divided into three different levels (low demand, normal, over saturated).

2. *Phase and sequence level*; the main goal of this level is to maximize the capacity by minimizing intergreen times.
3. *Green ending level or extension level*; the main goal of this level is to determine the first moment to terminate a signal group.

The rules of two lower levels are discussed in paper [21].

The main goal of fuzzy rules of the third level is to determine the first moment to terminate a signal group. The basic idea is not to terminate during queue discharge, thus enabling each vehicle to stop only once at each intersection. The input parameters in use are:

W(red)	=	weight of red signal groups; low, medium, high
GRN	=	running time of green signal group; short, medium, long
GAP	=	last gap between two approaching vehicles according to detection at 100 m before stop line; small, medium, long
S	=	indicator of discharging queue, (minimum discharging gap at the stop line); low, high.

The total number of rules is 54 (3* 3 *3 *2), but if *S* is high (queue is discharging) then we extend definitely. This means that we have 27 extension rules. The basic idea of output is that the consequences have some kind of uncertainty. Therefore the final consequence of each rule is a fuzzy number, like terminate certainly (0.00), terminate probably (0.25), either terminate or extend (0.50), extend probably (0.75) and extend certainly (1.00). The fuzzy rules are

if *W(red)* is low and *GRN* is short and *GAP* is long then extend probably (1) or ...
if *W(red)* is high and *GRN* is long and *GAP* is small then terminate certainly (27).

In the defuzzification part, the rules are located in five groups. The maximum value of the membership function of each group will be selected. The COG(center of gravity)-method calculates our extension ratio, which will be compared to the extension criteria (for instance 0.50).

4.5 Traffic Signal Control in High-Class Roads

The fuzzy control method for an arterial road is developed from functions of the Swedish vehicle actuated LHOVRA-control [41,11]. LHOVRA contains six different (crisp) control functions referred to (in Swedish) by the letters of the acronym "LHOVRA". In our application, four functions from the LHOVRA-control (HOVR) have been selected and connected to the fuzzy functions developed earlier in FUSICO-project. These functions, selecting the most suitable next phase and the termination moment of the current green phase, have already been tested in simulations and now form the basis of the control. The fuzzy H, O, V and R -functions provide extensions to the green phase and other timing parameters of the signal group control.

H-function gives priority to the main flow and relates closely to the selection of the green phase ending moment, which has been discussed earlier. The fuzzy variables of the H-function are: the discharging time gap, the time gap between arriving vehicles, the weight of the cars waiting for green phase and the running green time. In practice, the H-function, with some modifications for arterial roads, is the same function that is used to determine the green extensions in fuzzy multi-phase control (*See Section 4.4 Multi-Phase Control*). Arriving platoons are detected by means of the time gaps between approaching vehicles. By giving a green extension to the arriving platoons, the number of stopped cars can be decreased. Green phases that would be too long are avoided by the running green time variable. The final output of the fuzzy H-function is the probability of terminating the current green phase.

The O-function concentrates on safety aspects. The most common type of accident at signal controlled intersections is rear-end collision, which usually is a consequence of different stopping decisions of following vehicles. When the signal group turns to amber, the first driver may decide to stop while the second driver decides to drive through the intersection, resulting in a rear-end collision. The option zone is the area upstream from the stopping line where two successive drivers are likely to make different decisions. To avoid these situations it is important to clear the zone or at least minimize the number of vehicles in the zone when terminating the green phase. This is also the objective of the O-function, especially essential on arterial roads where speeds are high.

The fuzzy inputs of the safety function O are the number of unsafe vehicles in the option zone and the running green time. The outputs are crisp, and either *E*(xtend) or *T*(erminate) the green phase. The green phase is extended if the number of cars in the option zone is considered unsafe, and if the green time has not been running for too long time. In the O-function, our new defuzzification method, called total fuzzy similarity, is used. More information about total fuzzy similarity -method can be found in [38].

The other two LHOVRA-functions selected in our application, V- and R-functions, are basically fine-tuning control instruments that do not affect the green phase terminating moment. The V-function has only one input, the number of cars between the stopping line and the end of option zone (critical vehicles). The output is an extension of the amber time (0, 1.5 or 3 s). If the current traffic situation allows (i.e., if there are no cars between stopping line and option zone), the amber time can be set to minimum and some time saving is achieved without questioning safety. The R-function has two inputs, *time* between the beginning of the red phase and detected red-violence-driving, and the *number* of vehicles driving through the red light. The R-function lengthens the intergreen time between green phases if red-violence driving is detected, in order to reduce severe collisions.

4.6 Public Transport Priority

The main goal of the fuzzy rule base of public transport priority is to give correct priority as a function of the request moment and prevailing status of the signal control. Public transport priorities are tailor-made using some special algorithms. The most important functions of these algorithms are phase extension, phase recall, extra phase and rapid cycle (Fig. 3).

Phase extension is a function that extends the green phase if a public transport vehicle is approaching the intersection. The priority lasts until the public transport vehicle has passed the intersection or until some pre-set maximum value of the controller is exceeded. If an exit detector at the stop line for buses is used, the extension can be finished exactly at the right moment. Otherwise, some pre-programmed settings for the extension period during peak and off-peak hours should be used.

Phase recall is a function that cancels conflicting green phases and starts the green phase for the approach with public transport detection. This function should pay regard to some minimum green restrictions to avoid green times being too short. If there is traffic demand in those approaches where the green time is to be cancelled, the minimum green time should always be longer than normal minimum green time.

Extra phase is a function that is used only at multi-phase intersections. If the normal phase sequence in three-phase control is A-B-C-A-B-C-...(A being the phase with public transport), the priority call of phase A during the phase B forces the phase sequence to an extra phase. After that, the phase order is A-B-A-C instead of the normal A-B-C. *Extra phase* is very effective, especially when the it is very short (4–6 s). The extra phase -function can always be used in isolated signals. In coordinated signals, especially with a short cycle, *rapid cycle*-function is more useful. It keeps the length of all conflicting phases as short as possible until the priority phase has green again.

The multi-phase control is more complicated because there exists the possibility to affect both on phase and sequence order. The decision rules for priority algorithms are based on the moment (phase) of public transport detection (crisp) and the fuzzy parameter of next phase (*W(next)*). If the selected control algorithm will be phase extension or phase recall, then we can evaluate traffic situation using the phase extension and phase recall rules. If extra phase or rapid cycle will be selected, then the control algorithm works like in the conventional control. See also [30].

It is noteworthy that the fuzzy public transport priority rule base can also be implemented as a part of the basic rules. This kind of structural rule base is easy to develop and to use.

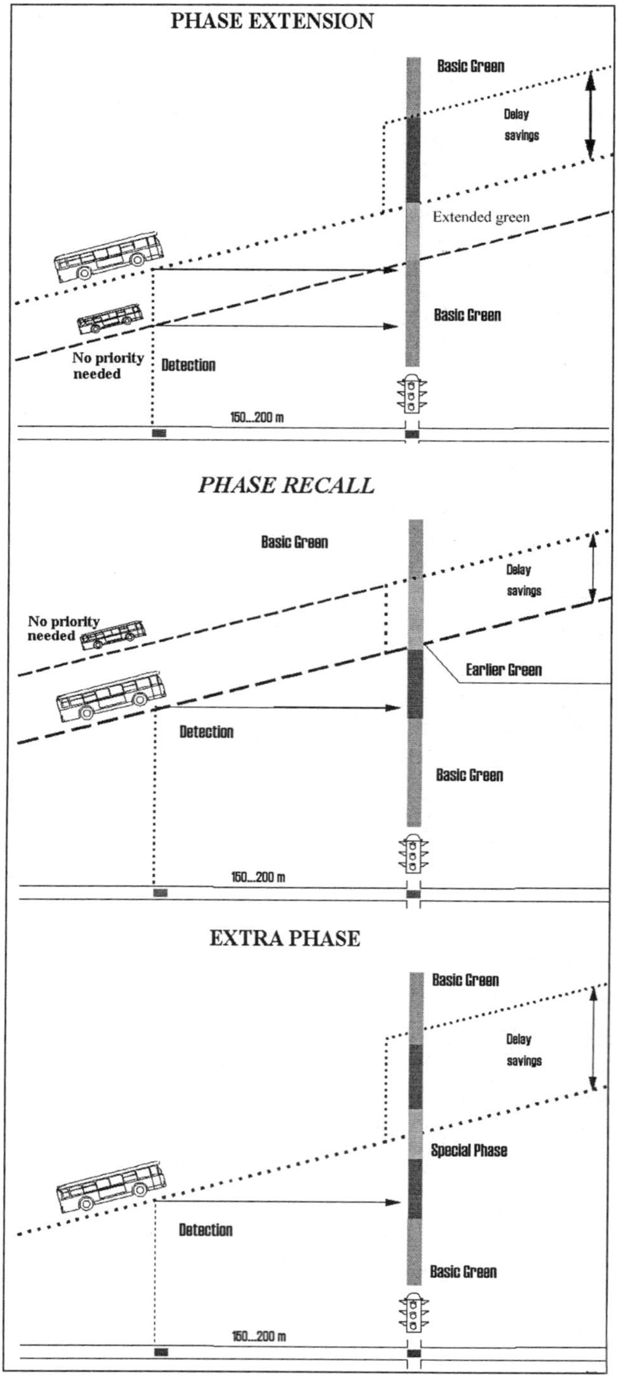

Fig. 3. Public transport priority functions [34].

4.7 Areal Urban Traffic Control

The main idea of the method is to adjust the cycle time to satisfy the network traffic demand according to the degrees of saturation observed at all intersections in the area. The areal fuzzy signal control works at two control levels: the areal main controller adjusts the cycle times of every local intersection signal controller and the local controllers decide individually the green split. The system structure of the areal control method is presented in Fig. 4.

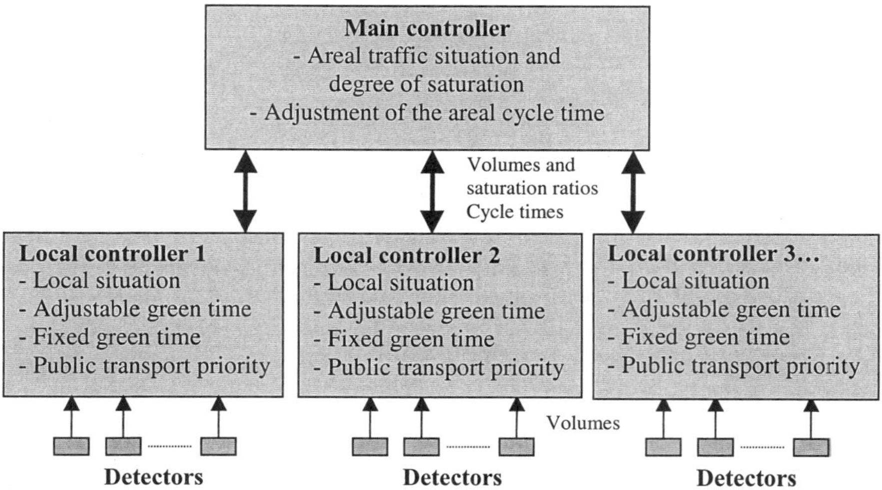

Fig. 4. The system structure of areal fuzzy signal control.

The local controllers collect information of volume and duration of every green phase. After the critical lane group of every phase of the last cycle is found, the local controllers send the traffic volumes of these critical lane groups to the main controller. After receiving the data, the main controller calculates the total degrees of saturation of every intersection; these are first compared individually one to each other and the highest degree of saturation is chosen. After the selection, this maximum is compared to the favorable degree of saturation. The cycle time is changed according to the difference between the favorable and the maximum.

When the fuzzy decision is made, the new cycle time is set in all local controllers in the area. The local controllers at each intersection decide the green split within the cycle time given by the main controller. The green split is made in two steps: first, a suitable share of the available green time is reserved for actuated green time (extensions) to balance the random changes in traffic, second, the remaining green time is divided into the phases according to the differences between saturation degrees in each phase.

This kind of areal control method is to be tested in the field in the near future.

5 Performance of Fuzzy Control: Results of Simulation Studies and Field Experiments

5.1 Signalized Pedestrian Crossing

The performances of the three controls (traditional vehicle-actuated, modified vehicle-actuated, and fuzzy) were compared with respect to three measures (pedestrian delay, average vehicle delay and percentage of stopped vehicles) for these three pedestrian volumes: 15 pph, 50 pph and 150 pph. The results are based on a simulation of more than 120 hours [19].

Low pedestrian volume (15 pph): The results show a comparison of the three controls in terms of the three performance measures. The average pedestrian delay is found to be significantly smaller for fuzzy control than that for the conventional vehicle-actuated control. However, it is slightly higher than that of the modified control, also called 'fast pedestrian green control'. The vehicle delays are similar in all cases. The modified vehicle-actuated controller causes slightly longer vehicle delays than the others, while the fuzzy controller causes the shortest delays. The average delay per vehicle is between 2 and 5 s. The percentage of vehicles stopped is smallest for the fuzzy control.

Medium pedestrian volume (50 pph): Here the results show that vehicle delays and the percentage of vehicles stopped are higher than in the case of the low pedestrian volume. Pedestrian delay, on the other hand, is nearly the same as in the case of low pedestrian volume for all three controls.

High pedestrian volume (150 pph): The results show that vehicle delay and the percentage of stopped vehicles are greater than in the previous two cases. These findings can be expected: the higher the pedestrian volume the greater the chance of interruption to the flow of traffic. The pedestrian delay remains nearly unchanged from the above two cases (low and medium pedestrian volumes).

Niittymäki and Turunen (2000) have compared different defuzzification methods in signalized pedestrian crossings (fuzzy similarity [38] vs. Mamdani-type COG). The fuzzy similarity gives a larger capacity, but Mamdani-type COG is more pedestrian friendly. Statistical analysis is shown in [29].

5.2 Two-Phase Control

5.2.1 Simulation Study

The efficiency of our FUSICO-algorithm was compared with the traditional vehicle-actuated control (called extension principle [7]). The results indicated that the extension principle is a reasonably good traffic signal control mode in the area of very low traffic volumes. However, the results also indicated that the application area of fuzzy control is very wide. This result is the same as that in Pursula and Niittymäki 1996 [33].

The FUSICO-control was also compared to fuzzy control as developed by Pappis and Mamdani. The biggest difference found between our FUSICO-algorithm and the fuzzy control algorithm of Pappis-Mamdani was the fact that FUSICO-control gained smaller stopping percentages than traditional vehicle-actuated control, while Pappis-Mamdami caused larger stopping percentages compared to vehicle-actuated control. In our case, the delays were also slightly shorter (–10…–20 % in the tested area 100–1200 vph) than in the Pappis-Mamdani case [28].

To compare the two control methods (vehicle-actuated, VA, and fuzzy, FU), identical simulations were run on both controllers. The testing was conducted using the HUTSIM – microscopic simulation program. The vehicles were generated simultaneously in both simulation runs, and any change in their behavior resulted only from a change in control. The simulation time of each traffic volume was 32 * 900 s (=8 h), while the simulated intersection was an isolated intersection of two one-way streets with 2 + 2 lanes. The speed distribution was Gaussian with a mean value of about 40 km/h. The average delays of both control methods are shown in Fig. 5.

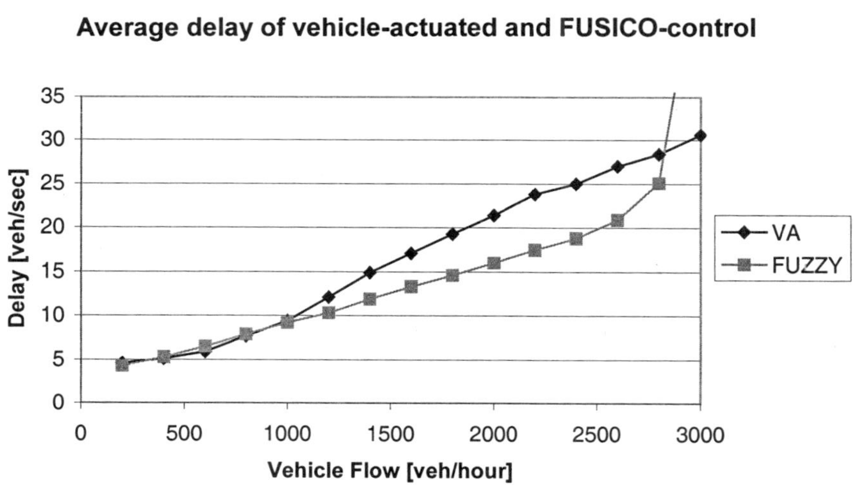

Fig. 5. Average delays of fuzzy and vehicle-actuated two-phase control in simulations [28].

Since the observations of delay in these two simulation runs were paired, a t-test on paired observations was used to compare the means of the delays in the simulation runs of the two-phase control. The results proved that the extension principle (VA) is a good traffic signal control mode in the area of low traffic volumes (no statistical difference), but fuzzy control (FUZZY) is a competitive alternative for isolated signal control, especially at medium volumes (statistical difference). The capacity of vehicle-actuated control seems to be slightly higher than the capacity of fuzzy control.

5.2.2 Field Tests

The two-phase fuzzy signal control of FUSICO-project has also been tested with field experiments. The first test intersection in Oulunkylä, Helsinki, was chosen on the basis of the following criteria: the intersection was isolated, the traffic volumes during peak hours were high, there were four approaches, bus-traffic, pedestrian crossings, a suburban location and two-phase control.

The results of field measurements show that the fuzzy control algorithm works better than vehicle-actuated control in most cases. Compared to vehicle-actuated control, the average travel times of FUSICO-control are approximately 4–10 s shorter, the percentages of stops are 2–12 % lower, the bus delays are shorter in 8 of 9 cases, and there are good savings in fuel and emissions based on simulations. All these results prove that the fuzzy control algorithm can be successfully used to control traffic at real intersections. However, the better traffic fluency is only one advantage. Pedestrians can also benefit. The cycle times are on average 8 seconds shorter, which means shorter waiting times for pedestrians because of shorter green times of conflicting vehicle signal groups. This also results in the maximum vehicle green being smaller if the traffic situation is not over saturated. In our experiment, the pedestrian signal groups were requested secondarily, which means that they appeared in every cycle at the same time as the non-conflicting vehicle groups [20].

5.3 Multi-Phase Control

The efficiency of our fuzzy algorithm was compared with the traditional vehicle-actuated control, called extension principle, which is actually used in real intersections. The aim was to ensure that the algorithm used for comparison was as effective as possible. The simulation time of each case was 3600 s, and the simulated traffic situation was exactly the same in both cases. As in the study of two-phase control, the results prove that the extension principle is the better traffic signal control mode in the area of low traffic volumes. However, the results also indicate that the application area of fuzzy control is available. If the major traffic flow is more than 500 veh/h, the results of fuzzy control are at least as good as the results of traditional control. According to the field measurements of the test intersection, the real traffic volume of major flow varies between 600–900 veh/h and the minor/major-ratio is approximately 1:5. Based on this, we can say that fuzzy control principles are competitive, in isolated multi-phase traffic signal control in this case. However, our simulations were based on the decision making of the signal group extension only; better results can be achieved for low traffic volumes also by using the second level fuzzy decision making (see *Section 4.4 Multi-Phase Control*) with the fuzzy phase selector. The phase selector is a fuzzy function that selects the most suitable next phase to the prevailing traffic situation by deciding the need to change the basic phase sequence in multi-phased control. The decision is based on the cumulative delays of vehicles waiting for green phase. The main goal of the fuzzy phase selector is to maximize the capacity by minimizing intergreen times [21].

5.4 Traffic Signal Control in Major Arterials

5.4.1 Simulation Study at Huddinge Test Intersection

The test simulations of fuzzy control on major arterials were completed using a HUTSIM-model of a real test intersection in a suburb of Stockholm called Huddinge. The fuzzy method was compared to a vehicle-actuated control and to pre-set fixed time control. The vehicle-actuated control was adjusted to each traffic combination by modifying manually the maximum green times. Also, the cycle time of the fixed time control was optimized separately for each simulation situation with the Webster algorithm [42], which minimizes the total delays for the pre-known traffic volumes. The optimum cycle time was shared to green times according to the occupancy levels of each phase. In comparison, the fuzzy control algorithm was not changed in any way during the simulations. Additionally, the fuzzy control used only the fuzzy functions of selecting the green phase termination moment and the most suitable next phase (i.e., the fuzzy LHOVRA-functions were not included) [26].

Tested traffic situations were formed from all combinations of main flow volumes 200, 500, 1 000, 1 500 and 2 000 veh/h and minor flow volumes 200, 500 and 800 veh/h. The origin-destination distribution was estimated from the real traffic distribution in the Huddinge test intersection.

The capacities of all the compared control methods were exceeded when traffic volumes were largest. In other traffic situations, the control methods caused approximately the same amount of delays. The fuzzy logic controller clearly provided the best stopping percentage results, except in the traffic situations with highest main, and lowest minor, flow volumes. The main reasons for these results are the poor stopping percentages (almost 100 %) of all control methods with the highest volumes, and the tendency of the fuzzy control to maintain shorter cycle times than other methods with low volumes (leading unavoidably to higher number of stopped vehicles). Nevertheless, the stopping percentages of the fuzzy controller were still competitive even with these volumes [26].

In addition to the basic study, the fuzzy control and vehicle-actuated control were also compared to the Swedish SOS-control method [12], which was in test use at the Huddinge test intersection at the time of study. SOS-method uses mathematical optimization in signal controlling. The optimization is based on minimizing the costs of green phase termination/extension–decision consequences. The costs of the current green phase are compared to the corresponding costs of the red phases waiting for the green phase. If continuing the current phase provides lower costs than termination at some point in the near future (0.5-20 s), the phase will not be terminated. The calculations and traffic predictions are updated twice a second. After the optimum phase termination moment is found, the algorithm gives 12 seconds to the incident-reduction function to find a safe final termination moment. In over saturated conditions, the algorithm tries to balance the waiting times of each direction [12]. See also [15,40].

The comparisons of fuzzy and vehicle actuated control to the SOS-control were made by simulating real traffic situations observed at the Huddinge test site during morning and evening rush hours and daytime traffic. However, the control algorithm of the SOS-method was not available for SOS-simulations and the comparison had to be made based on SOS field measurements taken at the Huddinge test site. This notwithstanding, the results still indicate that the simple fuzzy control method works at the same level as the complex SOS-method with daytime traffic. During the rush hours, SOS-method seems to give much better results. The traditional vehicle actuated method was not competitive in any situation when compared to SOS or fuzzy control. However, because of the different test environments and counting methods of the delays and stopping percentages, accurate conclusions concerning performance differences and their reasons cannot be drawn from these trend-setting results.

5.4.2 Pre-Study of Upcoming Field Experiment

The promising results of the fuzzy control algorithm on arterial roads have encouraged us to move to a field installation. An intersection on a road coming in to the city of Tampere has been chosen to be our test site. Tampere is one of the largest cities in Finland with some 200 000 inhabitants (year 2000). The fuzzy algorithm, which is to be installed at Tampere during year 2001, will include the fuzzy applications of LHOVRA-control and the functions that select the green phase termination moment and the most suitable next phase.

A simulation study was made to obtain a preliminary view of the fuzzy control's performance at the Tampere test site. The intersection modeling and simulation were carried out using HUTSIM-simulation software. As in the Huddinge study, the preliminary simulation study did not include all the fuzzy functions that will be installed on the test site. The fuzzy controller used only the fuzzy function that selects the green phase terminating moment. The fuzzy control was compared to a traditional vehicle actuated control (VA) operating with the ordinary green extension principle, in which each detection from every detector causes an individual, pre-set extension to the current green phase.

The simulated traffic situations were defined from the saturation degree of the test intersection to form a more general view of fuzzy control performance. Five separate degrees of saturation, 0.2, 0.4, 0.6, 0.8 and 0.9, were studied. The degree of saturation was calculated according to Finnish instructions [8], which are derived from Webster's theory [42]. The volume of each approach was iterated to fulfill the degree of saturation of each test situation. The daytime directional distribution of the Tampere test site was used in the iteration and in the simulations.

Fig. 6 presents the average delay of the whole test intersection and separate delays of main and minor approaches with both control methods. As can be seen, original VA-control has shorter delays with low saturation degrees, while the fuzzy controller has better results in more saturated traffic conditions.

On main flow, the performance of fuzzy controller exceeds the VA-control with saturation degrees 0.8 and higher. On minor flow, fuzzy control gives the best results even with low demand conditions. The performance gap increases as demand increases, which results in lower total delays on the intersection when the saturation degree exceeds 0.6…0.7. In all, already the non-complete fuzzy control tuned for high-class roads improves the capacity of the whole intersection and decreases delays on both main and minor flows during high traffic demand situations.

The fuzzy H-function, missing from the preliminary simulation study, will, in comparison to the VA-control, improve the fuzzy controller's performance on main flow, and also during low demand. This is important, because main flow priority is one objective of traffic signal control on an arterial road. Even though the main flow priorities will increase the minor flow delays, they should still remain lower than in traditional VA-control. This is enabled by the fuzzy reasoning and by correctly set extension criteria on main flow, which pay attention to the whole traffic situation instead of monitoring only separate vehicles approaching the green phase. In addition, the implementation of the fuzzy phase selector (*See Section 5.3 Multi-Phase Control*) will have an even bigger delay-reducing effect on both main and minor flows.

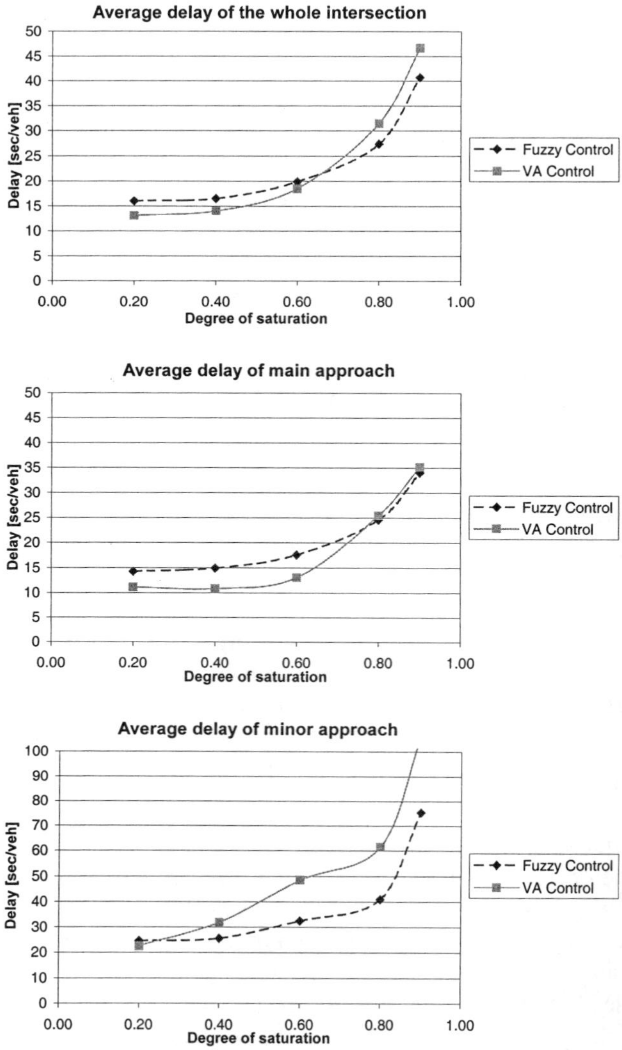

Fig. 6. Average delays of Tampere preliminary study.

5.5 Public Transport Priorities

The Finnish national TETRA-research program included three different test cases of fuzzy public transport priorities. In these three cases, there was not the possibility of using exit detectors for buses. Instead, a constant extension time based on the average driving time from requesting the detector to the stop line was used.

The test intersection in Vantaa is a four-leg intersection. There are three phases, and all phases include bus traffic. If no buses are detected, the fuzzy controller makes the decisions concerning green extensions according to the fuzzy rule base and membership functions. The bus detected first gets priority first. All three ways to give the priority are available: phase extension, phase recall and extra phase.

In Vantaa, the travel times of buses were measured in two directions with video cameras: buses approaching from the major street and either driving straight or turning left in their own phase. Before the fuzzy system was installed, the average travel times were 29.7 s (straight) and 41.4 s (turning left). After the installation, travel times were 27.1 s and 31.3 s, respectively. The average travel time of left turning buses especially, decreased considerably, by 10 s.

The test case in Lahti is also a four-leg intersection with three phases. The intersection is located in the center of the city. Buses approach the intersection from two directions, and their routes cross each other. The constant cycle time, 100 s, is one factor used to achieve coordination in the intersection.

When a bus is detected, the fuzzy signal controller decides according to the fuzzy rule base, whether a priority is given or not. Phase extension, phase recall and extra phase are the alternatives that give priority (*See Section 4.6 Public Transport Priority*). In situations where two or more buses are approaching from different directions, the priority is first given to the bus detected first.

In Lahti, the average travel times in both approaching directions were, before installation, 34.4 s (major street) and 38.3 s (minor street). After installation they were 36.6 s (major) and 35.7 s (minor). The average time for buses approaching from the major street increased 2.2 s, because some buses did not have time to pass the intersection within the green extension or extra phase, and had to wait through the whole cycle time again. The other reason for some fairly high travel times is that some buses were not detected due to their low speeds at the inductive loop. The same happened in the other direction as well, but not so often. If we leave those buses out of the measurement data, the average travel time for major street buses was 32.1 s. Some changes to the membership functions and the traffic situation model have been made after taking the measurements.

In *Jyväskylä*, the test intersection is a T-intersection with three phases, and with a fairly small volume of traffic arriving from the minor street (10–20 % of major street). Buses approach the intersection from both major street directions and drive straight through. The fuzzy signal controller gives priorities to buses if the current traffic situation allows. As in Vantaa, the fuzzy signal controller makes the decisions concerning green lengths also if no buses are detected. Bus priority can be phase extension, phase recall and extra phase. In Jyväskylä, the average travel time of buses was before the installation 22.7 s, and afterwards, 20.1 s.

More information about the simulations, the field measurement results, and statistical tests, can be found in [16] and [30].

6 Technical Properties of Fuzzy Signal Controller

A picture of a fuzzy controller in the field is shown in Fig. 7.

Fig. 7. The fuzzy controlling unit (center) installed beside the signal controller.

A normal PC computer with FUSICO-software that includes both the fuzzy algorithm and HUTSIM-simulator with the intersection model is installed beside the original vehicle-actuated signal controller of the intersection. The PC-card in use is the Octagon-systems PC-510, which runs at a speed of 133 MHz. Its operating temperature with adequate airflow ranges from –40 ˚C to +70 ˚C. The card has 1 MB on board memory and a 4 MB optional RAM-chip (required for running simulation software). Because of variations of temperature and humidity are very high, hard disks are not used. The software is stored on a flash RAM-chip (2 MB).

A simple parallel interface is used to send detector pulses from the controller to FUSICO, and control orders from FUSICO to the controller. The system structure is presented in [25]. In this arrangement, the original controller is a "slave" for the FUSICO-control algorithm [13,14].

The software used is HUTSIM, which is modified for real-time use [10]. When a vehicle passes a detector, the vehicle is generated in the simulation model. After that, any additional information about the vehicle remains uncollected. This idea leads to one key benefit of the system: the whole intersection can be handled with very few traffic detectors. The detectors of incoming traffic in the fuzzy control system are normally located 40–160 m before stop line, while the stop-line detectors and pedestrian push buttons are not in use in the fuzzy control system as in the vehicle-actuated control.

7 Conclusions and Discussion

Many problems in complex transportation systems are non-linear. Fuzzy logic can be used successfully to model situations in which people make decisions in an environment that is so complex that developing a mathematical model is very difficult. Our belief, encouraged by the research results of international studies as well as our own, is that the fuzzy methods and approximate reasoning are successful concepts in dealing with a complicated system like traffic signal control. In addition to the issues mentioned above, the uncertainty of input information to signal control favors approximate reasoning.

The problems in the present traffic signal control are:

- the critical need for input (detector) data in on-line mathematical optimization,
- the identification of the overall traffic situation through detector inputs,
- the vagueness of the definition and treatment of the multiple and partly conflicting goals of optimization, and
- the problems in the use of system-wide priority measures for public transport and pedestrians.

Increasing traffic demand in urban road networks has already promoted many traffic-actuated control concepts in the last twenty years. The main result of our work is a new systematic approach to traffic signal planning. By developing the general approach for traffic signal control planning, the study promotes a more efficient and goal-driven signal control planning in general. The results can be used in day-to-day planning and design of signal control in practice, too. New artificial intelligence methods, especially fuzzy logic, together with the development of traffic simulation, have opened totally new opportunities to solve traffic signal problems.

In this study, the theoretical framework for signal control has been constructed at a general and abstract level without strict ties to fuzzy control. Constructed in this way, it can promote the analysis and development of all kinds of signal control strategies and principles, including methods based on traditional on-line optimization.

We have tested several different control strategies in different isolated control environments:

1. The results of signalized pedestrian crossing indicated that the fuzzy control provides pedestrian friendly control keeping vehicle delay smaller than the conventional control. In other words, the fuzzy controller provided a compromise between the two opposing objectives, minimum pedestrian delay and minimum vehicle delay, in accordance with the level of the pedestrian and vehicular volumes.

2. The results of the comparison of the Pappis-Mamdani fuzzy control algorithm to traditional vehicle-actuated control indicated that fuzzy control is suitable for many other control applications. The long cycle times of the traditional extension principle indicated the weakness of this kind of gap-seeking control. On the other hand, the number of stopped vehicles in Pappis-Mamdani control was higher than the number in the case of traditional vehicle-actuated control. Based on these experiences, we developed our own control algorithm for two-phase vehicle control. The main goals of this control were to adjust the cycle time and to divide the cycle into the green times of phases. According to the results of our FUSICO-approach, the application area of fuzzy control is very wide. The difference between our FUSICO-control algorithm and Pappis-Mamdani control algorithm is that our FUSICO-control algorithm gives a smaller number of stops than the traditional extension principle or Pappis-Mamdani control. Basically, this means that the FUSICO-algorithm should also result in lower fuel consumption and better traffic safety than the traditional vehicle-actuated control algorithms or Pappis-Mamdani control algorithm.

3. The results of multi-phase control prove that the extension principle is the better traffic signal control mode in the area of low traffic volumes. However, the results indicate that the application area of fuzzy control exists. When the major traffic flow was more than 500 veh/h, the results of fuzzy control were at least as good as the results of traditional control.

4. The results of fuzzy signal control in major arterials have also been promising. The fuzzy logic controller provides clearly the best results except in the cases of the highest main flow and lowest minor flow volumes. Even with these volumes, the stopping percentages of the fuzzy controller are competitive.

5. The results of fuzzy public transport priorities have been at least promising, and the planners of Finnish cities involved in the study have been satisfied with the results of fuzzy control.

The preliminary comparison of fuzzy control and mathematical optimization was also conducted during the project.

The post-study strives to create a systematic methodological framework for traffic signal control from isolated signals to coordinated and area (urban) traffic control systems [18,22,23,27]. In the study we will try to develop a systematic approach to the goals, problems and methods of signal control. The analysis of the needs of input data, the formulation of goals and objectives in a general form, the discussion of different control strategies and of the consequences of alternative decisions give new input into the overall design and analysis of signal control systems. At the same time, they form a basis for the formulation and testing of a new fuzzy control system framework.

One very important sub-goal was to develop a real fuzzy controller and test it at real intersections. The results of field tests in Helsinki, Vantaa, Lahti and Jyväskylä have been good, and we can say that the fuzzy signal control is a part of reality. The number of fuzzy traffic signals is increasing and we plan to install 10–15 fuzzy traffic signals in the near future. At the moment, fuzzy applications are not yet commercialized (at least not in Finland), but the good performance and usability of fuzzy systems, as well as the satisfaction of traffic planners involved in the development projects, provide the opportunity to commercial activity in the future. Nevertheless, commercialization still requires careful documentation of the control method, comprehensive manuals for control tuning and installation, and overall approval and knowledge of fuzzy systems in the field of traffic control.

In general, the main goal is to continue the theoretical work of developing fuzzy control, to achieve something theoretically new in this field, and to apply it in practice. In all, this project will increase the theoretical competence and widen the level of practical know-how in the area of traffic signal control in Finland and the wider world. See also [2,24].

References

1. Bell M.G.H (1992), Future directions in traffic signal control, *Transportation Research*, **26A**, no. 4. pp. 303-313.
2. Bingham E. (1998), *Neurofuzzy traffic signal control*, Master's Thesis, Helsinki University of Technology, Department of Engineering Physics and Mathematics, Espoo, Finland.
3. Bång K.-L. (1976), Optimal control of isolated traffic signals, *Traffic Engineering and Control,* July 1976, pp. 288-292.

4. Chiu S. (1992), Adaptive traffic signal control using fuzzy logic, *Proceedings of the Intelligent Vehicles Symposium '92*, Detroit, MI, pp. 98-107.
5. Evans J.B. (1980), 'Structure' in intersection controllers. *Traffic Engineering and Control*, April 1980, pp. 178-181.
6. Favilla J., Machion A., and Gomide F. (1993), Fuzzy traffic control: Adaptive strategies, *Proceedings of the IEEE International Conference on Fuzzy Systems*, San Francisco, CA, pp. I-506-I-511.
7. FHWA (1985), *Traffic Control Systems Handbook*, Report FHWA-IP-85-12, Federal Highway Administration, Washington, DC.
8. Finnish National Road Administration (1996), *LIVASU 95, Traffic Signal Control Handbook*, Helsinki, Finland, (in Finnish).
9. Kim S. (1994), *Applications of Petri networks and fuzzy logic to advanced traffic management systems*, Ph.D.–thesis, Polytechnic University, USA.
10. Kosonen I. (1999), *HUTSIM – urban traffic simulation and control model: Principles and Applications*, Publication 100, Helsinki University of Technology, Transportation Engineering, Espoo, Finland.
11. Kronborg P. (1992), *MOVA and LHOVRA – traffic signal control for isolated Intersections*, TFK Report 1992:4E, Transport Research Institute, Stockholm, Sweden.
12. Kronborg P., Davidsson F., and Edholm J. (1997), *SOS – self optimising signal control, development and field trials of the SOS algorithm for self optimising signal control at isolated intersections,* TFK Report 1997:2E, Transport Research Institute, Stockholm, Sweden.
13. Könönen V. (1999), *New methods for traffic signal control - development of fuzzy controller*, Master's Thesis, Lappeenranta University of Technology, Department of Information Technology, Lappeenranta, Finland.
14. Könönen V. and Niittymäki J. (2000), New methods for traffic signal control - development of fuzzy controller, *Proceedings of the 9th IEEE International Conference on Fuzzy Systems*, San Antonio, TX, pp. I-358-I-362.
15. Miller A.J. (1963), A computer control system for traffic networks, *Proceedings of the 2nd International Symposium on the Theory of Traffic Flow*, Paris, France, pp. 202-206.
16. Mäenpää M. (2000), *Public transport priorities using fuzzy logic*, Master's Thesis, Helsinki University of Technology, Transportation Engineering, Espoo, Finland, (in Finnish).
17. Nakatsuyama M., Nakahashi H., and Nishizuka N. (1984), Fuzzy logic phase controller for traffic junctions in the one-way arterial road, *Proceedings of the IFAC 9th Triennal World Congress*, Pergamon Press, Oxford, UK, pp. 2865-2870.
18. Nevala R., Könönen V., Niittymäki J., and Kosonen I. (2000), Fuzzy platoon recognition for coordinated intersections, *Proceedings of the 5th On-Line World Conference on Soft Computing in Industrial Applications*, On the Internet, http://wsc-virtual.hut.fi (now closed).
19. Niittymäki J. and Kikuchi S. (1998), Application of fuzzy logic to the control of a pedestrian crossing signal, *Transportation Research Record,* no. 1651, Transportation Research Board, Washington, DC, pp. 30–38.
20. Niittymäki J. (1999), Installation and experiences of field test of fuzzy signal controller, *Proceedings of the 11th Mini-EURO Conference on Artificial Intelligence in Transportation Systems and Science, and the 7th EURO Working Group Meeting on Transportation*, Helsinki University of Technology, Transportation Engineering, Publication 98, Espoo, Finland, 5 pp.

21. Niittymäki J. (1999), Using fuzzy logic to control traffic signals at multi-phase intersections, *Proceedings of the 6th Fuzzy Days in Dortmund*, Springer, Berlin-Heidelberg, Germany, pp. 342–362.
22. Niittymäki J. (1999), Fuzzy logic two-phase traffic signal control for coordinated one-way streets, *Proceedings of the IEEE Midnight-Sun Workshop on Soft Computing Methods in Industrial Applications*, Kuusamo, Finland. pp. 69-74.
23. Niittymäki J. and Nevala R. (1999), Fuzzy logic traffic signal control for coordinated streets. *Proceedings of the 4th On-line World Conference on Soft Computing in Industrial Applications,* On the Internet, http//yukar.cc.muroran-it.ac.jp/papers/paper-I-45/index.html (now closed).
24. Niittymäki J. and Granberg M. (2000), Use of genetic algorithms in fuzzy traffic signal control, *Proceedings of the Rome Jubilee 2000 Conference, 8th Meeting of the EURO Working Group on Transportation*, University of La Sapienza, National Research Council of Italy, Rome, Italy, pp. 409–412.
25. Niittymäki J. and Könönen V. (2000), Traffic signal controller based on fuzzy logic, *Proceedings of the IEEE International Conference on Systems, Man & Cybernetics,* Nashville, TN. pp. 3578-3581.
26. Niittymäki J. and Nevala R. (2000). Multi-level and multi-objective traffic signal control using fuzzy methods, *Proceedings of the 6th AATT Conference*, Singapore, in CD-ROM format.
27. Niittymäki J. and Nevala R. (2000), Fuzzy-logic two-phase traffic signal control for coordinated one-way streets, In *Soft Computing in Industrial Applications,* Suzuki Y., Ovaska S., Furuhashi T., Roy R, and Dote Y. (Eds.), Springer-Verlag, London, UK, pp. 107–116.
28. Niittymäki J. and Pursula M. (2000), Signal-group control using fuzzy logic, *Fuzzy Sets and Systems, International Journal of Soft Computing,* **116**, no.1, pp. 11-22.
29. Niittymäki J. and Turunen E. (2000), Traffic signal control on total fuzzy similarity based reasoning, *Paper sent for publication in the Journal of Fuzzy Sets and Systems after revision.*
30. Niittymäki J. and Mäenpää M. (2001), The role of fuzzy logic public transport priority in traffic signal control, *Traffic Engineering & Control,* **42**, no.1, pp. 22-26.
31. Orcutt F.L. (1993), *The Traffic Signal Book,* Prentice Hall, Upper Saddle River, NJ, USA.
32. Pappis C. and Mamdani E. (1977), A fuzzy logic controller for a traffic junction, *IEEE Transactions on Systems, Man and Cybernetics*, **SMC-7**, no.10, pp. 707-717.
33. Pursula M. and Niittymäki J. (1996), Evaluation of traffic signal control with simulation - a comparison of the Pappis-Mamdani control vs. vehicle actuation with extension principle, *Proceedings of the 4th Meeting of the EURO Working Group on Transportation*, University of Newcastle, Newcastle upon Tyne, UK, 4 + 14 pp.
34. Sane K. (1997), Systematics of traffic signal control planning, *Working Report of FUSICO-project*, Helsinki University of Technology, Espoo, Finland.
35. Sayers T., Bell M.G.H., Mieden T., and Busch F. (1995). Improving the traffic responsiveness of signal controllers using fuzzy logic. *Presented at the IEE symposium on Urban Congestion Management*, University of Newcastle upon Tyne, UK.
36. Sayers T., Anderson J., and Bell M. (1998), Traffic control system optimisation: A multiobjective approach, In *Mathematics in Transport Planning and Control, Proceedings of the 3rd IMA International Conference on Mathematics in Transport Planning and Control*, Griffits J. D. (Ed.), Pergamon, Elsevier Science, UK, pp. 37-46.

37. Trabia M.B., Kaseko M.S., and Andel M. (1999), A two-stage fuzzy logic controller for traffic signals, *Transportation Research Part C*, **7**, Elsevier Science, UK, pp. 353-367.
38. Turunen E. (1999), Mathematics behind Fuzzy Logic, Physica-Verlag, Heidelberg, Germany.
39. Tzes W., McShane W.R., and Kim S. (1995). Expert fuzzy logic traffic signal control for transportation networks, *Proceedings of the Institute of Transportation Engineers 65th Annual meeting*, Washington, DC, pp. 154-158.
40. Vincent R.A., and Peirce J.R. (1988), *MOVA: Traffic responsive, self-optimising signal control for isolated intersections*, Research Report 170, TRRL, Growthorne, UK.
41. Vägverket (1983), *Signalreglering med LHOVRA-teknik*, Projekteringshandbook, Borlänge, Sweden (in Swedish).
42. Webster F. (1958), *Traffic signal settings*, Road Research Technical Paper no. 56, Road Research Laboratory, London, UK.

On Evolvable Hardware

Timothy G. W. Gordon and Peter J. Bentley

University College London

Keywords: *evolvable hardware, evolutionary electronics, evolutionary computation, genetic algorithms, automatic design, innovation, evolvability, generalization, intrinsic circuit evolution, FPGAs.*

1 Introduction

Electronic circuit production is a significant industry. Ever more complex behaviors are being demanded from electronic circuits, fuelled by relentless improvements in circuit embodiment technologies. Consequently a bottleneck is developing at the point of circuit design. Traditional circuit design methodologies rely on rules that have been developed over many decades. However the need for human input to the increasingly complex design process means that modern circuit production takes one of two paths. The first is to employ more designers with greater expertise. This is expensive. The second is to simplify circuit design by imposing greater and greater abstraction to the design space. An example of this is the use of hardware description languages. This results in mounting waste of potential circuit behavior.

Recently a new field applying evolutionary techniques to hardware design and synthesis has emerged. These techniques give us a third option - to use evolution to design the circuits for us. This field has been coined *evolutionary electronics*, *hardware evolution* and *evolvable hardware* amongst others. Here it will be referred to as evolvable hardware

The field of evolvable hardware draws inspiration from a range of other fields. The most important are shown in Fig. 1. For many years computer scientists have used ideas from biology to develop algorithms for soft computing. We have seen the advent of the artificial neural network (ANN) [73] and more recently the development of evolutionary computation, the field of problem solving using algorithms inspired by evolution [10]. Ideas from nature have also been used in electronic engineering for many years. An example of this is simulated annealing which is used in many partitioning algorithms. This algorithm is based on the physical phenomenon of annealing in cooling metals. Recently the field of bio-inspired hardware has also developed, using ideas from biology to explore methods of fault tolerance and reconfigurability in modern hardware designs.

There is much interchange of ideas between the fields of evolvable hardware and bio-inspired hardware, but in this work we focus on the field of evolvable hardware, which lies at the crossroads between all three of these major sciences.

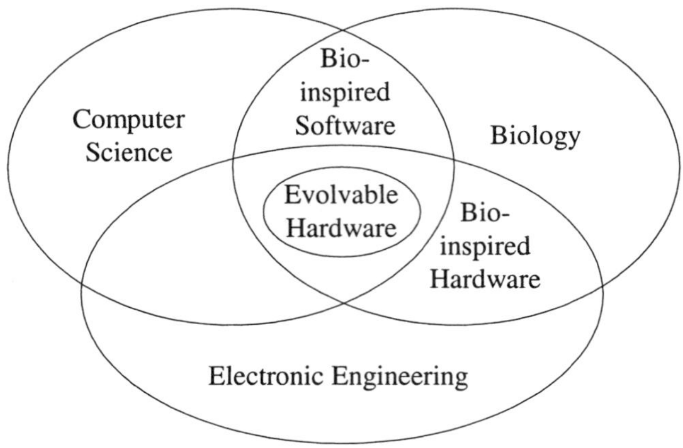

Fig. 1. The field of evolvable hardware originates from the intersection of three sciences.

The interrelationships between areas of hardware design and synthesis, and evolutionary computation are shown below in Fig. 2. Digital hardware synthesis is traditionally a combination of two processes. First a human-designed circuit specification is mapped to a logical representation through the process of logic synthesis. This is represented as the lower left-hand set in Fig. 2. This netlist then undergoes further combinatorially complex optimization processes in order to place and route the circuit to the target technology. This area is represented as the lower left-hand set in Fig. 2. Many modern EDA[1] tools use intelligent techniques in these optimization algorithms, and research into use of evolution for these purposes abounds [54, 14]. Hence we see a set representing evolutionary design optimization intersect with that of technology mapping, placement and routing in Fig. 2 to yield evolutionary mapping, placement and routing. However circuit design, along with some optimization decisions during the synthesis process are still in the domain of the human designer. It is only recently that significant interest has developed in implementing evolutionary techniques higher up the VLSI design flow at circuit design, a move that can allow evolution to generate creative designs that can rival or improve on human ones. The most widespread examples of this have been to use evolution for the design of logic, as represented by the intersection of the areas of creative evolutionary design and logic synthesis in Fig. 2. Some of the work in this intersection falls into the field of evolvable hardware. However much work at the logical level is carried out in the spirit of

[1] Electronic Design Automation

evolving programs or other forms of logic, and so is not specifically considered here.

The field of evolvable hardware is still in its infancy, and there are many problems that must be tackled before we will see large-scale industrial use of the techniques. Hence there is much active research associated with improving the algorithms used to evolve hardware. Following a brief discussion of applications of evolvable hardware in section 2, this research will be reviewed in section 3, using level of abstraction, learning bias and evolving platform as the main features to map out the area. The areas of research under review are innovation, generalization, and evolvability.

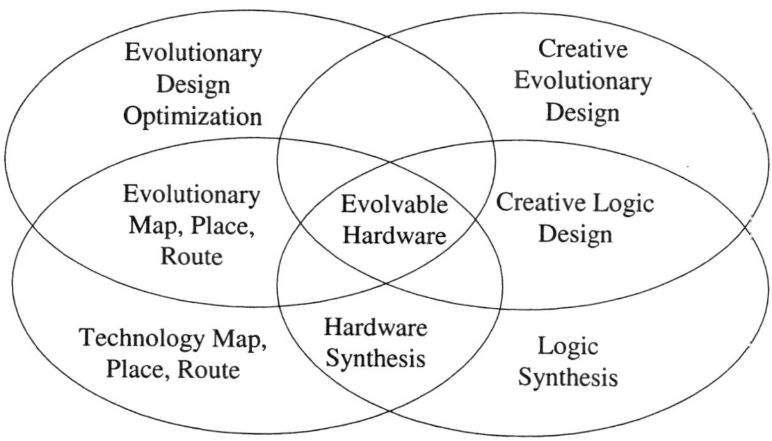

Fig. 2. Evolvable hardware can include aspects of design, optimization and traditional hardware development techniques.

A particularly exciting branch of evolvable hardware research is that of *intrinsic evolution*, where evolved circuit designs are synthesized and evaluated directly on programmable logic devices (PLDs). Here the success of evolution can depend on the choice of platform. Platforms that have been used for evolvable hardware will be appraised at the end of section 3, including those that are commercially available and those developed as research tools. Two case studies will then presented in section 4. One of these is carried out in simulation and one on a commercially available platform. Finally, a summary will be given in section 5.

2 Applications of Evolvable Hardware

Evolutionary Computation is the field of solving problems using learning algorithms inspired by biological evolution. These algorithms are collectively known as *evolutionary algorithms*, and model the cycle of selection, recombination and reproduction that biological organisms undergo. Typically they work on a population of prospective solutions at any one time. Each member of the population is evaluated according to a problem-specific *fitness function*, which tests how well the trial solution performs the required task. A *selection* operator then probabilistically chooses the solutions within the population that the algorithm will subsequently focus on, based on the fitness function evaluation. The selected solutions are recombined and mutated in order to search new but related areas of the problem space. Finally they are reinserted into the population, and the process iterates. The evolutionary algorithm most commonly used to evolve hardware designs is the *genetic algorithm* [15], where each trial circuit design is encoded as a bitstring. Recombination, or *crossover*, is achieved by the probabilistic exchange of bits between individuals, and mutation by the probabilistic toggling of bits in each individual, normally according to predefined rates. Thus the algorithm explicitly separates the genetic information that is recombined and mutated (the *genotype*) from the actual circuit that is evaluated (the *phenotype*). Another evolutionary algorithm commonly used is *genetic programming* [3], where an individual solution is a computer program typically represented by a tree, with no explicit mapping between genotype and phenotype. Here crossover and mutation typically operate on branches of the trees. Variations of these and other evolutionary algorithms have been also used to evolve hardware, but for the purposes of this paper the differences between any of these are not important.

Using evolution to design circuits brings a number of important benefits to electronics, allowing design automation and innovation for an increasing range of applications. Some of the more important areas where evolvable hardware can be applied include:

- Automatic design of low cost hardware;
- Coping with poorly specified problems;
- Creation of adaptive systems;
- Creation of fault tolerant systems; and
- Innovation in poorly understood design spaces.

The remainder of this section will explore these benefits in a little more detail.

2.1 Automatic Design of Low Cost Hardware

Automation of circuit synthesis has been with us for many years. Traditional digital design involves the mapping of an abstract human-designed circuit to a specific technology through the application of simple minimization, placement and routing rules. As our capability for synthesizing more complex circuits has grown, so has the need for more resourceful processes to handle the combinatorially complex mapping procedures. Intelligent techniques such as simulated annealing [76] and ANNs [96] have been routinely used to search these exploding spaces of mappings and minimizations for some time. More recently so has evolution [54, 14].

Useful though this work on searching mapping spaces is, it is not the focus of our discussion here. Instead we are interested in automating generation of the input to this process - the circuit design - from the specified behaviors of the circuit. The required behavior may be presented in the form of pairs of input/output response required to carry out a specified computation, or some other representation of circuit behavior in cases where this is more convenient. The important point here is that to allow evolution to design, a circuit is best evaluated as a black box: according to what it does, not how it does it.

These ideas of design automation can be of significant benefit to hardware that requires a low cost per unit. One example of this is low volume hardware. Low cost reconfigurable hardware can be used to embody evolved designs. For low volume designs this reduces cost by avoiding the need for a VLSI fabrication process. Use of reconfigurable hardware also allows changes in specification to be applied not only to new applications of a design, but also to hardware already in use, thus avoiding replacement costs. Risk, and its associated cost may also be reduced, as design faults could be corrected, either by hand or through further evolution.

Evolutionary automation can even make realistic the prospect of evolving hardware designs to suit an individual. Many medical applications have not been suitable for hardware solutions owing to the expense of personalization. Evolvable hardware allows cheap fast solutions to such medical applications. For example, a system has been developed to control a prosthetic hand by recognizing patterns of myoelectric signals in a user's arm [35]. The implementation was an entirely hardware based solution with reconfigurable logic, a hardware genetic algorithm unit, a CPU core for evaluation, a chromosome memory and a random number generator implemented on the same integrated chip.

A related application is the use of evolution to tune reconfigurable analog circuits. Often the components in analog circuits differ slightly from their original specification due to variations during fabrication. It has been shown that such discrepancies can be corrected using evolution to guide the circuit to its desired behavior [66]. In this case the solution was developed for tuning intermediate frequency filters for mobile telephones to allow the use of cheaper components, and increase yields.

2.2 Poorly Specified Problems

It is difficult to specify the functionality of some problems. In these cases design automation may allow solutions to be generated from a behavioral description of the problem. Evolution is one of a range of soft computing techniques that can be used to handle poor specifications. For instance ANNs have been applied to problems such as noisy pattern recognition for many years [73]. Evolvable hardware techniques have similarities with and advantages over ANNs, as noted by Yao and Higuchi [95]. Both can be feed-forward networks, and both can learn non-linear functions successfully. But in addition hardware is by nature a fast medium and in many cases such as when restricted to feed-forward networks, evolved hardware designs are more easily understood than ANNs. Therefore it is often suited to problems usually tackled with ANNs, but which require fast operation and good solution tractability. Evolvable hardware suitable for such purposes has already been developed for industrial use [68].

One problem where evolved hardware can rival ANNs is pattern recognition. For example high-speed robust classifiers have been developed by Higuchi et al. [24, 32]. One of the advantages of evolutionary systems is the ease with which learning biases can be incorporated. Here, robust generalization characteristics were incorporated into the solutions by specification of a bias towards short description lengths, a recommendation that results from the application of the Minimum Description Length principle to hypothesis representation in a Bayesian context [64, chapter 6].

The ability to generate solutions to poorly specified problems can be considered as a form of creativity [72]. Creativity and innovation are important features of evolutionary processes, and will be discussed later.

2.3 Adaptive Systems

With sufficient automation (i.e. real-time synthesis provided by PLDs), evolvable hardware has the potential to adapt autonomously to changes in its environment. This can be very useful for situations where real-time control over systems by humans is not possible, such as on deep space missions. It could be particularly useful when harsh or unexpected conditions are encountered.

Stoica et al. have noted that current lack of validation for on-line evolutionary systems mean that critical spacecraft control systems cannot realistically be evolved on-line [77]. This follows for other mission-critical problem control systems. However, sensors and sensory information control systems are not so critical, and their work has focused on these. Much of the data that spacecraft sensors acquire, process and transmit is highly specialized, but perhaps unknown until it is encountered. Hence systems designed to capture, process and transmit such data are suitable for evolutionary design. For instance an example of evolved hardware compression for space images is presented in [12]. Here a function is regressed for each image that is to be transmitted using genetic programming.

Other adaptive hardware compression systems have also been developed. Two systems have been developed at the Electrotechnical Lab (ETL) in Japan, both using predictive coding. The first breaks images into smaller sections and uses evolution to model a function for each section [74]. They also suggested that a similar system could be used for real-time adaptive compression of video signals. The other approach at ETL also aims to predict each pixel from a subset of surrounding pixels. The particular pixels that are chosen for the job are collectively known as the pixel template. This time a fixed function is used for prediction, but for each new image or image fragment, a genetic algorithm searches for the overall best pixel template to input to the prediction function. This works particularly well when the type of image to be compressed does not exhibit close correlation between neighboring pixels, but between wider ranging pixels. Hence this technique was used in [75] to compress bi-level images for high precision electrophotographic printers, images that exhibit this quality. The images were broken into smaller sections, and pixel templates to be used by a standard encoding engine (a QM-Coder) were successfully evolved. These templates and the compressed images were transmitted for decoding. It was found that this system could outperform JBIG, the ISO standard for bi-level image compression, by an average of around 50%.

Adaptive control systems have also been developed with evolvable hardware. Most commonly, this has been for robot control. For example, see [39, 70]. Industrial real-time control applications may also be suitable for this approach.

Damiani et al. have developed an on-line adaptive hashing system [8]. They have proposed that this could be used to map cache blocks to cache tags dependent on the access patterns of the data over time.

2.4 Fault Tolerant Systems

Another practical class of adaptive system is one that can adapt to faults in its own hardware, thereby implementing a level of fault-tolerance. Higuchi et al. [24] developed an adaptive hardware system that learned the behavior of an expert robot controller by example using a genetic algorithm. It could then be used as a backup controller if the expert controller failed.

On-line autonomous hardware fault detection and repair mechanisms have been developed [13, 69]. But although these architectures are examples of bio-inspired hardware and have been proposed as a platform for evolutionary experiments, they do not use evolution as an adaptive repair mechanism. Off-line systems can also be evolved to provide fault tolerance, as first shown by Thompson [81]. Thompson also showed that evolution may also generate fault tolerant solutions implicitly through the incremental nature of the evolutionary design process. Fault tolerance can exhibit itself at a population level as well as at an individual level. This is discussed in section 3.

2.5 Design Innovation in Poorly Understood Design Spaces

The design space of all circuits contains an infinitely large number of components that can be wired together in an infinite number of ways. In order to find useful circuits, human designers need to reduce this search space to a manageable size. To do this, they work in a space of lower dimensionality in which they are skilled in searching. For instance, some designers treat all components as perfect digital gates, when the components used in the embodied design are in fact high gain analog devices. The evolutionary approach may allow us to search spaces with a lower (or different) abstraction. This means that exploration of designs from the much larger, and often richer, solution space beyond the realms of the traditional hardware search spaces is possible, resulting in novel designs.

Such innovative solutions are needed when we do not have a good understanding of the design space. For instance, when compared to the logic design space, analog design is less well understood with no formal methods of abstracting the design process. Hence circuit design in this domain requires more expert knowledge. Much work has gone into using evolutionary algorithms to produce human-competitive (and better) analog circuit designs [41].

A good deal of this work centers around the optimization of parameters for current designs, and as such fits more with the field of evolutionary optimization, rather than evolutionary circuit design [2, 66]. Work in less abstract search spaces more relevant to this discussion was also been carried out. Grimbleby developed a hybrid genetic algorithm / numerical search method, using the genetic algorithm to search netlist topologies, and a numerical design optimization method to select parameter values for the evolved topologies [18]. Koza et al. as well as Lohn and Columbano have both evolved circuit designs at the analog netlist level [41, 49]. Handling the increase in search space when moving from optimization to design may require additional techniques which will be discussed later.

Developments in electronic engineering are beginning to generate new kinds of circuit. The design spaces of new technologies such as these are often very poorly understood. In these cases evolution can prove a useful technique in searching for innovative designs. An example of this is the field of nanoelectonics, where Thompson and Wasshuber have successfully evolved innovative (but at this stage not particularly useful) single electron NOR gates [87].

Although the digital search space is much better mapped than the spaces mentioned above, traditional logic synthesis techniques such as Karnaugh maps and the Quine McCluskey procedure are best suited to generating sum-of-products solutions. If there is a requirement to design a different representation of the circuit, for instance if it is required to optimize the map to a technology including XOR gate or multiplexers in addition to designing the logic, then the evolutionary approach can work with a design abstraction more related to the technology and potentially search areas of space that a human designer would miss if using the techniques above. This possibility has been demonstrated in work where evolutionary algorithms have been used to discover more parsimonious circuits for representations, for instance those with multiplexer and XOR gate primitives [56].

3 Research in Evolvable Hardware

Having discussed the benefits of evolvable hardware, and some of the applications that these benefits allow, this section reviews the main thrusts of research in this field. The research is decomposed into three areas - innovation, generalization and evolvability. Before we review this work, however, it is instructional to classify the research according to a number of other features.

3.1 Classifying Evolvable Hardware

A number of schemes have been developed for classifying work on evolvable hardware [97, 25, 1, 89]. We focus on three features that have been used by some of these classifications - level of abstraction, bias implementation and hardware evaluation process. Classification by level of abstraction was introduced by Hirst [25], and has been used by both Andersen [1] and Torresen [89] in different guises, using different levels of abstraction. Here we combine the levels of abstraction used by these three reviews to produce a more comprehensive means of highlighting the scale of behaviors open to the evolutionary process. We introduce a second feature - the bias implementation. With this addition we can see not only what abstractions researchers have found useful, but also how these abstractions can be imposed on an evolving system through the learning algorithm bias. This reveals a developing trend of movement from the imposition of static biases towards ones that alter throughout an evolutionary run. Finally, hardware evaluation process was a feature used by Hirst, Andersen, Torresen and Zebulum [97]. Although this classification is usually tightly linked to the level of abstraction, it will become clear that the evaluation process can lead to some important implementation choices and so is worth discussing in its own right.

Level of Abstraction

As we have already mentioned, the level of abstraction employed by evolvable hardware systems is often important. Hirst identified a number of stages within the design and synthesis lifecycle of reconfigurable hardware where the problem representation could be used as a genotype for an evolutionary algorithm [25]. This has been used as the basis for the classification of design abstraction used here, although categories that have not been applied to date have not been included, and additional levels have been added where other work in the field requires it. A diagram of the levels used here is shown in Fig. 3.

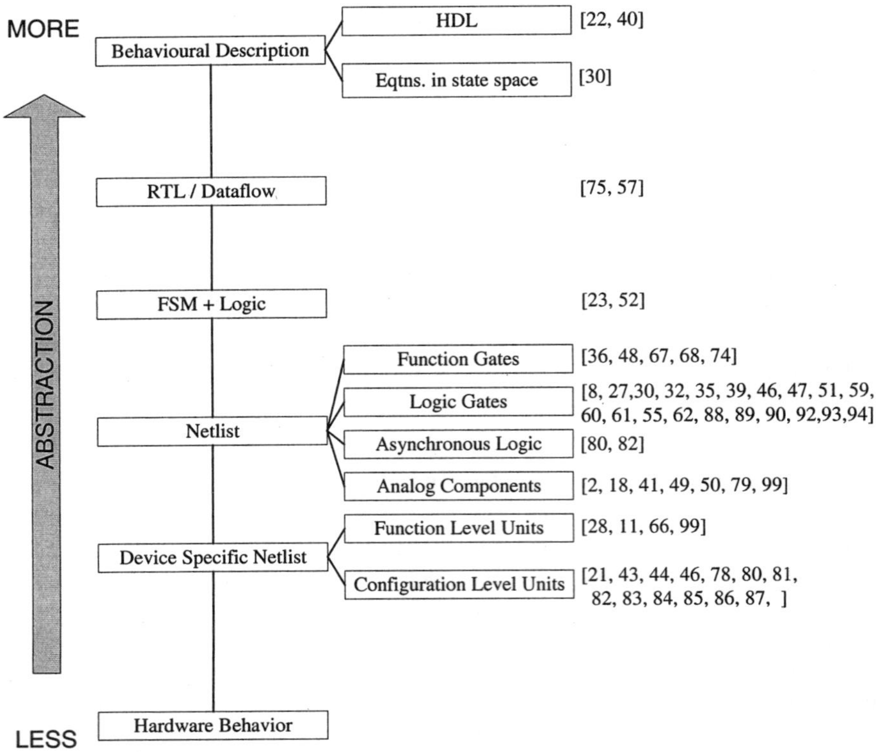

Fig. 3. Levels of abstraction needed to characterize behavior of evolved circuits.

Torresen [89] partitioned levels of abstraction into the digital and analog paradigms that are commonly used in the evolvable hardware literature, and are included here. Anderson [1] further divided the netlist level into netlists of function-level units and netlists of logical gates. We also make this differentiation. Note that the resultant levels are arbitrary levels that happen to fit most of the work in the field. Many variations of the abstractions shown here could and have been used to describe the field. However in the context of our discussion, grouping work into one of these levels allows the reader to understand the essence of work at different levels. The levels used throughout section three are displayed in Fig. 3 along with references to work that can be envisaged as evolving at each particular abstraction.

Bias Implementation

Implementation of a design abstraction is not only brought about through genotype constraint. All learning algorithms can be characterized in terms of a bias that guides the system though a space of possible solutions. These biases can be defined according to a concept learning framework introduced by Rendell [71] and clarified by Gordon and des Jardins [17], which is illustrated in Fig. 4.

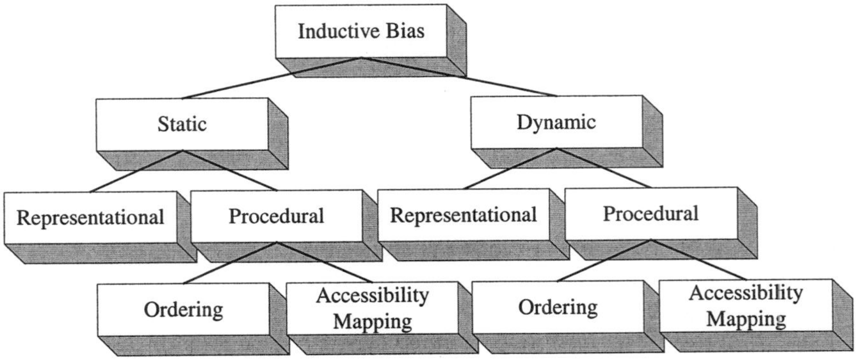

Fig. 4. Learning biases.

In this framework, the bias of a learning algorithm is defined in terms of functions operating on a solution space. Bias is separated into representational bias and procedural bias. The representational bias is the language that defines the search space within the entire space of possible solutions. We term constraints applied to the system by using such a bias as strong biases, because these biases are always obeyed. An example of this is the bias imposed by the genotype representation in a standard genetic algorithm. For the same algorithm we can make a further distinction between the representational bias that limits the search space and the representational bias that maps the search space to the solution space.

The procedural bias determines how the algorithm moves through the search space. The canonical genetic algorithm has three operators – selection, crossover and mutation, each of which exerts a different procedural bias on the search. Usually procedural biases consist of two components. The first is an accessibility mapping, which is used to map one point in search space to another. Selection and crossover have complex dynamic procedural accessibility mappings dependent on the genetic information present in the population at a given time. Mutation exerts no accessibility mapping bias, as any point in the population can be reached from any other. The second type of procedural bias accounts for mutation's effect. This is a partial ordering which is used to structure the candidate solutions in such a way that an order for search space traversal is defined. For mutation the ordering bias is that states close in space to the current state are preferred, the extent of which is determined by the mutation rate. Crossover points are chosen at random, so crossover has no ordering bias. Selection has an ordering bias determined by the fitness score awarded to each candidate solution by the evaluation function.

Dynamic representational biases can allow us to search a space of strong biases, and can easily be incorporated into evolutionary algorithms. For instance the Messy Genetic Algorithm implements such a search through new operators [16]. Another approach is that of developmental genotype-phenotype mappings [4].

Hardware Evaluation Process

Fitness values for the evaluation step of the evolutionary algorithm must be calculated for each member of the evolving population. To do this, early evolvable hardware experiments used simulations of the hardware that each member of the population specified. The main reason for this was because the production of hardware was impossible within the time-scale needed to use it for the evaluation stage of an evolutionary algorithm. Hardware designs evolved using simulation for evaluation was labelled *extrinsic* by de Garis [9], who also noted that developments in reconfigurable hardware technology could lead to the possibility of implementing solutions fast enough to evaluate real hardware within an evolutionary algorithm framework. This he called *intrinsic* evolvable hardware. Such reconfigurable hardware has been available for some time, most commonly in the form of field programmable gate arrays (FPGAs) although other devices are available. It is important to realize that the choice of an intrinsic or extrinsic system rarely determines the type of the circuit that can be evolved - it is the total design abstraction that imposes this limitation. As any abstraction can be modelled by simulation on any platform, it is possible to represent any kind of circuit if enough time and care are taken. However this may not be a feasible or sensible approach in terms of time and resources. Hence the intrinsic / extrinsic distinction is often useful as it tends to be well correlated with the use of high / low levels of abstraction.

With the three classification features of level of abstraction, bias implementation and hardware evaluation process in mind, we will now continue to review research in the field of evolvable hardware in three major threads - innovation, generalization and evolvability.

3.2 Innovation

The set of all possible circuits is a large space indeed. When humans come to design circuits, they do not consider this space. Rather, they simplify it so that the space of designs that they have to consider is much smaller. They do this by applying constraints to the search space, commonly termed design rules. This process is called abstraction.

Two common abstractions for circuits are analog and digital. Analog circuit design usually requires highly skilled practitioners who have developed sets of rules of such complexity that many see the field as an art. The need for the designer to understand the interactions within these designs eventually limit circuit complexity, as with increasing circuit complexity the interactions they must consider to search their design space effectively rapidly become unmanageable.

Designers of synchronous digital circuits use a much more constrained set of rules, which ideally reduces them to considering a sequence of circuits which can be fully specified logically. Successive applications of design rules can eventually reduce the search space to a single circuit configuration, or a very small choice of circuits. Because of the more restrictive abstraction of digital over analog behavior, humans can discover circuits of greater computational complexity within a digital search space. This however comes at a cost - for instance digital circuits tend to require more components than their analog counterparts to achieve a computation. As a result of such trade-offs, design abstractions tend to be chosen to suit the application, and possibly the designers available.

Evolution searches large spaces objectively - given a method of evaluating design performance, it can use this to guide its search of the space. Because evolution searches the design space in a different way to humans, the abstractions that human designers rely on are no longer necessary, and in many only serve to restrict evolution from finding novel designs.

3.2.1 Relaxation of Abstractions

Seminal work in this idea was carried out by Thompson. He first set out to show that evolution could successfully manipulate the dynamics and structure of circuits when the dynamical and structural constraints that human designers depend on heavily on had been relaxed. In [84] he demonstrated this by evolving a complex recurrent network of high-speed gates at a netlist level abstraction to behave as a low frequency oscillator. Fitness was measured as an average error based on the sum of the differences between desired and measured transition periods. Circuits were evaluated in simulation using an asynchronous digital abstraction. If we interpret this in terms of bias, we see a static representational bias was used to ensure only the space of recurrent netlists of logic gates was searched. The simulator imposed a static procedural ordering bias on the selection operator through the evaluation function. Hence the space of any behavior of electronic circuits not modelled by the simulator was not searched. On the other hand, the selection operator could explore the asynchronous dynamics afforded by the simulation, free to make use of any such behavior or ignore it as it saw fit.

The required behavior of the circuit was successfully evolved, showing that it is possible for evolution to search without the constraints usually needed by human designers. Further, a graph-partitioning algorithm showed the structure of the circuit contained no significant structural modules as would be seen through the successive abstraction approach of a top-down human approach. In addition circuit behavior relied on methods that would not have been used by human designers. So not only had evolution found a solution by searching the space beyond conventional circuit design space, but also it had found one that that actually lay in this space.

Thompson went on to show evolution with relaxed restrictions on circuit dynamics was possible in real hardware. The hardware was a finite state machine for a robot controller. However whether the states were controlled synchronously by a given clock or not was under genetic control. The evolved robot controller used a mixture of synchronous and asynchronous behavior, and interacted with the environment in a complex dynamical manner to produce behavior that would not have been possible using the finite state machine abstraction with such limited resources. Again evolution had found a solution through relaxation of a human design constraint, in this case the constraint being the synchrony imposed on the finite state machine abstraction. But in addition evolution has found a circuit that uses the rich dynamics that can arise by relaxing design constraints, demonstrating that such dynamics technique can be useful.

Thompson also carried out the first intrinsic evolution of a circuit evaluated on an FPGA. A 10x10 area of a Xilinx 6126 bitstream was evolved. Almost all bits in the bitstream corresponding to this area were evolved, directly as the bits of the chromosome of a genetic algorithm [80]. Thereby Thompson set about evolving a circuit at the lowest level of abstraction possible - that of the physical behavior of the target technology. The task was to evolve a circuit do discriminate between 1 kHz and 10 kHz signals. Fitness was calculated by subjecting each circuit to five 500 ms bursts of each signal in a random order, and awarding high fitness to circuits with a large difference between the average voltage of the output during these bursts. The average voltages were measured with an analog integrator. The only input to the circuit was the 1 kHz / 10 kHz signal - no clock was given. Again, a highly innovative circuit was evolved that used a fraction of the resources that a human designer would need. Following months of analysis Thompson and Layzell described the functionality of the circuit as 'bizarre' and to date the nature of some of the mechanisms it uses are not completely understood [85].

Thompson and Layzell carried out a similar experiment, this time providing the circuit with a 6 MHz oscillator signal, which could be used or ignored as evolution required [86]. The prime motivation for the experiment was to investigate robustness, and so evaluation was carried out under a range of conditions specified by the operational envelope detailed in section 3.3. Hence the constraints to the system were the same as before except that a weak bias towards robust behavior had been added. However an additional dynamical resource had been provided. The resulting circuit made use of the clock, and the design was simulated using the PSpice digital simulator. The simulated design behaved exactly as that of the real circuit, showing that evolution had found a solution within the digital design abstraction, even through the constraints did not require that. However, analysis of the simulation waveforms showed a large number of transient signals. This led Thompson to hypothesize that evolution can find circuits within the digital abstraction, with all the benefits of robustness and technology insensitivity that this entails, but that these circuits may lie outside the scope of the human digital design space. To put it another way, evolution can find innovative digital designs by searching a bigger space than human designers, but by using a different mapping though space.

3.2.2 Combining Common Abstractions

Miller et al. have also conducted research into the discovery of innovative circuits, one of their main motivations being the derivation of new design principles. In [55] they note that Boolean or other algebraic rules can map from a truth table of required circuit behavior to an expression in terms of that algebra. They then suggest that a bottom-up evolutionary approach can search not just the class of expressions that the algebraic rules map to, but a larger space of logical representations, beyond commonly used algebras.

In an aim to demonstrate this they successfully evolved one and two bit adders based on the ripple adder principle using a feed-forward netlist representation of AND, OR, NOT, XOR and MUX gates. This space lies beyond the commonly used Boolean and Reed-Muller algebra spaces, but is of interest as the multiplexer is available as a basic unit in many technologies. Hence this approach is very similar to Thompson's approach in principle - that the discovery of innovative circuits can be facilitated through the modification of design abstractions implemented through representational biases.

Many of the circuits reported in this and other work [56, 62] were unusual but interesting because of their efficiency in terms of gate count. They lay in the space of circuits making use of multiplexers and XOR gates, outside the space of traditional atomic Boolean logic units. They argued that these circuits were unlikely to be found using traditional algebraic methods, and so evolutionary "assemble-and-test" is a useful way that such a space can be explored. The work continued with the evolution of two bit and three bit multipliers. All work was carried out in simulation. Similar work has been carried out with multiple valued algebras [36].

Another aspect of this group's work is the contention that design principles could be discovered by searching for patterns in evolved circuits. In particular they hypothesis that by evolving a series of modules of increasing size, design principles that they have in common may be extracted from them. In [55] and [62] they evolved many one and two bit adders, and by inspection deduced the principle of the ripple adder. Although this principle is known, they went on to argue that evolution discovered and made use of it with no prior knowledge or explicit bias. As the design principle could be extracted from comparing one and two bit adders that had evolved to use the principle, they asserted that evolution could be used as a method of design principle discovery.

Their recent work in this area has concentrated on developing an automatic method of principle detection [57]. Having successfully evolved two and three bit multipliers that are much more compact than those of traditional design, they have integrated a data mining procedure [33] to search for design principles. The learning algorithm used for the data mining process is an instance based learning technique called Case Based Reasoning [64, chapter 8].

3.3 Generalization

Inductive learners such as evolutionary algorithms infer hypotheses from observed training examples. If it is infeasible for all possible training examples to be observed by the learner, it must use an inductive bias to generalize beyond the cases it has observed. This is a very common situation. For example, a combinational circuit with eighty input pins has 2^{80} possible sets of inputs. Even if we observe each of these training examples at the rate of one million examples a second, observing all the examples only once would take over three times the age of the universe. Choosing a good inductive bias is therefore crucial to the success of our learner.

Two approaches to generalization can be found in the evolvable hardware literature:

(a) Introduce knowledge about the *nature* of circuits that have the generalization characteristics required, perhaps in the form of a heuristic.

(b) Introduce knowledge about the *behavior* of circuits that have the generalization characteristics required, and make evolution learn about the nature of such circuits in addition to the primary task.

Both of these can be applied either as a strong bias in which case a circuit with the generalization capability follows deductively, or a weak bias in which case the other biases innate in the evolutionary process must make up the shortfall in knowledge.

3.3.1 Generalization Across Inputs

Iwata et al. have successfully managed to improve the generalization abilities of evolved pattern recognizers in the manner of case (a) above [32]. They introduced a heuristic commonly used in the machine learning literature to improve generalization. The heuristic results from the application of the Minimum Description Length (MDL) principle to the discovery of maximum a posteriori hypotheses in Bayesian settings, and biases the search towards small circuits. For details of this interpretation of MDL see [64, chapter 6]. Miller and Thomson investigated the generalization abilities of a system evolving a three bit multiplier with respect to the size of its input training set [60]. The task was to evolve a functional circuit from a subset of the truth table. They found that if evolution was presented with a subset of training cases throughout the entire evolutionary run it was not able to produce general solutions. This suggests that in the setting of this problem and algorithm there was no implicit bias towards generality. They also reported that even when evolution was provided with a new set of training cases randomly drawn from the truth table every generation, general solutions were still not found. This is an example of case (b) above, and would suggest that evolution had little memory in the context of this problem.

Miller and Thomson also investigated the evolution of square root functions [61]. In these cases, they discovered that some acceptable solutions were generated when evolution was limited to an incomplete training set. These cases occurred when the missing training cases tested low-order bits, which contributed less to the fitness. Hence they concluded that real-valued functions could be approximated.

Imamura, Foster and Krings have also considered the generalization problem [31], and concluded that evolving fully correct circuits to many problems was difficult. They pointed out that the problem was exacerbated in functions where each test vector contained equal amounts of information relevant to the problem, such as the case of the three bit multiplier studied by Miller and Thomson. However they suggested that in cases where the data contained a large amount of 'don't care' values evolvable hardware could be successful using a smaller test vector. They suggested suitable applications might be feature extraction, data mining and data cleaning.

3.3.2 Generalization Across Operating Environments

It is unrealistic for the algorithm to train from every conceivable circuit input. It is also unrealistic to train under every conceivable operating environment. Operating environments might include a range of technologies or platforms on which the designed circuit should operate, and a range of conditions that the embodied circuit may be subjected to.

Human designers often manage such generalization by imposing strong biases on the nature of the circuit. These biases are representational abstractions that are known to produce behavior common across all necessary operating environments. The abstractions are then mirrored on the physical hardware through some constraint on its behavior. A circuit that behaves correctly in all necessary conditions should then follow. For instance, the digital design abstraction requires that the physical gates of the target technology behave as perfect logic gates. In reality, they are physical devices that behave as high gain amplifiers. Timing constraints and operating environment constraints specified by the manufacturer of the physical device are imposed on the real hardware. This ensures that when an abstract computation takes place the voltages of the gates have saturated, any transient behavior generated before saturation has dissipated. From this point, their outputs can be treated as logical values. In synchronous systems, these constraints are usually imposed with respect to a clock. The manufacturer will then guarantee that for a range of operating conditions, the device will behave as it appeared to within the design abstraction. The design is then portable across a range of devices and operating conditions.

When automation is a more important requirement than innovation, evolutionary circuit design often takes a similar approach to the human design process. Consequently, human design abstractions are often used by evolutionary algorithms to ensure that certain familiar behaviors are embedded in the design. This is most easily done by imposing a representational bias. For example the Electrotechnical Lab in Tsukuba has centered on a netlist level design abstraction, implemented through a static representational bias. Early experiments were designed around the AND-OR networks of a Lattice GAL16V8 PLA. Test circuits were evaluated using a logic simulator, as this device could only be reprogrammed in the order of 10^4 times before failing. A circuits behavior in simulation is its behavior in one operating environment. In order for the behavior of the simulation to generalize to any GAL16V8 operating under standard conditions, a further representational bias was imposed such that no feedback was permitted. This prevented the search including any circuits that would rely on timing-specific features of the gates in simulation, and so the behavior of the circuit in both hardware and simulation would be identical. Note that in practice a real implementation of the evolved circuits would have a physical delay before the desired output signal reached the output. This would be dependent on the delays of the actual gates on that particular chip and the operating environment. Initially a 6-mulitplexer was evolved [23].

Work with the GAL16V8 at ETL broadened to use of a finite state machine (FSM) abstraction. First the same netlist representation was used to evolve the state transition functions of a counter given the counter inputs [23]. This was extended to the successful evolution of a four state Mealy FSM with one input and one output. The FSM abstraction was applied through the fitness evaluation, a static procedural ordering bias. This is a weak bias. Consequently circuits corresponding to FSM, and thereby providing the generalized states were not the only points searched. In fact function FSMs only appeared towards the end of the evolutionary runs. It is not clear whether this additional exploration was advantageous. Similar work has been carried out by Manovit et al. [52].

3.3.3 Generalization Across Operating Environments Using Behavior

In cases where no knowledge about the nature of solutions that generalize across all operating environments is available, the only solution is for evolution to infer this information from examples.

Early work with intrinsically evolved circuits by Thompson focused on design innovation through relaxation of constraints [80]. He successfully evolved an innovative circuit to distinguish between two frequencies, using a Xilinx 6200 FPGA. However he then went on to note the lack of robustness to environmental conditions such as temperature, electronic surroundings, and power supply may occur. It was also noted that the design was not portable, not only when moved to a different FPGA, but also a different area of the same FPGA. Similar results have been reported by Masner et al. [53]. We can think of Thompson's solution as including the bounds of the operating requirements as a procedural ordering bias, although not one specified directly by the fitness function [83]. He took a previously evolved FPGA circuit that discriminated between two tones. He then specified a number of parameters for an operational envelope which when varied affected the performance of this circuit: temperature, power supply, fabrication variations, packaging, electronic surroundings, output load and circuit position on the FPGA. The final population from the previous experiment was then allowed to evolve further, this time on a five different FPGAs maintained at the limits of environmental conditions specified by the operational envelope parameters. Although there was no guarantee that the circuit would generalize to behave robustly to all environmental conditions within the envelope, Thompson found a level of robustness evolved in four out of five cases. Hence, it appears that the biases of the evolutionary algorithm and the 6200 architecture promoted good operating condition generalization characteristics. It is interesting to note that incorporating a procedural bias towards generality in this way was successful when applied to operating conditions, but not in the case of Miller's input test cases discussed in section 3.3.1.

Another example of this is the portability problem of evolving analog circuits extrinsically. Analog circuit simulators tend to simulate circuit behavior very closely. Hence we would expect extrinsically evolved circuits to generalize well to the real circuit. However this does not happen in practice. One issue is that some behaviors that simulate according to the physics programmed into the simulator may not be feasible in the chosen implementation technology. A common example is that simulators often allow the use of extremely high currents. Koza et al. have evolved many circuits extrinsically at an analog abstraction using the Berkeley SPICE simulator [41], but have not been able to build them in real life because of such behaviors. Additionally analog simulators use very precise operating conditions. The circuits of Koza et al. are evolved to operate at 27 °C, and so there is no explicit bias towards generalization across a range of temperatures.

Portability both between simulated and real environments and between changing real environments is problem for robot controllers. In order to develop robustness in both cases Keymeulen et al. used a combination of both generalization techniques - they introduced the robot to as many simulated environments as possible and also introduced mutation operators designed to generalize the evolved circuit model.

Stoica et al. have evolved networks of transistors intrinsically and suffered from the reverse problem - circuits evolved intrinsically operate well in hardware, but may not in software [78]. Their solution to the problem is to impose a procedural ordering bias towards working in simulation and in hardware by evaluating some circuits of each generation intrinsically, and some extrinsically. This they term mixtrinsic evolution [79]. They also suggested that another use of mixtrinsic evolution would be to reward solutions which operate differently in simulation than when instantiated in a physical circuit. This would place a procedural ordering bias towards innovative behavior not captured by simulation.

3.3.4 Inherent Generalization

One other possibility is that the biases of the evolutionary algorithm have an inherent tendency to generate solutions that generalize across certain conditions. Thereby evolved circuits would exhibit robustness to changes in those particular conditions "for free".

One example of such robustness is robustness to faults. The obvious method of evolving such robustness is to include a requirement to robustness in the fitness function, thereby altering the procedural ordering bias of the selection operator [83]. This method of generalization has already been discussed.

Thompson has also postulated that evolved circuits may be inherently robust to some types of fault. He observed that an evolutionary algorithm will by nature be drawn to optima surrounded by areas of high fitness, and suggested that as a result, a single bit mutation from such an optimum will also tend to have a high fitness. He then conducted experiments on an artificial NK landscape to demonstrate this. For details of this type of landscape see [38]. He then proposed that such an effect could have beneficial engineering consequences if a mutation causes a change in the circuit that is similar to a fault - namely that the evolved system is likely to be inherently robust to such faults. He went on to highlight this using the evolution of a state machine for a robot controller as an example. The state machine used a RAM to hold a lookup table of state transitions. Each bit of the RAM was directly encoded in the chromosome, and so mutation of one of these bits had a similar effect to a 'single stuck at' (SSA) fault. Examination of the effect of SSA faults on a previously evolved state machine revealed that it was quite robust to faults. However as state machines for this problem with similar fitness could not be easily generated by any means other than evolution, statistical tests of the evolved machine's resilience to faults could not be carried out. Even so, the idea of a general characteristic developing inherently as a result of a mapping bias inducing the same move through space as a change in environment seems reasonable, and in some cases may be easier than specifying a generalization requirement directly through the procedural ordering bias.

Following this, Masner et al. [53] have carried out studies of the effect of representational bias on the robustness of evolved sorting networks to a range of faults. The aim of the work was to explore the relationship between size and robustness of sorting networks using two representations - tree and linear. They noted that robustness first increases and then decreases with size, and is therefore not due purely to the existence of redundant non-functional gates in the sorting networks. They also noted that the linear representation tended to decrease in robustness with respect to size faster than the tree representation. Again this demonstrates that robustness is a feature that can be evolved through the proper selection of procedural ordering and representational biases.

Layzell has suggested that robustness of solutions can also be generated at the level of populations [44]. In particular he was interested in the ability of another member of the population to be robust with respect to a fault that causes the original best solution to fail. This he called populational fault tolerance (PFT). He went on to demonstrate that PFT is inherent in certain classes of evolved circuit, and test various hypotheses that could explain its nature [45]. As with Masner et al. he noted that fault tolerance did not seem to be a result of non-functional redundancy in the current design. Instead he showed that descendants of a previously best and inherently different design were still present in redundant genes in the members of the population. It was these individuals that resulted in PFT.

3.4 Evolvability

The evolutionary paradigm is no panacea for searching the extremely large spaces that we have been discussing. Researchers have found evolution of small low complexity circuits such as a three bit multiplier difficult [60]. For evolution to search for useful (i.e. large and/or complex) designs over large spaces, we need to make the search for incremental improvements by the evolutionary algorithm easier.

Traditional theory in evolutionary computation suggests that evolution cannot continue to search effectively after a population has converged [15, 63]. Hence from this point of view, making evolution's job as easy as possible improves the quality of the search, as well as the speed.

Intuitively, one would expect evolvability to be improved by limiting the search to a smaller space than the space of all circuits possible in a target technology. However we do not have to limit the space using human design rules - we could apply new abstractions, which assist the method of search that evolution employs. These abstractions may reduce the size of the space that evolution searches, or they may transform the space into something of the same size (or bigger) that is more tractable to evolution.

Finding useful abstractions may be very hard. However we could allow evolution itself to search for abstractions that it finds useful. Also, these abstractions no longer need to be hard constraints that must be adhered to, but may be soft constraints used to bias the search without forcing it to avoid potentially useful areas of space. The flip-side of this is that evolution is given more work to do.

3.4.1 Static Representational Biases

Choosing a good representational bias is very important to the success of the algorithm. As discussed in the section on innovation, we must ensure that solutions we are interested in finding can be represented. However there is another issue. Because the directions of many of the dynamic biases within a genetic algorithm depend on the space around them, the choice of representation will affect the efficiency of the algorithm. Work has been carried out on how static representational biases can affect the evolvability of logic netlist design spaces. Miller and Thomson explored how changes in circuit geometry [60], and how functionality-to-routing ratio affected evolvability [59]. Both appeared to be important. In the case of geometry little else was concluded but for functionality/routing more could be said. They found that forcing differentiated routing and functional units for evolution was important, but functional resources were more important than routing. They also noted that the importance of tuning the average number of neighbors to each cell. Kalganova et al. have analyzed how representational biases can affect the geometry both of multiple valued logic netlists [36] and Boolean logic netlists [37]. However all these studies are likely to be dependent on the problem and the other biases employed, making it difficult to draw general conclusions.

3.4.2 Function Level Evolution

The function level approach to improving evolvability was developed at ETL, and has been adopted by many others. Early evolvable hardware work at ETL used a combinational digital netlist representation that could be easily mapped to hardware. For harder problems they suggested that the size of the chromosome limited the speed of evolution. They suggested two solutions - one was to use more abstract structures in the representation, thereby reducing the search space. This was dubbed function-level evolvable hardware [48, 67]. The difficulty here is in choosing the correct structures to use in the representation. Any abstraction made makes assumptions about the type of problem. Therefore problem-dependent functions would have to be developed for each class of problem. Once this trade-off has been made, evolution is now limited to search the space of this abstraction, and any innovative solutions at the lower abstraction will be unattainable.

In response to the work on function level evolvable hardware Thompson argued that such course-grained representations could reduce the evolvability of a hardware design space [82], noting that as the control evolution has over an evolving platform becomes coarser, so can the fitness landscape. This can result in less evolvable landscapes, at the limit reducing evolutionary search to random search. Instead, Thompson argues that traditional evolution has the capability to search larger spaces than are advocated by those at ETL in [48, 57]. In particular he suggests that there may be features of many hardware design landscapes that allow us to search large spaces beyond the point where the evolving population has converged in fitness. Such a feature, he suggested, was the neutral network.

3.4.3 Neutral Networks

If the genotype search space is considered as a high dimensional fitness landscape, neutral networks can be conceived as pathways or networks of genotypes whose phenotypes share the same fitness. It has been suggested [29] that genetic drift along such networks can allow evolution to escape local optima they would otherwise be anchored to. The idea of neutral mutations has been recognized in the field of evolutionary biology for some time but has only recently been used as a paradigm for search in evolutionary computation. Taking advantage of such a paradigm requires the evolutionary algorithm to be modified to include this knowledge. To this end Harvey [20] developed the Species Adaptation Genetic Algorithm (SAGA), which advocates incremental changes in genotype length and a much greater mutation rate than is common for genetic algorithms. Thompson however used a fixed length genetic algorithm with a SAGA-style mutation rate to search an incredibly large circuit design space (2^{1800}) for good solutions. This he succeeded in doing, and when the algorithm was stopped owing to time constraints, fitness was still increasing even though the population had converged long before [21]. Analysis of the evolutionary process did indeed reveal that a converged population had drifted along neutral networks to more fruitful areas of the search space. He put much of this behavior down to the increased mutation rate, a change to the static procedural mapping of the algorithm. In this paradigm, mutation usurps the role of the genetic algorithm's primary variation operator from crossover. He noted that the nature of the solution representation space was also important – without the existence of neutral networks in the solution space would not be possible, and speculated that neutral networks might be a feature of a great deal of design spaces including many hardware design spaces.

Vassiliev and Miller also endorse the neutral network theory. Their work on neutrality in the three bit multiplier logic netlist space [93] suggests that neutral changes at the start of an evolutionary run occur because of high redundancy in the genotype. As the run continues and fitness becomes higher, redundancy is reduced. However the number of neutral changes does not drop as quickly, suggesting that selection actually promotes neutral changes in order to search the design space. They then went on to show that when neutral mutations were forbidden, the evolvability of the landscape was reduced. Comparisons between the three bit multiplier space and other problem spaces with the same representational bias suggested that the nature of the spaces was similar, and so the results from this space may well hold for others. They have also shown that the search for innovation may be assisted by using current designs as a starting point for evolution, and proposed that neutral bridge could be used to lead us from conventional design space to areas beyond [92].

Much of the work on neutrality uses evolutionary strategies as opposed to the more traditional genetic algorithm. Evolutionary strategies do not use the crossover operator. Because of this their use in studies of neutral mutations, the driving force of evolution in the neutral network paradigm, simplifies analysis.

3.4.4 Dynamic Representational Biases

The second proposal from ETL to improve the speed of evolution was to use a variable length representation. This imposes a dynamic representational bias – it allows evolution to searches the space of representations in addition to the problem space. If successful, a representation useful to the algorithm can be found. This is a good idea, as the search is directed by the procedural ordering bias imposed by the problem, rather than a bias of the algorithm, meaning the technique could be portable to a range of problems. Applied to a pattern recognition problem, the results were greatly improved over an algorithm that did not search the bias space, both in terms of solution parsimony and efficacy [34]. However, the evaluation of the chromosome was such that the representational space was still always limited to a feed-forward network of Boolean gates.

A similar approach was taken by Zebulum in an experiment to evolve Boolean functions using a chromosome of product terms that were summed by the fitness function [98]. However the search order of representation space differed from the ETL experiments. Inspired by the observation that complex organisms have evolved from simpler ones, the population was seeded with short chromosomes. A new operator was introduced to increase chromosome length, under the control of a fixed parameter. Hence a fixed dynamic bias to move from short representations to long ones was set. It was found that a low rate of increase allowed fully functional, but more parsimonious solutions to be found over a larger rate.

An interesting approach was that of the Adaptive Architecture Methodology (AdAM) system [22]. Here a system was developed to evolve a hardware description language (HDL). These are high-level behavioral languages, but also allow register transfer level constructs to be embedded within the code. The approach taken here was to evolve a tree, each node of which referred to a rewriting rule from a set that effectively make up the grammar of the HDL. As the mapping between genotype and phenotype depends on the structure of the genotype itself, this becomes a method of searching representation space - it is a dynamic representational bias. This allowed evolution to find representations in which problem-dependent structures or modules could be better represented.

Modularity has been recognized as important in evolutionary systems for some time, particularly in genetic programming [3, 41] and more recently in evolvable hardware [57]. For this purpose, Kitano has proposed a similar rule-based system that promotes large-scale modularity [40].

The representation used by Koza has similar properties. It is a tree of circuit modifying functions that act on an embryonic circuit. It is this mapping between the embryo and the phenotype that is evolved, again allowing search of the representational bias space. The rules are chosen to avoid creation of invalid circuits [41]. Lohn and Columbano have used a similar approach, but with a linear mapping representation which is applied to an embryonic circuit in an unfolding manner, rather than a circuit modifying one. Although its representational power is limited, it has been found to be effective [49].

3.4.5 Dynamic Procedural Biases

Rather than finding new ways in which to express the solution, we can search the space of ways to express the problem.

Torresen recognized both the benefits of modularity and the advantages of allowing evolution to search for its own modules. He has suggested using what we can interpret as dynamic procedural ordering biases to search for them. The first is to partition the training vector for a problem by hand and evolve solutions to each separate section. He applied this to a character recognition problem and gained promising results [88]. The second method he suggested was inspired by the concept of incremental learning introduced by Brooks [5]. This bias advocates presentation of training vectors representative of an increasingly large set of behaviors to a single learner over time. This way, more complex behavior can be learned over time. Both methods provide a way of generating problem modularity, but they both require intervention from the designer in order to determine how to break up the problem into sets of behaviors. Torresen showed this technique improved evolvability for a real-world image recognition problem [90].

Lohn et al. have also worked on dynamic procedural biases. They compared three dynamic fitness functions against a static one [50]. The dynamic fitness functions increased in difficulty during an evolutionary run, directly altering the procedural ordering bias. One had a fixed increase in difficulty, one had a simple adaptive increase based on the best fitness within the population, and one put the level of difficulty was under genetic control by co-evolving the problem and the solution. In the former cases the newly introduced biases are new procedural ordering biases. In the latter, a complex interaction between the procedural ordering and mapping biases of two evolving systems is set up. The results showed that static and co-evolutionary biases performed best on an amplifier design problem.

The evolvability of two bit multipliers has also been studied with respect to the evolutionary operator bias, a procedural ordering bias [94]. Uniform crossover and point mutation were studied. Analysis of the ruggedness of the landscape showed that uniform crossover generated a large amount of ruggedness and was therefore unsuitable for searching this space. Point mutation fared much better.

3.5 Platform Research

We have now reviewed most of the research into evolvable hardware. We have seen that many researchers believe that working at low levels of abstraction can have advantages. We have also seen mechanisms to deal with the increased size of the search space, if needed at all, are being actively investigated. What we have not considered is the availability of platforms for low abstraction hardware evolution.

In this section, we cover the platforms that have been reported in the evolvable hardware literature. Some are commercially available, and some have been developed by researchers. Of these, few have been developed with evolvable hardware as a primary goal, and so suffer from various shortcomings. Some have been developed expressly for evolvable hardware. However, industry is already interested in evolvable hardware [24, 66], and as research tools all of these solutions are unlikely to be available to industry for real-world experiments. Even if they are, the cost associated with low-volume research platforms may be prohibitive.

Other devices are commercial, and have not been designed with evolvable hardware in mind. Because of this, most struggle to compete with dedicated evolvable hardware on performance, versatility and ease of use for our purposes. However they have the advantages of availability and cost, and so are a more likely candidate for future industrial applications.

3.5.1 Criteria for Successful Evolutionary Platforms

In [82] Thompson listed a number of criteria for intrinsic circuit evolution platforms. These are discussed below:

Reconfigurable an unlimited number of times - Many field programmable devices are designed to be programmed only once. Others are designed to be programmed a small number of times, but repeated configuration can eventually cause damage. Evolutionary experiments can require millions of evaluations, and so devices for intrinsic experiments should be able to be reconfigured infinitely.

Fast and / or partial reconfiguration - If millions of evaluations are needed, the evaluation process should be fast. Modern programmable devices have millions of configurable transistors and consequently have large configuration bitstreams. This can mean that downloading the configuration becomes the bottleneck of the evolutionary process. The brute force solution to this problem is to use devices with high bandwidth configuration ports. Another solution is to evaluate many individuals at once, as proposed by Higuchi amongst others [28]. Batch evaluation limits the type of evolutionary algorithm to those with large populations, ruling out the use of steady state genetic algorithms, or low-population evolutionary strategies. A more elegant solution is that of partial reconfiguration, where only the changes from the current configuration need to be uploaded. This yields similar bandwidth use with no constraints on the learning algorithm.

Indestructibility or validity checking - In conventional CMOS technologies, a wire driven from two sources can result a short circuit if one drives the wire to a different voltage level than another. The high currents generated from such an event are extremely undesirable, as they can damage the device, and so should be prevented by hard constraints, rather than the softer ones advocated so far. Some hardware platforms are designed around an architecture with which contention is impossible. For those that are not, there are two options - either an abstract architecture can be imposed on top of the real hardware, or circuits can be tested for contention before they are synthesized, and evaluated by an alternative means if such a condition is detected.

Fine grain reconfigurability - In order to allow evolution the ability to innovate, evolution must be able to manipulate candidate circuits at a low level of abstraction. Hence a good platform needs fine-grain control over the evolving platform. Thompson also points out the distinction between fine grain architectures and fine grain reconfigurability - namely that although a device's architecture may be based on repeated large units, if these can be reconfigured at a finer level then this criterion will be met.

Flexible I/O - The method of supplying input and retrieving output from an evolved circuit can affect the feasibility of successful evolution, and so a platform that allows experimentation with this is useful.

Low cost - This is of particular importance when the motive behind using evolution is to lower costs through design automation.

Observability - In order to analyze how evolved circuits work, their internal signals need to be probed. Although when working with low design abstractions it may be impossible to avert the potential of signal probes to change the behavior of the circuit, and the probed signal, architectures should be chosen with this as a consideration.

3.5.2 Platforms

Whilst bearing these criteria in mind, the platforms that have been used or proposed for use for evolvable hardware experiments are now considered briefly. These can be classified into three groups - commercial digital, commercial analog and research platforms, and are tabulated below.

Commercial Analog Platforms
Zetex TRAC [11] - Based around two pipelines of op. amps. Linear and non-linear functions successfully evolved. Large grained reconfigurability and limited topology limit worth for evolution. Has been used with evolvable motherboard to provide external components.
Motorola MPAA020 [99] - 20 cells containing an op. amp, comparator, transistors, capacitors and SRAM. Range of circuits have been evolved. Much of the bitstream is proprietary. Geared towards circuits based around the op. amp.

Commercial Digital Platforms
Xilinx 6200 [80, 86, 41] - Developed for dynamic reconfig. apps. Fast and infinite reconfig., fully or partially. Homogenous fine-grained architecture of MUXes. All configurations valid. Good I/O. Expensive, no longer produced.
Xilinx XC4000 [46] - Low cost, infinite but slow reconfig. SRAM LUT based architecture. Damaged by invalid configurations. Parts of bitstream proprietary and undisclosed. Reconfigurable at resource level using Xilinx JBits software.
Xilinx Virtex [27, 47] - Medium cost. Can be reconfigured infinitely and quickly, fully and partially. Can be damaged by invalid configurations. Some of the bitstream is proprietary and undisclosed, but most hardware resources can be reconfigured using Xilinx JBits software. Widely available.

Research Platforms
Field Programmable Transistor Arrays [78, 79] - Reconfigurable at transistor level, additionally supporting capacitors and multiple I/O points. Programmable voltages control resistances of connecting switches, hence they act as additional transistors. Flexible enough to evolve both filters and amplifiers. Fits criteria for evolvable hardware well.
Field Programmable Processor Arrays [13] - Bio-inspired fault tolerant architecture. Early prototypes used a multiplexer as the reconfigurable unit. New revisions increased the granularity of the unit under genetic control through decision tree machine to RISC processor. Now better suited to evolution of more abstract structures, e.g. genetic programming.
Palmo [19] - Based around array of integrators. Signals encoded using PWM. All configurations valid. Integrator unit is useful for evolution of analog and mixed signal processing circuits. Beyond this, not so versatile.
Programmable Transistor Array [42] - 16x16 array of programmable PMOS and NMOS transistor cells. Each cell contains 20 transistors of varying channel height and width, allowing great flexibility. Transistors can be connected in parallel to approximate other channel widths. Fast configuration, good I/O.
Evolvable Motherboard [43] - Array of analog switches, connected to six interchangeable evolvable units. Evolution of gates, amplifiers and oscillators demonstrated using bipolar transistors as evolvable units. Good I/O. Board-based architecture is not suitable for real world problems due to size, cost and number of evolvable units.
FIPSOC [65] - Complete evolutionary system aimed at mixed signal environments. Analog and digital units that can undergo evolution. CPU and memory to encode evolutionary algorithm. Analog units based around amplifiers. Digital units based on LUTs and flipflops. Context-based dynamic reconfiguration suitable for real-time adaptive systems.
Complete Hardware Evolution [91] - FPGA-independent evolutionary system on a chip consisting of a genetic algorithm pipeline including evaluation, functional design and storage for population. Limited to small populations of small chromosomes. Large FPGAs required.

4 Case Studies

The final section of this paper presents two case studies – one evolved at a logic netlist level and evaluated extrinsically, and one evolved at a device specific netlist level, evaluated intrinsically. The problem selected for both was the evolution of a two bit adder. This problem has been well studied in the past, initially by Louis and Rawlins [51] and more recently by Miller, Thomson and Fogarty [62], Coello and Aguirre [7] and Hollingworth and Tyrell [29].

4.1 Case Study 1 – Logic Level Evolution

4.1.1 Phenotype Abstraction

In this experiment we evolve the adder at a reasonably high level of abstraction. The representation is restricted to combinational digital circuits only. The minimum reconfigurable structural units are at the level of gates - we allowed AND, OR , NOT, XOR, and multiplexer gates. The selection of these resources was based loosely on the resources available in the Xilinx 6200 series FPGA architecture. This means we have abstracted away any of the physics in the device and consider only logical solutions, which limits the opportunity for innovation. However we have made the search space a lot smaller than if such features were included. This abstraction allows fast simulation of the evaluation function. If mapping rules are followed properly and the circuit is operated within the constraints set by the manufacturer of the technology it is mapped onto, it will function as it did in the abstracted simulation.

The circuit abstraction used in the first case study is based on that used by Miller et al. [62] to investigate the evolution of combinational circuits. The circuit is represented by a numbered rectangle of cells. They are indexed from the top left cell, row-wise then column-wise. Each cell has two inputs and one logic function. The function may either be a two input logic gate or a multiplexer. Inputs to a cell can be either from other cells, or the circuit inputs. The circuit inputs are the test input vector, the inverted test input vector, logic 0 or logic 1. To avoid feedback, each input must be from a cell with a lower number than the cell itself. The circuit outputs required are restricted to the top and right hand side of the cell array.

4.1.2 Genetic Algorithm and Genotype

The circuit is presented to the algorithm as an integer string. There is a triplet of integers for each cell, representing the sources of the two cell inputs and the cell function, with the triplet locus mapping to the cell index. If the cell function is a logic gate, the function allele represents a specific logic gate. If the cell function is a multiplexer, then the allele represents the multiplexer control signal source, which can be either the output of another cell or a circuit input. The list of functions used is shown below. Cell outputs are represented by an integer each, the allele referring to the output cell index.

The genetic algorithm used was a standard population-based linear genetic algorithm with selection, crossover and mutation operators, after Goldberg [15]. However the allele range of each gene varied. Therefore it was required that the cardinality of each gene can be set independently of the others. On application to a gene the mutation operator selected a random integer between zero and the cardinality. No bias was placed on selecting integers similar to the current value, as in most cases the relationship between the alleles in the context of this problem was not clear, and so the ordering of the alleles was arbitrary.

Fitness was measured by subjecting each candidate circuit to a test vector containing the complete two bit adder truth table. One fitness point was awarded for each correct bit in the output sequence, giving a total of 96 for maximum fitness.

The circuit inputs were A0, A1, B0, B1, Carry In, !A0, !A1, !B0, !B1, !Carry In, 0, and 1. The outputs were S0, S1, and Carry Out.

Allele	Function	Allele	Function
0	A . B	7	!A
1	A . !B	8	!A +B
2	!A . B	9	!B
3	A ⊕ B	10	!A + B
4	A + B	11	!A + !B
5	!A . !B	12....	!C . A + C . B, C = circuit input 0
6	!A ⊕ B	...(n)	!C . A + C . B, C = input (n-12)

Table 1. List of function to allele mappings used.

Following informal experiments the genetic parameters were mostly chosen by selecting those that Miller had found useful in [62]. Uniform crossover was used with breeding rate at 100%. The mutation rate was set to 5% of all genes in the population. The population size was set to 30. Qualitative examination of early runs led us to believe that the suggested 40,000 - 80,000 generations were not necessary to achieve good results, with 20,000 generations sufficing. Two-member tournament selection was used. A tournament selection pressure as described by Miller was also introduced, and set to 0.7, meaning that the winner of each tournament was selected only with 70% probability. The first generation of each run was randomly generated.

4.1.3 Results

10 runs were each made for cell array sizes of 3x3, 3 wide by 4 high, and 4x4. Overall results are shown below, and relate well to the results presented by Miller. Fitnesses and deviations have been scaled to 100. Each evolutionary run took around four minutes to complete on a 433 MHz Celeron CPU.

Array Size	Mean fitness of Best Solutions	Std. Dev. of Best Solutions	% of Runs with Perfect Solutions
3x3	96.25	4.99	50%
4x3	96.88	3.71	50%
4x4	97.50	4.03	60%

Table 2. Results from 10 runs of 2 bit adder evolution across a range of array sizes.

A series of 10 random searches were also carried out on the 3x3 array search space, using the same number of evaluations as the evolutionary runs. These are shown in Fig. 5, along with the best fitnesses of each generation of the 3x3 array evolutionary runs.

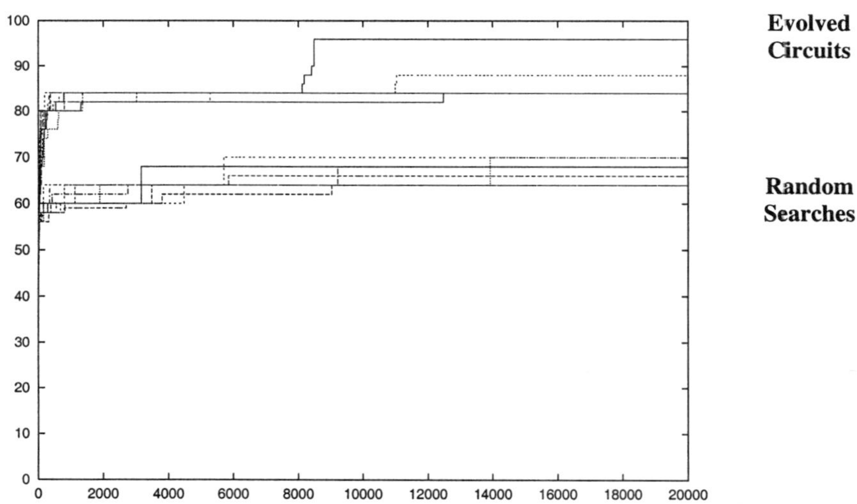

Fig. 5. Best fitness against generations of an evolved 2-bit adder on a 3x3 array along with 10 random searches of the same length.

4.1.4 Discussion

The results achieved from these experiments broadly agree with Miller's results. For example, for a 3x3 array, Miller reported a mean fitness of 96.14 with a standard deviation of 4.52, and 50% of cases perfect.

An example of a two bit adder evolved on the 3x3 array is show n in Fig. 6. It can be seen that this is a minor variation on a traditional two bit ripple-carry adder [26].

Note that this design uses XOR gates to generate the sum signals, and XOR and multiplexer gates to generate both carry signals. We have already noted that it is not trivial for human designers to develop efficient circuits that use only XOR and multiplexer gates. Thus evolution is demonstrated as a means for successfully searching a poorly understood design space for an efficient (in terms of gate count) circuit. Miller et al. have presented similar circuit designs.

The results display a trend towards higher mean fitnesses for larger arrays. This suggests that it is easier to find perfect solutions in larger arrays. However the solutions found within these arrays tended to involve more gates and connections than those discovered on smaller arrays, and so in this respect are less efficient. This is not surprising as no bias was used to search for parsimonious solutions. Miller et al. also noted such trends.

There were some minor differences between the algorithm and that reported by Miller. Miller used a "levels back" parameter to limit the length of the routing connections to a certain number of columns to the left of the current cell. When using such small arrays, we felt this was not necessary. Miller also ran his algorithm for more generations. Informal examination of our early results showed no or little improvement from running more than 20,000 generations. Miller also used a slightly different scheme to map multiplexer behavior to the chromosome.

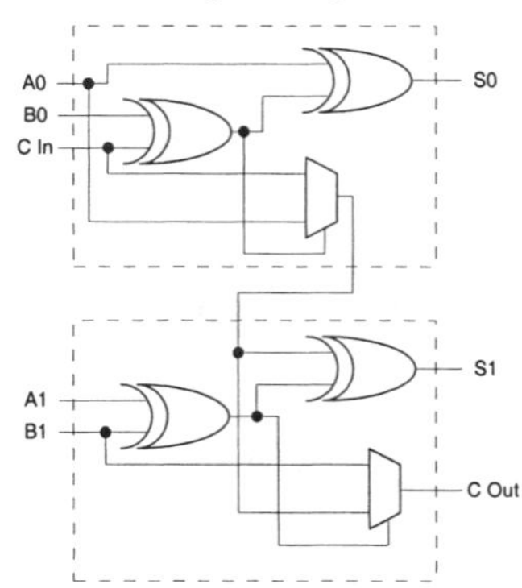

Fig. 6. Example of an adder evolved on a 3x3 grid.

4.2 Case Study 2: Device Specific Netlist Level Evolvable Hardware

For our intrinsic case study we selected the Xilinx Virtex FPGA. Of current commercial FPGA architectures, it fitted the criteria for a general evolvable hardware platform specified above better than any others. It is also a relatively new architecture with a long life ahead of it, so architecture-specific findings will be of interest for some time to come.

4.2.1 The Virtex Architecture

The SRAM-based Virtex can be reconfigured infinitely. With up to 60 MBs^{-1} of configuration bandwidth, it can also be configured quickly. Although some of the bitstream is proprietary and undisclosed, most hardware resources can be configured in Java with the Xilinx JBits software. It allows partial reconfiguration to reduce configuration time, and is widely available at reasonable cost.

A simplified diagram of the Virtex architecture is shown in Fig. 7. It is arranged as an array of configurable logic blocks (CLBs). There are three categories of routing between CLBs - single lines that connect CLBs to their neighbors, hex lines that connect a CLB to CLBs six blocks away, and long lines. I/O blocks providing logic and drivers for several I/O standards, surround the CLB array.

Each CLB consists of two slices, which each contain two SRAM-based four input function lookup tables (LUTs), some fast carry logic, and two flip-flops. The inputs to each CLB slice consist of thirteen inputs. These are the inputs to both LUTs, and flip-flop clock, data, enable and set/reset lines. Outputs from each CLB can be drawn from a range of internal signals, including the lookup tables, the carry logic and the flip-flops. Configurable multiplexers select the connection between the routing and the CLB inputs and outputs.

The main problem with Virtex is that it can be damaged by invalid configurations. Each CLB is driven independently, so it is possible for contention to arise between two drivers if two output multiplexers from two different CLBs drive the same line.

Experiments showing the feasibility of Virtex as a reconfigurable platform were first presented by Hollingworth et al. [27]. JBits was used to map circuit designs to the configuration bitstream. Contention has been avoided by using fixed, handcrafted routing between the CLBs, initially only in a feed forward structure, but more recently by imposing a 6200-like sea of gates structure where one routing signal for each neighbor is allowed in and out of each CLB [28]. Levi has also used Virtex for evolutionary experiments [47], but the method of contention avoidance has not been reported.

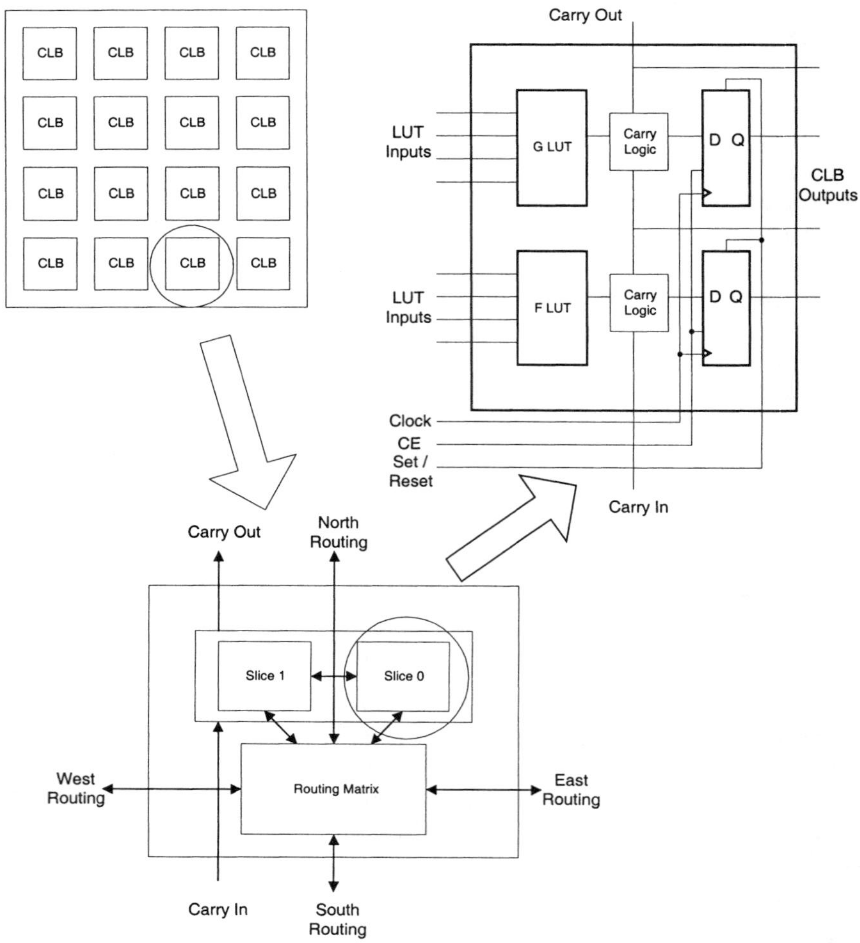

Fig. 7. A simplified diagram of the Virtex architecture.

4.2.2 Genotype Abstraction

As discussed earlier, the most innovative design space of circuits that can be represented using an FPGA is that provided by the configuration bitstream itself – the finest level of reconfigurability of the FPGA. Although the Virtex architecture is course grained it can be configured at a fine level of detail. Unfortunately a good deal of the bitstream-to-architecture mapping is proprietary knowledge. However Xilinx JBits software allows us to reconfigure at a fairly low level of detail, that of a netlist specific to the Virtex device with timing, routing and geometry included in the abstraction. With this we were able to encode almost all the resources available on the hardware, thereby allowing evolution to search a large space for innovative solutions.

To capture the netlist level abstraction correctly in the genotype, we opted to avoid using a binary representation. This was mainly due to our desire to include as many of the resources available to us as possible. First, if all possible configurations were encoded as a series of unique binary strings, arbitrary bias would be introduced against resources that spanned several bits. Alternatively, if each fine-grain resource were separately encoded as a number of bits, arbitrary redundancy would be introduced when the number of states that resource could assume (i.e. number of alleles) is not a power of two. This would also introduce unknown biases.

For these reasons it was decided that each resource that could be modified by the JBits API would be encoded as a separate integer gene. The exception to this was the LUT configurations, which were encoded as sixteen bits.

Routing representation was also an issue. It was so important to avoid damage of the device due to contention that we restricted the representation to avoid these areas of space. In more detail, only the single (nearest neighbor) wires were evolved. It was noted that although CLB input multiplexers can connect to a wide range of single lines, the connections between the output multiplexers and singles are sparse, and few can connect to any that their neighbors can. In fact, only eight of the possible forty-eight connections had to be prohibited to prevent any possible contention arising. The connection points between routing wires were not evolved. An overview of the chromosome structure for one CLB and its associated routing is shown in Table 3.

Number of Genes	Type of Gene	Alleles / Gene (Cardinality)
16	LUT Input MUXes	27
2	Clock Input MUXes	11
2	SR Input MUXes	10
2	Clock Enable Input MUXes	11
4	Other Input MUXes	4
64	LUT Configuration	2
48	Other CLB Logic	1-3
8	Output signal MUX	13
40	Output MUX to single switches	2

Table 3. Overview of the genotype for one CLB.

A small rectangle of the chip was selected for evolution. The cells on the edges of this area were further restricted. They were only allowed to use the routing connections to the rest of the evolved area or circuit input wires, and not outside. Inputs were fed directly into LUTs on the west edge. For instance all four LUTs on the southwest corner were restricted to using the A0 and B0 inputs, the carry input and the remaining input for each LUT was restricted to west and north connections. All other inputs to the CLB were also restricted to the west and north. The output of the evolved area was taken from hex lines on the west edge, which were not evolved. This led to a chromosome of 604 genes for a 2x2 evolved area. A diagram of the inputs to and outputs from the evolved area is shown in Fig. 8.

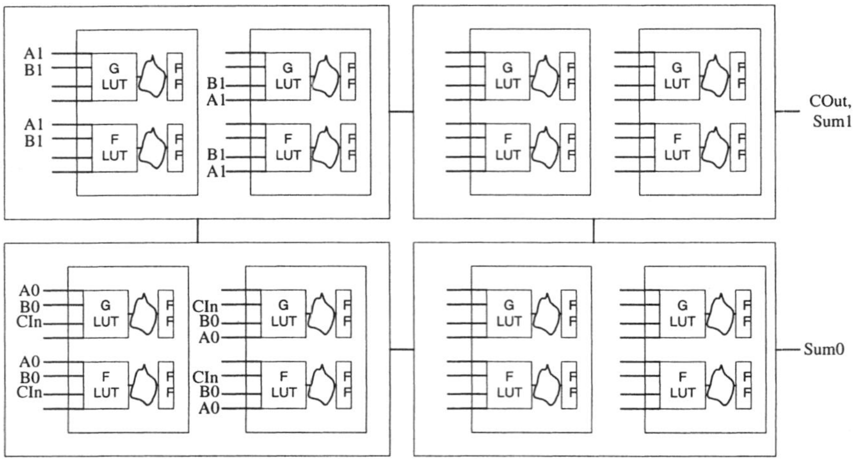

Fig. 8. Inputs and outputs to the evolved area. Unlabelled LUT inputs were completely under evolutionary control.

One-point crossover was used as opposed to uniform, following recent work by Vassilev and Miller [94] on the suitability of uniform crossover to three bit multiplier landscapes. The population size was also increased to 50. All other parameters were set as in the first experiment.

Unlike the experiment in simulation described in the first case study, the solutions were not constrained to combinational circuits, as the representation allowed a feedback. Working at low abstractions it may be necessary to specify generalization that is normally hidden by higher abstractions, as discussed in section 3.3. In order to ensure generalization across any order of input sequences the order of presentation was randomized for each evaluation. The genetic representation allows for circuits that may not generate the same fitness when evaluated twice - the outputs may exhibit dynamical variations unrelated to the inputs. Although such circuits are not useful final solutions they may contain valuable information about how to solve part of the problem, or how to traverse the fitness landscape. Because of this each individual was evaluated five times, and its worst fitness selected rather than discarding these solutions.

Initial experiments suffered from the speed it takes to reconfigure the FPGA. With a bitstream of over 300k configuration of each individual took around 1 second. Consequently the number of evaluations that could be carried out in a reasonable timeframe were low.[2] The work in simulation by both Miller and us suggested that large numbers of evaluations are needed, and this was confirmed by failure to evolve fully functional two bit adders within 1000 generations of two test runs. As we were evolving only a small area of the FPGA we had two options open to us – either batch multiple evaluations together in one configuration or use partial reconfiguration to reduce the bus traffic. As discussed earlier the most elegant solution to this problem is the use of partial reconfiguration. This is the approach we took, giving us an evaluation speedup of an order of magnitude. Simulation in the much more abstract logical space used in the first case study was still faster to complete the genotype-phenotype map and evaluation of the intrinsic evolution.

4.2.3 Results

10 runs were carried out for a 2x2 cell. Overall results are shown below. Fitnesses and deviations have again been scaled to 100. The mean number of generations to find a perfect solution and the corresponding standard deviation ignore the run that did not find a perfect solution. The average time to find a perfect solution was around 3 hours and 45 minutes using a 433 MHz Celeron PC.

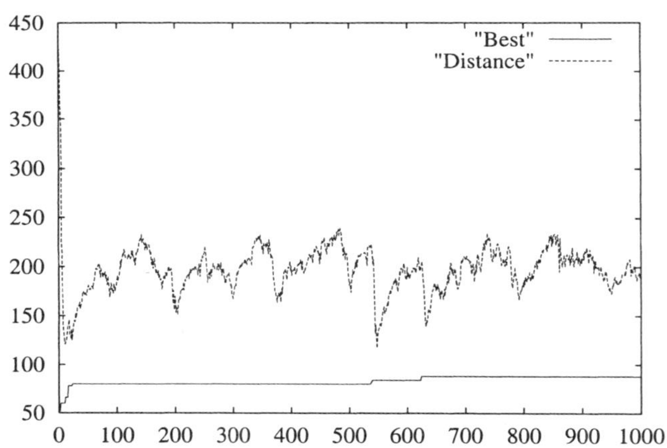

Fig. 9. Best fitness and average Euclidean distance between members of the population against generations for the first 1000 generations of the initial run.

---

[2] Note that this highlights the misconception that extrinsic evaluation is slower than intrinsic. If a logical abstraction is used, evaluation in simulation on a PC can be very fast as the microprocessor is designed to carry out fast logical calculations.

Mean Fitness of Best Solutions	Std. Dev. of Best Solutions	% of Runs With Perfect Solutions	Mean Gens to Find Perfect	St. Dev of Gens to Find Perfect
99.58	1.26	90	2661	2169

Table 4. Results from 10 runs of intrinsic 2-bit adder evolution.

4.2.4 Discussion

The results achieved from these experiments show an improvement in all statistics over the results carried out in simulation, even though the size of the space that the search was conducted across is much greater. Hence the combination of biases mean this space is more tractable to search. However it is difficult to say much more than this. The biases of the two experiments are very different - there were variations in representation, mapping and ordering biases. One point that should be made is that the Virtex architecture is rich in resources suitable for creating adders, which is likely to help evolvability. The LUT is a very useful primitive for generating logical behaviors, especially when the behavior is specified in truth tables. Additionally, logic is present within each Virtex CLB intended for performing logical carrys. In the spirit of working at a low level of abstraction that we have advocated during this discussion, our representation allows much of this logic to be manipulated by the algorithm.

Hollingworth that Tyrell have also evolved two bit adders (but without carry) intrinsically using Virtex [28]. In this case they used a fixed feed-forward routing structure and evolved only the bits of six LUTs. Such a representation cannot make use of the additional carry logic, which many of the two bit adders evolved here do. Hence we see that relaxing abstraction constraints may facilitate design innovation as discussed in section 3.2.

We should expect that such relaxations mean more work for the algorithm. Comparison with Hollingworth that Tyrell's results support this. They evolved a population of 100 chromosomes using a genetic algorithm with one point crossover and mutation. Their results showed a 100% success rate, and a mean and standard deviation of 808 and 259 generations respectively to find a perfect solution. What is most revealing is that such a huge increase in space does not appear to have been accompanied by a huge decrease in algorithm performance, rather only a moderate decrease. It was noted in section 3.4 that evolvability does not necessarily decrease with relaxation of constraints, and in some cases increases. So again this can be taken as a sign that working in low abstraction spaces should be advocated.

However it should be noted that adder circuits are not the best problems to demonstrate innovation over a sequential search space. Such a space is more suited to problems that make use of the range of behaviors over the time dimension opened by the relaxation in abstraction, such as Thompson's tone discriminator [80]. More complex problems such as this are also likely to benefit from the use of dynamic representational biases such as developmental processes.

This experiment was carried out using parameters in line with neutral network theory, which was discussed in section 3.4. It can be seen from Fig. 9 that evolution continues long after the initial convergence of the population, which in this example occurs within the first few generations. In addition, improvements take place after long periods of no improvement, which may correspond to periods when the population is moving across neutral planes or ridges. These improvements are also associated with a sudden convergence of the population, which then gradually diverges again. Our observations are in line with the hypothesis that neutral networks permeate this kind of search space.

5 Summary

The problems of electronic circuit design are increasing as demand for improvements increases. In this review we have introduced a promising new type of solution to these difficulties - evolvable hardware. This emerging field exists at the intersection of electronic engineering, computer science and biology. The benefits brought about by evolvable hardware are particularly suited to a number of applications, including the design of low cost hardware, poorly specified problems, creation of adaptive systems, fault tolerant systems and innovation.

As research in this field accelerates, new methods of classifying the many strands of research must be found. Here we identified three viewpoints: the level of abstraction, the bias implementation and the hardware evaluation process. With these in mind, current research trends in evolvable hardware were reviewed and analyzed. In particular, the research focusing on innovation, evolvability and platforms were described.

Finally, this work presented two case studies - one evolved at a logic netlist level and evaluated extrinsically, and one evolved at a device specific netlist level, evaluated intrinsically. The problem selected for both was the evolution of a two bit adder. These demonstrated the ability of evolution to successfully find solutions using a low abstraction and illustrated how such an approach can allow innovation in addition to automation.

Evolvable hardware is still a young field. It does not have all the answers to the problems of circuit design and there are still many difficulties to overcome. Nevertheless, these new ideas may be one of the brightest and best hopes for the future of electronics.

Acknowledgments

The authors would like to thank Peter Rounce for his insights and advice. Timothy G. W. Gordon is supported by an EPSRC studentship.

References

1. Andersen P. (1998) Evolvable hardware: Artificial evolution of hardware circuits in Simulation and Reality, M.Sc. Thesis, University of Aarhus, Denmark.
2. Arslan T. and Horrocks D.H. (1995), The design of analogue and digital filters using genetic algorithms, *Proc. of the 15th SARAGA Colloquium on Digital and Analogue Filters and Filtering Systems*, pp. 2/1 – 2/5, London, U.K.
3. Banzhaff W., Nordin P., Keller E., and Francone F.D. (1998), *Genetic Programming*, Morgan-Kaufmann, San Francisco, CA, U.S.A.
4. Bentley P.J. and Kumar S. (1999), Three ways to grow designs: A comparison of evolved embryogenies for a design problem, *Proc. of the Genetic and Evolutionary Computation Conf.*, Orlando, FL, U.S.A., pp.35-43.
5. Brooks R. A. (1991), Intelligence without representation, *Artificial Intelligence J.*, **47**, pp. 139-159.
6. Cliff D., Harvey I., and Husbands P. (1993), Explorations in evolutionary robotics, *Adaptive Behaviour*, **2**, 1, pp.73-110.
7. Coello Coello C.A., Christiansen A.D and Hernández Aguirre A. (2000), Using evolutionary techniques to automate the design of combinational circuits, *International Journal of Smart Engineering System Design*, **2**, no. 4, pp. 229-314.
8. Damiani E., Liberali V. and Tettamanzi A.G.B. (2000), Dynamic optimisation of non-linear feed-forward circuits, *Proc. of the 3rd Int. Conf. on Evolvable Systems*, Edinburgh, U.K., pp. 41-50.
9. de Garis H. (1994), An Artificial Brain: ATR's CAM-Brain Project Aims to Build/Evolve an Artificial Brain with a Million Neural Net Modules Inside a Trillion Cell Cellular Automata Machine, *New Generation Computing J.*, **12**, no. 2, pp. 215-221.
10. de Jong, K.A. (2001), *Evolutionary Computation*, MIT Press, Cambridge, MA, U.S.A.
11. Flockton S.J. and Sheehan K. (1999), A system for Intrinsic Evolution of Linear and Non-linear Filters, *Proc. of the 1st NASA/DoD Workshop on Evolvable Hardware*, Pasadena, CA, U.S.A, pp.93-100.
12. Fukunaga A. and Stechert A. (1998), Evolving nonlinear predictive models for lossless image compression with genetic programming, *Proc. of the 3rd Annual Genetic Programming Conf.*, Madison, WI, U.S.A, pp. 95-102.
13. Girau B., Marchal P., Nussbaum P., and Tisserand A. (1999), Evolvable platform for array processing: A one-chip approach, *Proc. of the 7th Int. Conf. on Microelectronics for Neural, Fuzzy and Bio-inspired Systems*, Granada, Spain, pp. 187-193.
14. Göckel N., Drechsler R. and Becker B. (1997), A multi-layer detailed routing approach based on evolutionary algorithms, *Proc. of the IEEE Int. Conf. on Evolutionary Computation*, Indianapolis, IN, U.S.A., pp. 557-562.
15. Goldberg D.E. (1989), *Genetic Algorithms in Search, Optimization and Machine Learning*, Addison-Wesley, Reading, MA, U.S.A.

16. Goldberg D.E., Deb K., and Korb B. (1991), Don't worry, be messy, *Proc. of the 4th Int. Conf. on Genetic Algorithms and their Applications*, San Diego, CA, U.S.A., pp. 24-30.
17. Gordon D.F. and des Jardins M. (1995), *Machine Learning J.*, **20**, pp. 1-17.
18. Grimbleby J.B. (2000), Automatic analogue circuit synthesis using genetic algorithms, *IEE Proc. - Circuits Devices, Systems*, **147**, no. 6, pp. 319-323.
19. Hamilton A., Papathanasiou K., Tamplin M., and Brandtner T. (1998), Palmo: Field programmable analogue and mixed-signal VLSI for evolvable hardware, *Proc. of the 2nd Int. Conf. on Evolvable Systems*, Lausanne, Switzerland, pp. 335-344.
20. Harvey I. (1991), Species adaptation genetic algorithms: The basis for a continuing SAGA, *Proc. of the 1st European Conf. on Artificial Life*, Paris, Franc, pp. 346-354.
21. Harvey I and Thompson A. (1996), Through the labyrinth evolution finds a way: A silicon ridge, *Proc. of the 1st Int. Conf. on Evolvable Systems*, Tsukuba, Japan, pp. 406-422.
22. Hemmi H., Mizoguchi J., and Shimohara K. (1996), Evolving large scale digital circuits, *Proc. of the 5th Int. Workshop on the Synthesis and Simulation of Living Systems*, Nara, Japan, pp. 168-173.
23. Higuchi T., Iba H., and Manderick B. (1994), Evolvable hardware, in *Massively Parallel Artifical Intelligence*, MIT Press, Cambridge, MA, U.S.A., pp. 398-421.
24. Higuchi T., Iwata M., Kajitani I., Iba H., Hirao Y., Manderick B., and Furuya T. (1996), Evolvable hardware and its applications to pattern recognition and fault-tolerant systems, in *Towards Evolvable Hardware: The Evolutionary Engineering Approach*, Sanchez E. and Tomassini M. (Eds.), Springer-Verlag, Berlin, Germany, pp. 118-135.
25. Hirst A.J. (1996), Notes on the evolution of adaptive hardware, *Proc. of Adaptive Computing in Engineering Design and Control*, Plymouth, UK, pp. 212-219.
26. Holdsworth B. (1993), *Digital Logic Design*, Butterworth-Heinemann, Oxford, UK.
27. Hollingworth G., Smith S., and Tyrrell A. (2000), The safe intrinsic evolution of Virtex devices, *Proc. of the 2nd NASA/DoD Workshop on Evolvable Hardware*, Palo Alto, CA, U.S.A.
28. Hollingworth G., Smith S. and Tyrell A. (2000), The intrinsic evolution of Virtex devices through Internet reconfigurable logic, *Proc. of the 3rd Int. Conf. on Evolvable Systems*, Edinburgh, UK, pp. 72-79,
29. Huynen M.A., Stadler P.F., and Fontana W. (1996), Smoothness within ruggedness: The role of neutrality in adaptation, *Proc. of the National Academy of Science*, **93**, 397-401.
30. Heyworth K. (1998), The "modeling clay" approach to bio-inspired electronic Hardware, *Proceedings of the Second International Conference on Evolvable Systems: From Biology to Hardware*, Lausanne, Switzerland, pp. 248-255.
31. Imamura K., Foster J. A. and Krings A.W. (2000), The test vector problem and limitations to evolving digital circuits, *Proc. of the 2nd NASA/DoD Workshop on Evolvable Hardware*, Palo Alto, CA, U.S.A, pp.75-79.
32. Iwata M., Kajitani I., Yamada H., Iba H., and Higuchi T. (1996), A pattern recognition system using evolvable hardware, *Proc. of the 4th Int. Conf. on Parallel Problem Solving from Nature*, Berlin, Germany, pp. 761-770.
33. Job D., Shankararaman V., and Miller J.F. (1999), Hybrid AI techniques for software design, *Proc. of the 11th Int. Conf. on Software Engineering and Knowledge Engineering*, Kaiserslautern, Germany, pp. 315-319.
34. Kajitani I., Hoshino T., Iwata M., and Higuchi T. (1996), Variable length chromosome GA for evolvable hardware, *Proc. of the 3rd Int. Conf. on Evolutionary Computation*, Nagoya, Japan, pp. 443-447.

35. Kajitani I., Hoshino T., Nishikawa D., Yokoi H., Nakaya S., Yamauchi T., Inuo T., Kajihara N., Iwata M., Keymeulen D. and Higuchi T. (1998), A gate-level EHW chip: implementing GA operations and reconfigurable hardware on a single LSI, *Proc. of the 2nd Int. Conf. on Evolvable Systems*, Lausanne, Switzerland, pp. 1-12.
36. Kalganova T., Miller J.F., and Lipnitskaya N., (1998), Multiple valued combinational Circuits Synthesised using Evolvable Hardware Approach, *Proc. of the 7th Workshop on Post-Binary Ultra Large Scale Integration Systems*, Fukuoka, Japan.
37. Kalganova T. and Miller J.F. (1999), Evolving more efficient digital circuits by allowing circuit layout evolution and multi-objective fitness, *Proc. of the 1st NASA/DoD Workshop on Evolvable Hardware*, Pasadena, CA, U.S.A, pp. 54-63.
38. Kauffman S. and Levin S. (1987), Towards a general theory of adaptive walks on rugged landscapes, *J. of Theoretical Biology*, **128**, pp.11-45.
39. Keymeulen D., Iwata M., Kuniyoshi Y. and Higuchi T. (1998), Comparison between an off-line model-free and an on-line model-based evolution applied to a robotics navigation system using evolvable hardware, *Proc. of the 6th Int. Conf. on Artificial Life*, Los Angeles, CA, U.S.A. pp.109-209.
40. Kitano H. (1998), Building complex systems using developmental process: An engineering approach, *Proc. of the 2nd Int. Conf. on Evolvable Systems*, Lausanne, Switzerland, pp. 218-229.
41. Koza J., Bennett F. H, III, Andre D., and Keane M.A. (1999), *Genetic Programming III*, Morgan-Kauffmann, San Francisco, CA, U.S.A.
42. Langeheine J., Folling S., Keir K., and Schemmel J. (2000), Towards a silicon primordial sourp: A fast approach to hardware evolution with a VLSI transistor array, *Proc. of the 3rd Int. Conf. on Evolvable Systems*, Edinburgh, UK, pp. 123-132.
43. Layzell P. (1999), Reducing hardware evolution's dependency on FPGAs, *Proc. of the 7th Int. Conf. on Microelectronics for Neural, Fuzzy and Bio-Inspired Systems*, Granada, Spain, pp. 171-178.
44. Layzell P. (1999), Inherent qualities of circuits designed by artificial evolution: A preliminary study of populational fault tolerance, *Proc. of the 1st NASA/DoD Workshop on Evolvable Hardware*, Pasadena, CA, U.S.A, pp. 85-86.
45. Layzell P. and Thompson A. (2000), Understanding inherent qualities of evolved circuits: Evolutionary history as a predictor of fault tolerance, *Proc. of the 3rd Int. Conf. on Evolvable Systems*, Edinburgh, UK, pp. 133-144.
46. Levi D. and Guccione S.A. (1999), GeneticFPGA: Evolving stable circuits on mainstream FPGA devices, *Proc. of the 1st NASA/DoD Workshop on Evolvable Hardware*, Pasadena, CA, U.S.A, pp. 12-17.
47. Levi D. (2000), HereBoy: A fast evolutionary algorithm, *Proc. of the 2nd NASA/DoD Workshop on Evolvable Hardware*, Palo Alto, CA, U.S.A, pp. 17-24.
48. Liu W., Murakawa M., and Higuchi T. (1996), ATM cell scheduling by function level evolvable hardware, *Proc. of the 1st Int. Conf. on Evolvable Systems*, Tsukuba, Japan, pp. 180-192.
49. Lohn J.D. and Columbano S.P. (1998), Automated analog circuit synthesis using a linear representation, *Proc. of the 2nd Int. Conf. on Evolvable Systems*, Lausanne, Switzerland, pp. 125-133.
50. Lohn J.D., Haith G.L., Colombano S.P. and Stassinopoulos D. (1999), A comparison of dynamic fitness schedules for evolutionary design of amplifiers, *Proc. of the 1st NASA/DoD Workshop on Evolvable Hardware*, Pasadena, CA, U.S.A, pp.87-92.
51. Louis S.J. and Rawlins G.J.E. (1991), Designer genetic algorithms: Genetic algorithms in structure design, *Proc. of the 4th Int. Conf. on Genetic Algorithms*, San Diego, CA, U.S.A., pp. 53-60.

52. Manovit C, Aporntewan C., and Chongstitvatana P. (1998), Synthesis of synchronous sequential logic circuits from partial input/output sequences, *Proc. of the 2nd Int. Conf. on Evolvable Systems*, Lausanne, Switzerland, pp.98-105.
53. Masner J., Cavalieri J., Frenzel J., and Foster J. (1999), Representation and robustness for evolved sorting networks, *Proc. of the 1st NASA/DoD Workshop on Evolvable Hardware*, Pasadena, CA, U.S.A, pp.255-261.
54. Mazumder P. and Rudnick E. M. (1999), *Genetic Algorithms for VLSI Design, Layout and Test Automation*, Prentice-Hall, Upper Saddle River, NJ, U.S.A.
55. Miller J.F., Kalganova T., Lipnitskaya N., and Job D. (1999), The genetic algorithm as a discovery engine: Strange circuits and new principles, *Proc. of the AISB Symposium on Creative Evolutionary Systems*, Edinburgh, UK, pp. 65-74.
56. Miller J.F., Job D., and Vassilev V. K. (2000), Principles in the evolutionary design of digital circuits – Part I, *Genetic Programming and Evolvable Machines*, **1**, no. 1/2, pp. 7-35.
57. Miller J.F., Job D., and Vassilev V. K. (2000), Principles in the evolutionary design of digital circuits – Part II, *Genetic Programming and Evolvable Machines*, **1**, no. 3, pp. 259-288.
58. Miller J.F. and Thomson P. (1995), Combinational and sequential logic optimisation using genetic algorithms, *Proc. of the 1st Int. Conf. on Genetic Algorithms in Engineering Systems: Innovations and Applications*, Sheffield, UK, pp. 34-38.
59. Miller J.F. and Thomson P. (1998), Aspects of digital evolution: Evolvability and architecture, *Proc. of the 5th Int. Conf. on Parallel Problem Solving in Nature*, Amsterdam, The Netherlands, pp. 927-936.
60. Miller J.F. and Thomson P. (1998), Aspects of digital evolution: Geometry and learning, *Proc. of the 2nd Int. Conf. on Evolvable Systems*, Lausanne, Switzerland, pp. 25-35.
61. Miller J.F. and Thomson P. (1998), Evolving digital electronic circuits for real-valued function generation using a genetic algorithm, *Proc. of the 3rd Annual Conf. on Genetic Programming*, San Francisco, CA, U.S.A, pp. 863-868.
62. Miller J.F., Thomson P., and Fogarty T.C. (1997), Designing electronic circuits using evolutionary algorithms. Arithmetic circuits: A case study, in *Genetic Algorithms and Evolution Strategies in Engineering and Computer Science: Recent Advancements and Industrial Applications*, Quagliarella D., Periaux J., Poloni C., and Winter G. (Eds.), John Wiley & Sons, London, UK.
63. Mitchell M. (1998), *An Introduction to Genetic Algorithms*, MIT Press, Cambridge MA, U.S.A.
64. Mitchell T.M. (1997), *Machine Learning*, McGraw-Hill, London, UK.
65. Moreno J.M., Madrenas J., Faura J., Canto E., Cabestany J., and Insenser J.M. (1998), Feasible evolutionary and self-repairing hardware by means of the dynamic reconfiguration capabilities of the FIPSOC devices, *Proc. of the 2nd Int. Conf. on Evolvable Systems*, Lausanne, Switzerland, pp. 345-355.
66. Murakawa M., Yoshizawa S., Adachi T., Suzuki S., Takasuka K., Iwata M., and Higuchi T. (1998), Analog EHW chip for intermediate frequency filters, *Proc. of the 2nd Int. Conf. on Evolvable Systems*, Lausanne, Switzerland, pp. 134-143.
67. Murakawa M., Yoshizawa S., Kajitani I., Furuya T., Iwata M., and Higuchi T. (1996), Hardware evolution at function level, *Proc. of the 4th Conf. on Parallel Problem Solving from Nature*, Berlin, Germany, pp. 62-71.
68. Murakawa M., Yoshizawa S., Kajitani I., Yao X., Kajihara N., Iwata M., and Higuchi T. (1999), The GRD chip: Genetic reconfiguration of DSPs for neural network processing, *IEEE Trans. on Computers*, **48**, no. 6, pp. 628-639.

69. Ortega C. and Tyrrell A. (1999), Biologically inspired fault-tolerant architectures for real-time control applications, *Control Engineering Practice*, **7**, no. 5, pp. 673-678.
70. Pollack J.B., Lipson H., Ficici S., Funes P., Hornby G., and Watson R. (2000), Evolutionary techniques in physical robotics, *Proc. of the 3rd Int. Conf. on Evolvable Systems*, Edinburgh, UK, pp. 175-186.
71. Rendell, L. (1987), Similarity-based learning and its extensions, *Computational Intelligence*, **3**, pp. 241-266.
72. Rosenman, M. (1997), The generation of form using an evolutionary approach, in *Evolutionary Algorithms in Engineering Applications,* Dasgupta D. and Michalewicz (Eds.), Springer-Verlag, pp. 69-86.
73. Rumelhart, D.E., Widrow B., and Lehr M. (1994), The basic ideas in neural networks, *Communications of the ACM*, **37**, no. 3, pp. 87-92.
74. Salami M., Murakawa M. and Higuchi T. (1996), Data compression based on evolvable hardware, *Proc. of the 1st Int. Conf. on Evolvable Systems*, Tsukuba, Japan, pp. 169-179.
75. Salami M., Sakanashi H., Tanaka M., Iwata M., Kurita T., and Higuchi T. (1998), On-line compression of high precision printer images by evolvable hardware, *Proc. of the Data Compression Conf.*, Los Alamitos, CA, U.S.A. pp. 219-228.
76. Sechen C. (1988), *VLSI Placement and Global Routing Using Simulated Annealing*, Kluwer Academic Publishers, Boston MA, U.S.A.
77. Stoica A., Fukunaga A., Hayworth K., and Salazar-Lazaro C. (1998), Evolvable hardware for space applications, *Proc. of the 2nd Int. Conf. on Evolvable Systems*, Lausanne, Switzerland, pp. 166-173.
78. Stoica A., Keymeulen D., Tawel R., Salazar-Lazaro C., and Li W. (1999), Evolutionary experiments with a fine-grained reconfigurable architecture for analog and digital CMOS Circuits, *Proc. of the 1st NASA/DoD Workshop on Evolvable Hardware*, Pasadena, CA, U.S.A, pp. 76-85.
79. Stoica A., Zebulum R., and Keymeulen D., (2000), Mixtrinsic evolution, *Proc. of the 3rd Int. Conf. on Evolvable Systems*, Edinburgh, UK, pp. 208-217.
80. Thompson A. (1996), Silicon evolution, *Proc. of the 1st Annual Conf. on Genetic Programming*, Stanford, CA, U.S.A., pp. 444-452.
81. Thompson A. (1997), Evolving inherently fault-tolerant systems, *Proc. of Institution of Mechanical Engineers*, **211**, Part I, pp. 365-371.
82. Thompson A. (1998), *Hardware Evolution*, Springer-Verlag, London, UK.
83. Thompson A. (1998), On the automatic design of robust electronics through artificial evolution, *Proc. of the 2nd Int. Conf. on Evolvable Systems*, Lausanne, Switzerland, pp. 13-25.
84. Thompson A., Harvey I. and Husbands P. (1996), Unconstrained evolution and hard consequences, in *Towards Evolvable Hardware: the evolutionary engineering approach*, Sanchez E. and Tomassini M. (Eds.), Springer-Verlag, Berlin, Germany, pp. 136-165.
85. Thompson A. and Layzell P. (1999), Analysis of unconventional evolved electronics, *Communications of the ACM*, **42**, no. 4, pp. 71-79.
86. Thompson A. and Layzell P. (2000), Evolution of robustness in an electronics design, *Proc. of the 3rd Int. Conf. on Evolvable Systems*, Edinburgh, UK, pp. 218-228.
87. Thompson A. and Wasshuber C. (2000), Design of single electron systems through artificial evolution, *Int. J. of Circuit Theory and Applications*, **28**, no. 6, pp. 585-599.
88. Torresen J. (1998), A divide and conquer approach to evolvable hardware, *Proc. of the 2nd Int. Conf. on Evolvable Systems*, Lausanne, Switzerland, pp.57-65.

89. Torresen J. (2000), Possibilities and limitations of applying evolvable hardware to real-world applications, *Proc. of the 10th Int. Conf. on Field Programmable Logic and Applications*, Villach, Austria, pp. 230-239.
90. Torresen J. (2000), Scalable evolvable hardware applied to road image recognition, *Proc. of the 2nd NASA/DoD Workshop on Evolvable Hardware*, Palo Alto, CA, U.S.A, pp. 245-252.
91. Tufte G. and Haddow P.C. (1999), Prototyping a GA pipeline for complete hardware evolution, *Proc. of the 1st NASA/DoD Workshop on Evolvable Hardware*, Pasadena, CA, U.S.A, pp. 18-25.
92. Vassilev V. and Miller J.F. (2000), Embedding landscape neutrality to build a bridge from the conventional to a more efficient three-bit multiplier circuit, *Proc. of the Genetic and Evolutionary Computation Conf.*, Las Vegas, NV, U.S.A.
93. Vassilev V. and Miller J.F. (2000), The advantages of landscape neutrality in digital circuit evolution, *Proc. of the 3rd Int. Conf. on Evolvable Systems*, Edinburgh, U.K., pp. 252-263.
94. Vassilev V., Miller J.F., and Fogarty T.C. (1999), On the nature of two-bit multiplier landscapes, *Proc. of the 1st NASA/DoD Workshop on Evolvable Hardware*, Pasadena, CA, U.S.A, pp. 36-45.
95. Yao X. and Higuchi T. (1996), Promises and challenges of evolvable hardware, *Proc. of the 1st Int. Conf. on Evolvable Systems*, Tsukuba, Japan, pp. 55-78.
96. Yih J.S. and Mazumder P. (1990), A neural network design for circuit partitioning, *IEEE Trans. on Computer Aided Design*, **9**, no. 10, pp. 1265-1271.
97. Zebulum R.S., Aurélio Pacheo M., and Vellasco M. (1996), Evolvable systems in hardware design: Taxonomy, survey and applications, *Proc. of the 1st Int. Conf. on Evolvable Systems*, Tsukuba, Japan, pp. 344-358.
98. Zebulum R.S., Aurélio Pacheo M., and Vellasco M. (1997) Increasing length genotypes in evolutionary electronics, *Proc. of the 7th Int. Conf. on Genetic Algorithms*, East Lansing, MI, U.S.A.
99. Zebulum R.S., Aurélio Pacheo M., and Vellasco M. (1998), Analog circuits evolution in extrinsic and intrinsic modes, *Proc. of the 2nd Int. Conf. on Evolvable Systems*, Lausanne, Switzerland pp. 154-165.

Category Classification Using Neural Networks

Toru Fujinaka, Michifumi Yoshioka, and Sigeru Omatu

Osaka Prefecture University

Keywords: *neural networks, learning vector quantization, layered neural networks, self-organization, remote sensing, land cover mapping, synthetic aperture radar, maximum likelihood method, competitive neural networks, Jeffries-Matusita distance.*

1 Introduction

Remote sensing data includes reflectional and radiational characteristics of naturally occurring features found on earth. It is possible to categorize the data by analyzing its spectral information. The categorization employs a pattern recognition method, where a category corresponds to a pattern. There are two types of recognition methods known, i.e., supervised and unsupervised methods. The former method picks up some training data from the remote sensing data in advance and defines a category for each of them. This method then classifies the whole data based on a specific feature of each category. The latter method does not require selection of the training data, and this method classifies the remote sensing data based on similarities among the pixels [1].

The maximum likelihood (ML) method has often been applied to the classification of remote sensing data as a supervised classification method [2], [3]. The ML method uses a statistical approach based on the Bayesian theorem. It can be applied theoretically to any probability distribution. In real data processing, however, the discrimination function works under the assumption that the probability distribution for each category is an n-dimensional normal distribution where n denotes the number of spectral bands. The categories are determined from the land cover information. The probability distribution of real data for each category does not always follow normal distribution, which is essential to the conventional ML method. Therefore, a non-parametric classification method becomes important for real data processing. Recently, a method using the multi-layered neural network has been applied to the analysis of remote sensing data [4], [5], [6]. The neural network is trained by the well-known back-propagation (BP) method [4], [7]. However, in the neural network approach to remote sensing analysis, the convergence of a learning process is sensitive to the selection of the training data set. Furthermore, the learning method often requires a large number of iterations and much computational time. Since the back-propagation method is a black-box approach, it is difficult to give physical meaning to the weights connecting the neurons.

The remote sensing techniques have various advantages in monitoring the global environments on the earth surface in a short time. However, they also have a disadvantage that a successful monitoring depends on the weather condition. Rainy days are unfavorable in general, which becomes especially serious in the analysis of a tropical rain forest. Recently, multi-frequency and multi-polarimetric Synthetic Aperture Radar (SAR) data can be obtained by using airborne or spaceborne SAR sensor systems. Calibration techniques for the SAR data have been established, so that it may be used for such applications as terrain detection and environmental research. With several basic approaches, the SAR data classification methods have been gradually developed. These methods classify the SAR data into specific categories. The data observed by the quadruple polarization, i.e., the combination of horizontal and vertical polarization in the transmitting and receiving signals, includes the polarization backscattering coefficients and relative phases between polarizations [8], [9]. Thus, better classification accuracy may be expected by using the polarimetric SAR data instead of mono-polarization data.

Supervised classification methods for the polarimetric SAR data can be divided into statistical and neural network approaches. The Bayesian methods reported in [10], [11], and [12] belong to the former approach. In recent years, the latter approach has been recognized for its classification performance. Algorithms using layered neural networks have been considered in [13], [14], and [15]. Algorithms using competitive neural networks have been considered in [2], [3], [16], and [17].

In general, the layered neural networks have a problem that they need much learning computation time in comparison with competitive neural networks. When adopting competitive neural networks to the classification of remote sensing data, it is difficult to accomplish sufficient classification accuracy using only basic learning algorithms. Hence, Hara et al. [2] proposed a classification method that incorporates both the neural network and the maximum likelihood methods. Ito and Omatu [17] use a classification method that integrates several competitive neural networks trained by a conscience algorithm. Although the classification accuracy is improved in these methods, they are more complex than the basic learning algorithms and considerable adjustments are often needed in real applications. Furthermore, a feature vector must be selected appropriately before the classification phase. Otherwise, satisfactory accuracy cannot be obtained even if all the observation values are applied. In Chen et al. [13], experiments using airborne SAR (AIRSAR) data for several types of feature vectors have been carried out, and their accuracies have been compared. Unfortunately, there exists no detailed discussion up to date concerning a method of selecting a suitable combination.

In Section 2, we discuss a new supervised classification method based on the self-organizing neural network in order to easily select a training data set, which will have reliable convergence with the neural network. Then we propose a category classification method, which uses a self-organizing neural network, and a k-nearest neighbor method. In Section 3, we explain a classification method using a competitive neural network trained by only two Learning Vector Quantization (LVQ) algorithms [18]–[20]. We discuss a method in which a suitable feature vector is selected using the JM (Jeffries-Matusita) distance [1]. Moreover, we introduce a pseudo-relative phase between polarimetries in order to obtain higher classification accuracy. The polarimetric SAR data observed by Shuttle Imaging Radar-C (SIR-C) is employed for the experiments [21]. The proposed and conventional methods are compared with respect to classification accuracy and quality of the resulting images. The neural network approach can be applied to more general area of image classification, while its effectiveness may depend on the specific features of the given data.

2 LANDSAT Data Classification by Neural Networks

This section is concerned with the LANDSAT data analysis where the data are measured using optical sensors.

2.1 Implementation of Neural Networks

The basic self-organizing neural network consists of two layers, i.e., an input layer and a competitive layer, both being completely interconnected as in Fig. 1 ([7], [18]–[20]). In this figure, w_{ij} denotes a connection weight from neuron j to neuron i, and o_i is the output of neuron i. The remote sensing data x_j is the input to neuron j in the input layer, and the output o_j of this neuron is distributed to all the neurons in the competitive layer. In the competitive layer, a neuron c is selected such that

$$E_c = \min_i E_i , \tag{1}$$

where

$$E_i = \sum_j (x_j - w_{ij})^2 . \tag{2}$$

The neuron c is called the winner neuron. After competing, the output from neuron c is one and the outputs from all the others are zero. We refer to the neuron c as the one that fired, and others that did not fire.

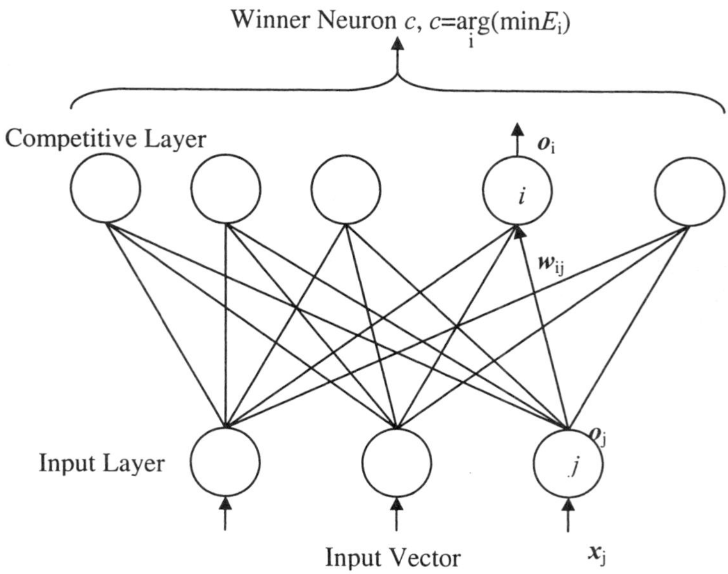

Fig. 1. Basic self-organizing neural network with two layers.

In competitive learning, we adjust connection weights w_{ci} (i=1, 2, ...) for a fixed c such that an error function $\sum_i E_i$ becomes smaller. This adjusting rule will be explained in detail in Section 2.2. The feature of a category ω_i (i=1, 2, ..., m) is represented by the neuron weights after competitive learning. The basic self-organizing neural network stated above is trained for a specific category, and the same training method will be applied to all the categories. Then we obtain a set of basic self-organizing neural networks whose number is equal to m. We propose a new type of neural network, whose structure is a combination of m basic self-organizing neural networks, for the purpose of category classification. The original remote-sensing data with unknown category is fed into the input layer of each block of the neural networks. After the neurons in all the competitive layers have competed with each other, we select k neurons in the order of minimum pattern distance between the input data and the neuron weights. The k neurons are referred to as the winner neurons, and this algorithm classifies the input data into one of the m categories where maximal number of the winner neurons are included.

2.2 Learning of the Network

The learning algorithm of the network explained above is based on a competitive learning rule of the self-organizing neural network. If this algorithm is applied to self-organization in the remote sensing data, what occurs is either only one specified neuron will fire or no reaction will take place at all. Therefore, we use the improved learning algorithm proposed by DeSieno [22]. This algorithm, which is regarded as the neuron's conscience, counts the firing frequency of each neuron and adjusts its firing rate. If reduces the firing rate of a neuron that fires often, while increasing the rate for less firing neurons. Network learning is achieved by feeding the training data whose category is known beforehand and by updating the neuron's weight. After completing the learning, neurons in the competitive layer for each block will fire with equal probability according to the input data.

Not only the weight vector $\mathbf{w}_i$ but also the firing frequency p_i is associated with the neuron i in the competitive layer. The firing frequency is updated according to the equation

$$p_i^{new} = p_i^{old} + B(o_i - p_i^{old}) \tag{3}$$

where the fired neuron has the value $o_i = 1$, the other resting neurons $o_i = 0$, and B is a constant value satisfying $0 < B \ll 1$. If the input training data range is [0,1], B will be a suitable small value in the order of 10^{-4}. Pattern distance d_i between input vector $\mathbf{x}$ and weight vector $\mathbf{w}_i$ in neuron i is

$$d_i = \|\mathbf{x} - \mathbf{w}_i\| - b_i \tag{4}$$

where b_i is called the conscience term related to firing frequency and is given by

$$b_i = C\left(\frac{1}{n} - p_i\right). \tag{5}$$

Here, n is the number of neurons in the competitive layer, and C is a positive constant. If p_i is smaller than the uniform firing probability $1/n$, b_i is positive and d_i becomes smaller than $\|\mathbf{x} - \mathbf{w}_i\|$. Thus, the conscience term b_i in (4) works in such a way that the firing frequency of the neurons in a category will become more even. After the neurons in the competitive layer compete with respect to i, a firing neuron c becomes

$$d_c \le d_i, \ ^{\forall}i. \tag{6}$$

The other neurons rest and the weight vector $\mathbf{w}_c$ associated with the neuron c is updated by

$$\mathbf{w}_c^{new} = \mathbf{w}_c^{old} + \alpha(\mathbf{x} - \mathbf{w}_c^{old}) \qquad (7)$$

for input $\mathbf{x}$, where α is a constant value, and its range is typically chosen in the range $[0.01, 0.3]$. The weight vectors in other resting neurons are not updated at all. The self-organizing type neural network is repeatedly trained by the same data until the weight vectors converge to some values.

2.3 The Classification Method

The classification algorithm is based on the k-nearest neighbor method in a hyper-spherical region. The distribution of reference vectors represents the feature of a category. Each input vectors is classified into a category represented by a reference vector.

The input vector should be classified into a category based on the highest likelihood among the reference vectors. However, it is difficult to directly estimate the likelihood for a category by computing the probability from the n-dimensional hyper-spherical region's size associated with a reference vector to which the input vector belongs. Hence, we suppose a classification method based on the pattern distance between the input vector and reference vectors instead of using the likelihood. For example, this method classifies an input vector into one of the two categories by computing the 2-dimensional distances between the input vector and each reference vector representing the categories. The input vector is classified into the category where the nearest reference vector to the input data belongs. In a real remote sensing data, many observation values are located close to category boundaries. However, successful classification is possible, if the probability distribution of each category is taken into consideration. Hence, we adopt the classification method based on k-nearest neighbor reference vectors for the remote sensing data. This method selects the k reference vectors in the order of minimum pattern distance and classifies an input data into a category to which most of the reference vectors belong.

If an input vector is located close to a category boundary, it becomes difficult to classify the input vector, as the winner neurons may be distributed in many categories. In such a case, the number k may be increased adaptively to admit more winner neurons. However, if k becomes larger than a predetermined value, the category for such an input data should be regarded as unknown [16], [17], [23], and [24].

2.4 Classification Results

The proposed classification method is applied to the LANDSAT Thematic Mapper (TM) data observed by LANDSAT 5 satellite on April 24, 1988. Study area is included in the observation image at path 110 and row 36. Classification analysis is carried out on a rectangular area (640×480 pixels) in Takamatsu, Kagawa Prefecture, Japan. Figure 2 shows the remote sensing image of the study area. Figure 3 shows the study map corresponding to the area of Fig. 2.

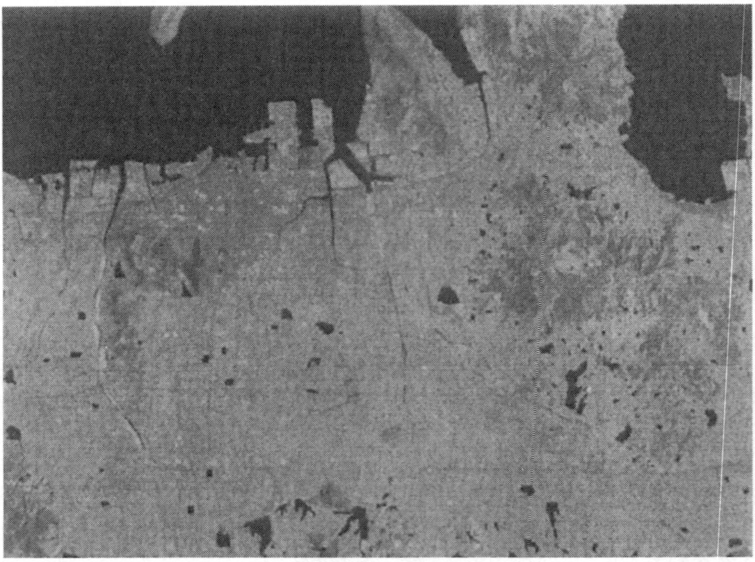

Fig. 2. Remote sensing image of LANDSAT-TM data (640×480 pixels).

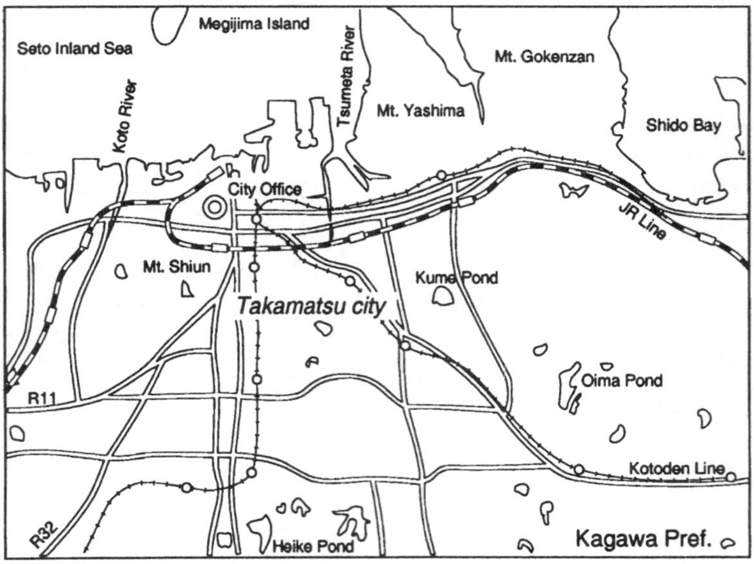

Fig. 3. Study map corresponding to the remote sensing image.

We define six categories, ω_1 : sea or river, ω_2 : pond, ω_3 : bare land, ω_4 : forest, ω_5 : farm, and ω_6 : density inhabited district. Taking the spectral feature of categories into consideration, we use three (Bands 3, 4, and 5) from seven bands of TM data. Next, the training data for each category has been picked up from the area where the correct category can be clearly distinguished. We have also generated an evaluation image by classifying all the pixels manually, using auxiliary information for actual land coverage.

Table 1. Parameters of combined self-organizing neural network.

Number of neurons in input layer	3
Number of neurons in competitive layers	15
Initial values of weight vectors	0.5
Learning iteration	10,000
Number of training data	150
Parameter α	0.03
Parameter B	0.0001
Parameter C	10.0

Table 1 shows experimental parameters of the combined self-organizing neural network. As these parameters are sensitive to the classification results, we must select them carefully. The learning iteration limit is less than 10,000 and the number of training data for each category is 150. A sub-network of each block has 3-input and 15-output neurons. The 3-input neurons correspond to the above three bands. The number of output neurons is n=15, which is equal to the number of training data divided by 10. This number was chosen from our experience. Consequently, the competitive layer has 90 neurons, which are 15 neurons times six categories. All the weight vectors are initialized to 0.5, which is selected in the range of [0.0, 1.0]. We set α, B, and C to constant values by trial and error.

In the learning stage, each sub-network is trained individually by using the training data. Initially, all neuron weights are concentrated to 0.5. After the 10,000 times of learning iterations, the weights showed convergence to various values, and they are used as reference vectors in the classification stage.

After completing the self-organization of each category, we evaluate the degree of vector quantization to investigate whether the appropriate reference vectors have been generated. As the degree of vector quantization, we adopt the entropy H defined by

$$H = -\sum_{i=1}^{n} q_i \log_2 q_i \tag{8}$$

where n is the number of neurons in the competitive layer per category and q_i is the firing probability of a neuron i. Therefore, the closer H is to maximum entropy $\log_2 n$, the better the vector quantization is achieved. The entropy for each category of $\omega_1 - \omega_6$ turned out to be 3.52, 3.78, 3.87, 3.88, 3.85, 3.84 for the training data, and 3.54, 3.82, 3.84, 3.88, 3.88, 3.83 for the test data, respectively. Since n = 15 and $\log_2 15 \approx 3.91$ in our case, the firing probabilities are close to uniform distribution, except for category ω_1. There is a shouldered-out neuron in the category ω_1, and it rarely fired.

In the next step, we classify the unknown pixels by using the learned neural network. Two types of k values are considered by trial and error, where k denotes the number of firing neurons in the competitive layer. One is to fix k=5 at the beginning, which is denoted by 5-NN. The other is to vary k adaptively within the range of three to six, which is denoted by A3-NN. For comparison with the above methods, we also apply two conventional methods, the usual ML method and the basic neural network method. Here, the latter is based on the back-propagation (BP) method of a three-layered (3-input, 6-hidden and 6-output) neural network. The same training data is used for the methods except for the BP method for the comparison of classification accuracy. For the BP method, the data was smoothed in order to achieve convergence. Figure 4 shows the result of classification using the A3-NN method, and Fig. 5 shows the corresponding result of using the BP method. The difference may not be obvious at a first glance, but the details are quite different, and it turns out that the A3-NN method gives better result in general. It has also been verified that the ponds (category ω_2) peculiar to this region are more correctly classified using the A3-NN method. Furthermore, the BP method suffers from more incorrectly classified pixels in the water area and shadow part of the mountain area.

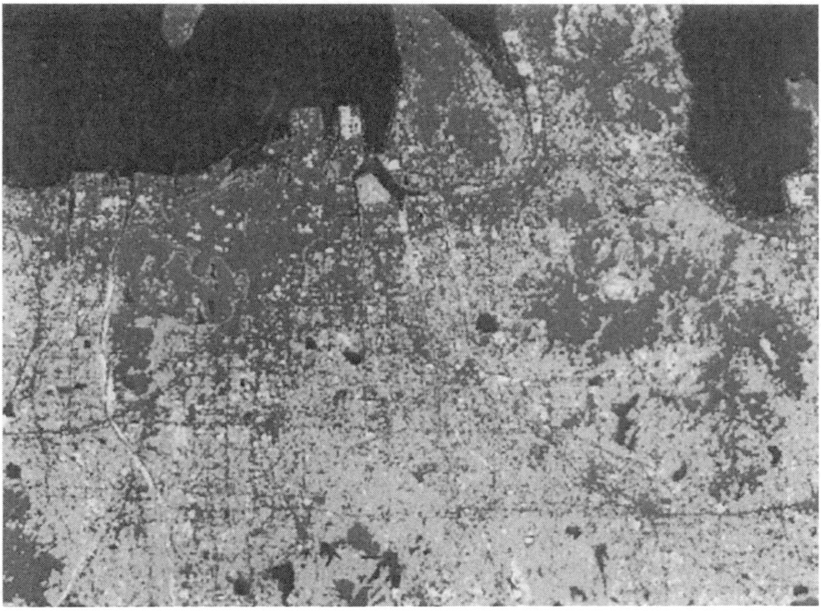

Fig. 4. Classification result using A3-NN type of self-organizing neural network.

Fig. 5. Classification result using 3-input, 6-hidden, 6-output BP method.

3 Polarimetric SAR Data Classification by Neural Networks

This section is concerned with the SAR data analysis in which the data are observed by measuring the active radar signals, while in Section 2 the passive sensing data were used. Here, the data belong to the range of microwave, and it is difficult to get the geographical scenary like a picture. Therefore, a kind of data mining approach should be used. In this section, we adopt various types of the LVQ methods to achieve this purpose.

3.1 The Classification Algorithm

We consider a classification algorithm using a competitive neural network similar to those discussed in Section 2. The competitive neural network is trained by the Learning Vector Quantization (LVQ) method [18]–[20]. The LVQ method will cyclically update the weight vectors so as to reward correct classifications and punish incorrect ones.

Kohonen [18]–[20] proposed the LVQ1, LVQ2.1, and OLVQ1 (Optimized-learning-rate LVQ1) algorithms for the LVQ method. We will first describe the LVQ1 learning algorithm. Let $\mathbf{w}_i$(t) denote a sequential value of $\mathbf{w}_i$ in the discrete-time, t=0, 1, 2, ..., where i=1, ... ,M. After the training data $\mathbf{x}$(t) is presented, competition occurs in the competitive layer where neuron c ultimately becomes the winner. Let $\mathbf{x}$(t) belong to a category ω_x and let neuron c be assigned to a category ω_c. The LVQ1 updates the weight vectors as follows:

i) $\omega_x = \omega_c$

$$\left.\begin{aligned} \mathbf{w}_c(t+1) &= \mathbf{w}_c(t) + \alpha(t)[\mathbf{x}(t) - \mathbf{w}_c(t)] \\ \mathbf{w}_i(t+1) &= \mathbf{w}_i(t), \quad i \neq c \end{aligned}\right\} \tag{9}$$

ii) $\omega_x \neq \omega_c$

$$\left.\begin{aligned} \mathbf{w}_c(t+1) &= \mathbf{w}_c(t) - \alpha(t)[\mathbf{x}(t) - \mathbf{w}_c(t)] \\ \mathbf{w}_i(t+1) &= \mathbf{w}_i(t), \quad i \neq c \end{aligned}\right\} \tag{10}$$

Here, $\alpha(t)$ denotes the learning rate such that $0 < \alpha(t) < 1$ and $\alpha(t) \to 0$ if $t \to \infty$. This learning algorithm tends to push weight vectors away from the decision surfaces of Bayes rule [25]. To accelerate the convergence of the network, the OLVQ1 updates the weight vectors by means of the individual learning rate in every neuron. Learning rate $\alpha_i(t)$ in a neuron i is updated as follows:

i) $\omega_x = \omega_c$

$$\left.\begin{aligned} \alpha_c(t+1) &= \frac{\alpha_c(t)}{1+\alpha_c(t)} \\ \alpha_i(t+1) &= \alpha_i(t), \quad i \neq c \end{aligned}\right\} \tag{11}$$

ii) $\omega_x \neq \omega_c$

$$\left.\begin{aligned} \alpha_c(t+1) &= \min\left(\frac{\alpha_c(t)}{1+\alpha_c(t)}, \alpha_0\right) \\ \alpha_i(t+1) &= \alpha_i(t), \quad i \neq c \end{aligned}\right\} \tag{12}$$

where the initial learning rate is given by $\alpha_i(0) = \alpha_0$, i=1, ... M. Thus, weight vectors in the center of the category will have rapidly decreasing learning rates and conversely those near category boundaries will have increasing rates.

The LVQ2.1 approximates the Bayes rule by adjusting category boundaries. It updates only the weight vectors in a window region. The window region is defined by a parameter w, $0 \leq w \leq 1$. E_u and E_v denote pattern distances from $\mathbf{x}$(t) to the nearest weight vector $\mathbf{w}_u$(t) and the second nearest weight vector $\mathbf{m}_v$(t), respectively. Either of the weight vectors, $\mathbf{m}_u$(t) or $\mathbf{m}_v$(t) belongs to the same category as $\mathbf{x}$(t) and the other belongs to a separate category. It is known that the window region is represented as

$$\min\left(\frac{E_u}{E_v}, \frac{E_v}{E_u}\right) > \frac{1-w}{1+w} \tag{13}$$

Then the LVQ2.1 updates the weight vectors as follows:

$$\left.\begin{aligned} \mathbf{m}_u(t+1) &= \mathbf{m}_u(t) + \alpha(t)[\mathbf{x}(t) - \mathbf{m}_u(t)] \\ \mathbf{m}_v(t+1) &= \mathbf{m}_v(t) + \alpha(t)[\mathbf{x}(t) - \mathbf{m}_v(t)] \end{aligned}\right\} \tag{14}$$

In this chapter, we employ the following learning approach:

[Step 1] Move the weight vectors roughly by LVQ1 or OLVQ1.

[Step 2] Tune up the weight vectors in category boundaries by LVQ2.1.

In Step 2, an improvement in the classification ability is expected by re-training the weight vectors using LVQ2.1. The proposed classification method is applied to the SAR data with unknown category to the trained neural network, and the data is classified to the category where the winner neuron belongs.

3.2 Feature Vector Selection

Next, we describe the selection method of an appropriate feature vector using the JM distance for the classification. The JM distance is a divergence which measures the separability of a pair of probability distributions. The multi-looking polarimetric SAR data has a nine-dimensional real vector **x** given by

$$\begin{aligned}\mathbf{x} = \big[&S_{hv}S_{hv}^{*},\ S_{vv}S_{vv}^{*},\ S_{hh}S_{hh}^{*},\ \Re\{S_{hh}S_{hv}^{*}\},\ \Im\{S_{hh}S_{hv}^{*}\},\\ &\Re\{S_{hh}S_{vv}^{*}\},\ \Im\{S_{hh}S_{vv}^{*}\},\ \Re\{S_{hv}S_{vv}^{*}\},\ \Im\{S_{hv}S_{vv}^{*}\}\big]^{T}.\end{aligned} \tag{15}$$

Here, S_{hv} denotes the complex scattering coefficient, where the transmitting and receiving polarization are vertical and receiving, respectively. The symbols $\Re$ and $\Im$ indicate the real and imaginary parts of a complex number, and * the complex conjugation. Note that **x** is derived from the Stokes matrix. The relative phase $\phi_{hhvv}^{(t)}$ between S_{hh} and S_{vv} is given by

$$\phi_{hhvv}^{(t)} = \tan^{-1}\frac{\Im\{S_{hh}S_{vv}^{*}\}}{\Re\{S_{hh}S_{vv}^{*}\}},\qquad 0^\circ \le \phi_{hhvv}^{(t)} < 360^\circ \tag{16}$$

Since divergences using $\phi_{hhvv}^{(t)}$ between categories may be small, we introduce pseudo-relative phase $\phi_{hhvv}^{(c)}$ defined by

$$\phi_{hhvv}^{(c)} = \cos^{-1}\frac{\Re\{S_{hh}S_{vv}^{*}\}}{\sqrt{S_{hh}S_{hh}^{*}S_{vv}S_{vv}^{*}}},\qquad 0^\circ \le \phi_{hhvv}^{(c)} < 360^\circ \tag{17}$$

Note that $\phi_{hhvv}^{(c)}$ suffers from a mirror effect such that values of $\phi_{hhvv}^{(t)}$ in $180^\circ \le \phi_{hhvv}^{(t)} < 360^\circ$ are reflected to the range of $0^\circ \le \phi_{hhvv}^{(c)} < 180^\circ$. Relative phases $\phi_{hhvv}^{(t)}$ and $\phi_{hhvv}^{(c)}$ between S_{hh} and S_{hv}, and $\phi_{hvvv}^{(t)}$ and $\phi_{hvvv}^{(c)}$ between S_{hv} and S_{vv} are defined in the same manner [16], [17], [23], and [24].

An appropriate feature vector is selected from the nine elements in (15) and six relative phase elements according to the following method. First, we estimate one-dimensional probability density function (PDF) for each category in every element based on data collected from test sites. JM distances between all category pairs for each element are computed and their averages are obtained. Thus, the feature vector elements can be selected in order of magnitudes of the average JM distance.

3.3 Experimental Results

In this section we will show some numerical results for the preceeding methods presented.

3.3.1 SAR Data and Study Area

The experimental study area is a rectangular stretch of land (approximately 8.7 km by 12.5 km) in Sakaide, Japan. Figure 6 illustrates the study map. The area includes vegetation, industrial area, and densely populated districts. We perform experiments using this area in particular, as it is possible to obtain training and test data easily. "Data take 61:40" indicated in Table 2 was acquired in quadruple polarization mode (both horizontal and transmission and reception polarizations) at C-band (5.7 cm) and L-band (24.0 cm) wavelengths. The study area is covered by the SIR-C data. A gray scale image with $696\times 1{,}000$ pixels corresponding to the study area is shown in Fig. 7.

In relatively high relief areas, foreshortening and layover effects may occur if side looking radar is used. Backscattering coefficients from these areas may be different from observation values estimated by target shapes and dielectric characteristics. Thus, in this chapter classification of only the smooth areas is considered. When referring to the topographical map corresponding to Fig. 6, the high relief areas in Fig. 7 are masked. These areas will not be classified but called masked areas. We define five categories (L=5), ω_1 : factory, ω_2 : golf course, ω_3 : vegetation, ω_4 : urban, and ω_5 : water, respectively. Therefore, the number of output neurons M is a multiple of five in the competitive neural network. In the experiments, the training and test data set with 500 samples for each category was used. Figure 8 shows a location diagram of the training and test data where the white regions represent unselected areas. The training data are denoted by ■, ▲, and ▼ where ■ and ▲ show 10 by 10 and 5 by 5 pixels areas, respectively and ▼ is a closed region of 25 pixels. The reason for using ▼ is that it shows an area of category ω_2 consisting of golf courses where the area shape is too complex to be described by squares. The training data consisting of 500 pixels for each category was obtained from a field survey. We adopt the entire part of ω_1 and ω_2 areas as the test data since there are not so many pixels. The remaining categories ω_3, ω_4, and ω_5 range over wider areas whose scales are different. To make the adopted test data equal in size, we pick up one pixel at intervals of 3, 4, and 32 pixels from the ω_3, ω_4, and ω_5 areas, respectively. Here, all classification methods employ the same training and test data in order to compare classification accuracies. The scoring matrix called confusion matrix [2], [12] is computed for each classification result. We will evaluate the classification methods using average accuracy defined by averaging the trace values of the confusion matrix. It is not meaningful to evaluate confusion matrices using the overall accuracy and miss-classification rate in this experiment, since the test data for some parts of the study area are missing, as can be seen in Fig. 8.

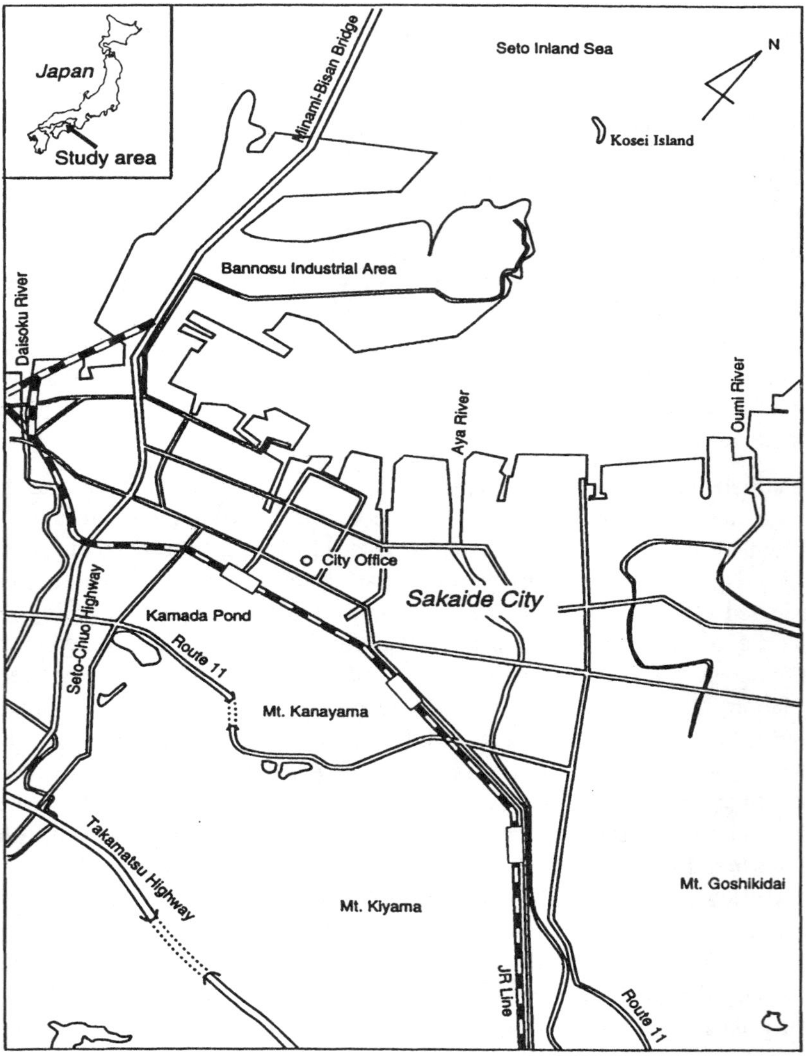

Fig. 6. Study area (Sakaide, Japan) for category classification.

Fig. 7. Gray scale image of SIR-C data for the study area.

Table 2. SIR-C data characteristics.

Data take No.	61:40
Date	April 13, 1994
Time	4:26:00.981 (GMT)
Wavelength	C-band (5.7cm), L-band (24.0cm)
Polarizations	HH, VV, HV, VH
Scene center	34°9.4´N, 133°58.6´E
Incidence angle range	26.3°–29.6°
Image size	Range 16.2 km/Azimuth 99.9 km
Digital image dimensions	1,296 pixels × 7,995 lines

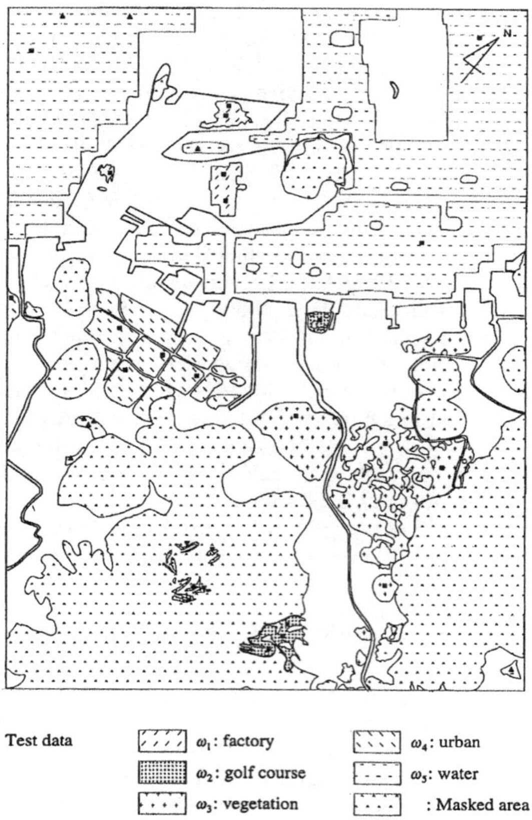

Fig. 8. Training and test data area.

3.3.2 Feature Vector

The JM distances for all category pairs are computed by using the test data of each element for appropriate selection of a feature vector. The average JM distances with all category pairs using C-band and L-band backscattering are calculated. The C-band and L-band backscattering data have a similar tendency. As the JM distances of the backscattering coefficients become larger, inclusion of these elements leads to greater effectiveness. Moreover, the JM distance of the pseudo-relative phase $\phi_{hhvv}^{(c)}$ is considerably larger than the other relative phases. Therefore, we select $[10\log_{10}(S_{hv}S_{hv}^{*})$, $10\log_{10}(S_{vv}S_{vv}^{*})$, $10\log_{10}(S_{hh}S_{hh}^{*})$, $\phi_{hhvv}^{(c)}]$ for both C-band and L-band as the best combination, i.e., there are eight elements in the selected feature vector. The input layer also has eight neurons (N=8).

Next, we note that $\phi_{hhvv}^{(t)}$ is inferior to $\phi_{hhvv}^{(c)}$ with respect to the divergence between categories. Investigation of the PDF of $\phi_{hhvv}^{(t)}$ for each category using the C-band revealed that the phase $\phi_{hhvv}^{(t)}$ for factory area (ω_1) differs from other categories by about $180°$. The phase $\phi_{hhvv}^{(t)}$ for water area (ω_5) has a concentration around $0°$, and $\phi_{hhvv}^{(t)}$ of ω_2, ω_3, and ω_4 vary with the PDF peaks. Hence, applying $\phi_{hhvv}^{(t)}$ to a classification method would be significant. The differences of the PDFs for $\phi_{hhvv}^{(c)}$ are significant among most of the categories compared with those of $\phi_{hhvv}^{(t)}$.

3.3.3 Classification Results

Table 3 indicates four kinds of learning algorithms applied to the LVQ methods in the experiments. Each algorithm includes either LVQ1 or OLVQ1, and the latter two algorithms incorporate LVQ2.1 as the second step of the algorithm. The number of training iterations for each step of the algorithm is determined by preliminary experiments. We set the initial learning rates to 0.03 for LVQ1 and LVQ2.1 and to 0.3 for OLVQ1. The window parameter is set to 0.3 in LVQ2.1.

Table 3. Training iterations.

Learning algorithms	Number of training iterations	
	Step1	Step 2
LVQ1	40,000	---
OLVQ1	10,000	---
LVQ1+LVQ2.1	40,000	40,000
OLVQ1+LVQ2.1	10,000	40,000

The LVQ methods are compared with the conventional methods of maximum likelihood (ML) method and the back-propagation (BP) method. The BP method employs three-layered neural network, where the number of neurons H in the hidden layer varies from 3 to 15. For the LVQ methods, five times as many, i.e., 15 to 75 neurons are used in the competitive layer for the LVQ methods. The selected eight-dimensional feature vectors are then adopted for each method and their test data classified.

Figure 9 shows this comparison with respect to the average accuracies. Improvement of classification accuracy by adding LVQ2.1 to LVQ1 is 1.95% on average as shown in Fig. 9(a). Similarly, improvement of classification accuracy by adding LVQ2.1 to OLVQ1 is 1.42% on average as shown in Fig. 9(b). LVQ1+LVQ2.1 can produce higher average accuracies than the ML method. However, OLVQ1+LVQ2.1 can only produce similar average accuracies to the ML method. Thus, LVQ1+LVQ2.1 having M ($35 \leq M \leq 70$) produces better results than other methods as shown in Fig. 9(c). Especially, when the number of output neurons M is 40, it yields a highest accuracy reading of 86.40%. In most cases, the LVQ methods outperform the BP method. The resultant image of LVQ1+LVQ2.1 turns out to be more distinguishable than those obtained by the ML method where comparisons of the image quality are made.

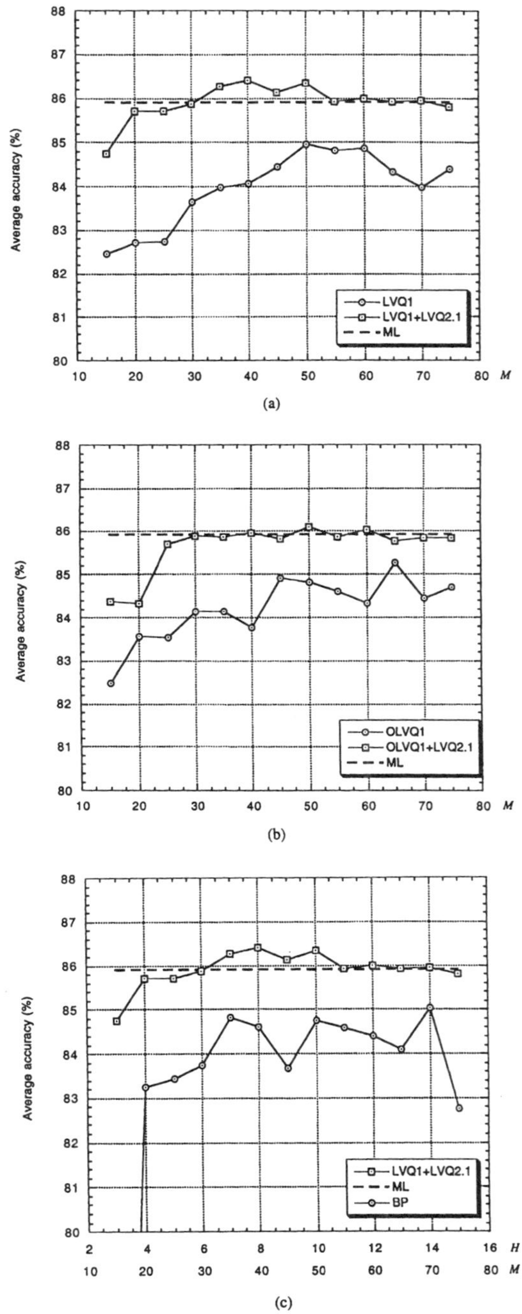

Fig. 9. Comparison of classification accuracy among LVQ, ML and BP methods.

4 Conclusions

We have proposed a new classification method of the remote sensing data, which uses the self-organizing neural network constructed from several neural network blocks, corresponding to the categories. In the experimental study, these blocks worked well as vector quantizers. The unknown pixels are successfully classified as well, based on the *k*-nearest neighbor rule. We could demonstrate that the proposed method generally yields higher classification accuracies compared to other methods. If all neurons in the competitive layer have equal firing probability, the network works well as the vector quantization of the training data.

Then we proposed a classification method that employs a competitive neural network and a procedure to select a suitable feature vector. After a preliminary experiment, three backscattering coefficients and a pseudo-relative phase between S_{hh} and S_{vv} are selected, using the JM distances. Among the methods considered, the neural network trained by LVQ1+LVQ2.1 produces the best classification results.

For future study, we will consider a classification algorithm which can classify the data into more detailed categories using multi-frequency and multi-polarimetric SAR data as well as physical backscattering models.

References

1. Richards J. A. (1993), *Remote Sensing Digital Image Analysis*, 2nd Edn. Springer-Verlag, Berlin, Germany, pp. 246-251
2. Hara Y., Atkins R. G., Yueh S. H., Shin R. T., and Kong J. A. (1994), Application of neural networks to radar image classification, *IEEE Transactions on Geoscience and Remote Sensing*, **32**, pp. 100-109.
3. Hara Y., Atkins R. G., Shin R. T., Kong J. A., Yueh S. H., and Kwok R. (1995), Application of neural networks for sea ice classification in polarimetric SAR images, *IEEE Transactions on Geoscience and Remote Sensing*, **33**, pp. 740-748.
4. Heermann P.D. and Khazenie N. (1992), Classification of multispectral remote sensing data using a back-propagation neural network, *IEEE Transactions on Geoscience and Remote Sensing*, **30**, pp. 81-88.
5. Yoshida T., Omatu S. and Teranishi M. (1991), Pattern classification for remote sensing data using neural network, *Transactions of the Institute of Systems, Control and Information Engineers*, **4**, pp. 11-20.
6. Yoshida T. and Omatu S. (1994), Neural network approach to land cover mapping, *IEEE Transactions on Geoscience and Remote Sensing*, **32**, pp. 1103-1109.
7. Hecht-Nielsen R. (1990), *Neurocomputing*, Addison-Wesley, New York, NY.

8. Ulaby F. T. and Elachi C. (1990), *Radar Polarimetry for Geoscience Applications*, Artech House, Norwood, MA.
9. Van Zyl J. J., Zebker H. A., and Elachi C. (1987), Imaging radar polarization signatures: Theory and observation, Radio Science, **22**, pp. 529-543.
10. Lim H. H., Swartz A. A., Yueh H. A., Kong J. A., Shin R. T., and Van Zyl J. J. (1989), Classification of earth terrain using polarimetric synthetic aperture radar images. *Journal of Geophysical Research*, **94**, pp. 7049-7057.
11. Van Zyl J. J. and Burnette C. F (1992), Bayesian classification of polarimetric SAR images using adaptive a priori probabilities, *International Journal of Remote Sensing*, **13**, pp. 835-840.
12. Yueh H. A., Swartz A. A., and Kong J. A. (1988), Bayes classification of terrain cover using normalized polarimetric data., *Journal of Geophysical Research*, **93**, pp. 15261-15267.
13. Chen K. S., Huang W. P., Tsay D. H., and Amar F. (1996), Classification of multifrequency polarimetric SAR imagery using a dynamic learning neural network, *IEEE Transactions on Geoscience and Remote Sensing*, **34**, pp. 814-820.
14. Decatur, S. E. (1989), Application of Neural Networks to Terrain Classification, *Proceedings of International Joint Conference on Neural Networks,* Washington, DC, pp. 283-288.
15. Kanellopoulos I., Varfis A., Wilkinson G. G., and Megier J. (1992), Land-cover discrimination in SPOT HRV imagery using an artificial neural network - a 20-class experiment, *International Journal of Remote Sensing*, **13**, pp. 917-924.
16. Ito Y. and Omatu S. (1996), Land cover mapping method using a competitive neural network, *Journal of the Japan Society of Photogrammetry and Remote Sensing*, **35**, pp. 5-17.
17. Ito Y. and Omatu S. (1997), Category classification method using a self-organizing neural network, *International Journal of Remote Sensing*, **18**, pp. 829-845.
18. Kohonen T. (1997), *Self-Organizing Maps*, Springer-Verlag, Berlin, Germany, pp. 203-220.
19. Kohonen T. (1984), *Self-Organization and Associative Memory*, Springer-Verlag, New York, NY.
20. Kohonen T. (1988), An introduction to neural computing, *Neural Networks*, **1**, pp. 3-16.
21. Stofan E. R., Evans D. L., Schmullius C., Holt B., Plaut J. J., Van Zyl J., Wall S. D., and Way J. (1995), Overview of results of spaceborne imaging radar-C, X-band synthetic aperture radar (SIR-C/X-SAR), *IEEE Transactions on Geoscience and Remote Sensing*, **33**, pp. 817-828.
22. DeSieno D. (1988), Adding a conscience to competitive learning, *Proceedings of 2nd Annual IEEE International Conference on Neural* Networks, San Diego, CA, pp. 117-124.
23. Ito Y. and Omatu S. (1994), Category classification using self-organization model, *Proceedings of International Symposium on Noise and Clutter Rejection in Radars and Imaging Sensors*, Kawasaki, Japan, pp. 537-542.
24. Ito Y. and Omatu S. (1997), Land cover mapping method for polarimetric SAR data, in *Algorithms for Synthetic Aperture Radar Imagery IV*, SPIE, Bellingham, MA, pp. 388-397.
25. Barndorff-Nielsen O. E., Jensen J. L., and Kendall W. S. (1993), *Networks and Chaos-Statistical and Probabilistic Aspects*, Chapman & Hall, London, UK, pp. 40-123.

List of Contributors

Xiao-Zhi Gao and
Seppo J. Ovaska
Helsinki University of Technology
Institute of Intelligent Power Electronics
Otakaari 5 A
FIN-02150 Espoo, Finland
gao@csc.fi
ovaska@ieee.org

Marian P. Kazmierkowski
Warsaw University of Technology
Department of Electrical Engineering
Institute of Control and Industrial Electronics
ul. Koszykowa 75, Gm. E.
00-662 Warsaw, Poland
mpk@isep.pw.edu.pl

Teresa Orlowska-Kowalska
Technical University of Wroclaw
Institute of Electric Machines and Drives
ul. Smoluchowskiego 19
50-372 Wroclaw, Poland
tok@imne.pwr.wroc.pl

Adrian David Cheok
National University of Singapore
Department of Electrical and Computer Engineering
Singapore 117576
adriancheok@nus.edu.sg

Bogdan Tomescu
Space Systems/Loral
3825 Fabian Way, M/S G23
Palo Alto, CA 94303, U.S.A.
tomescu.bogdan@ssd.loral.com

Hugh F. VanLandingham
Virginia Polytechnic Institute and State University
The Bradley Department of Electrical and Computer Engineering
340 Whittemore Hall
Blacksburg, VA 24061, U.S.A.
hughv@vt.edu

Akimoto Kamiya,
Masakazu Kato, and
Kazue Shimada
Toshiba Corporation
1, Toshiba-Cho, Fuchu-Shi
Tokyo 183-8511, Japan
akimoto.kamiya@toshiba.co.jp
masakazu.kato@toshiba.co.jp
kazue.shimada@toshiba.co.jp

Shigenobu Kobayashi
Tokyo Institute of Technology
4259, Midori-Ku, Nagatsuda
Yokohama 226-8502, Japan
kobayashi@dis.titech.ac.jp

Adrian Halinka,
Pawel Sowa, and
Michal Szewczyk
Technical University of Silesia
Institute of Electric Power Engineering
ul. B. Krywoustego 2
44-101 Gliwice, Poland
ahal@iesu.pols.gliwice.pl
pawels@eas.pols.gliwice.pl
michals@iesu.pols.gliwice.pl

Les M. Sztandera
Philadelphia University
School of Science and Health
Computer Science Department
Philadelphia, PA 19144, U.S.A.
sztanderal@philau.edu

Jarkko Niittymäki,
Riku Nevala, and
Marko Mäenpää
Helsinki University of Technology
Laboratory of Transportation Engineering
P.O. Box 2001
FIN-02015 HUT, Finland
jarkko.niittymaki@hut.fi
riku.nevala@hut.fi
marko.maenpaa@hut.fi

Timothy G.W. Gordon and
Peter J. Bentley
University College London
Department of Computer Science
Gower Street
London WC1E 6BT, United Kingdom
t.gordon@cs.ucl.ac.uk
p.bentley@cs.ucl.ac.uk

Toru Fujinaka,
Michifumi Yoshioka, and
Sigeru Omatu
Osaka Prefecture University
Graduate School of Engineering
Division of Computer and Systems Sciences
1-1, Gakuen-Cho, Sakai
Osaka 599-8531, Japan
fuji@cs.osakafu-u.ac.jp
yoshioka@cs.osakafu-u.ac.jp
omatu@cs.osakafu-u.ac.jp

About the Editors

Dr. Seppo J. Ovaska is a Professor of Industrial Electronics and Vice Chair of the Department of Electrical and Communications Engineering at the Helsinki University of Technology in Espoo, Finland. His international experience in industrial electronics spans 21 years and includes R&D contributions in elevator group dispatching, overlay modernization of high-rise elevator banks, and motion control. Dr. Ovaska is a prolific author in soft computing applications and predictive signal processing, having published 60 journal papers, as well as over 100 conference publications, and 2 book chapters. In addition, he has co-edited the book *Soft Computing in Industrial Applications* (Springer, 2000), and holds 9 patents in the area of elevator systems and control.

Dr. Ovaska received the D.Sc. degree in electrical engineering from the Tampere University of Technology, Finland, in 1989. He earned the Lic.Sc. degree in computer science and engineering from the Helsinki University of Technology in 1987, and the M.Sc. in electrical engineering from the Tampere University of Technology in 1980. Prior to joining the Helsinki University of Technology in 1996, Dr. Ovaska was a Professor of Electronics at the Lappeenranta University of Technology, Finland (1992–96). From 1980 to 1992, he held engineering, research, and R&D management positions with Kone Corporation and Nokia Research Center, both in Finland and Kentucky, U.S.A. He has taught university courses at the graduate and undergraduate level in soft computing, signal processing, real-time systems, microcomputer hardware design, and computer architectures. He serves as associate editor for the IEEE Transactions on Industrial Electronics. In addition, he is the coordinating chair of technical committees in the area of systems, and an elected member of the administrative committee, IEEE Systems, Man, and Cybernetics Society.

Dr. Ovaska served as a Visiting Scholar at the Muroran Institute of Technology, Japan, in the summer of 1999, and at the Virginia Polytechnic Institute and State University, U.S.A., in the summers of 2000 and 2001. He is a senior member of the IEEE, a member of the International Society for Computers and Their Applications, and a member of the American Society for Engineering Education. He was the founding general chair of the IEEE Midnight-Sun Workshop on Soft Computing Methods in Industrial Applications in 1999, and the general chair of the 5th Online World Conference on Soft Computing in Industrial Applications in 2000. Dr. Ovaska is a recipient of the Outstanding Contribution Award of the IEEE Systems, Man, and Cybernetics Society. He is listed in the Marquis Who's Who in Science and Engineering, Who's Who in Finance and Industry, and Who's Who in the World.

Dr. Les M. Sztandera is an Associate Professor of Computer Science and Head of the Computer Science Program at the Philadelphia University, Philadelphia, Pennsylvania, U.S.A.

He has been involved in soft computing teaching and research since 1987. Dr. Sztandera has 11 years of full time university teaching experience, and is a recipient of a Teaching Excellence Award. He developed a sequence of soft computing courses coupled with laboratory assignments in which students work with real life problems, such as detecting an industrial pollutant, predicting strength and density of materials, designing a medical expert system, simulating protective systems in complex power generating units, detecting carcinogenic dyes, or designing new drugs.

Complementary with his teaching effort, Dr. Sztandera has been involved in a variety of research activities. That has resulted in numerous research grants from the Department of Commerce, National Textile Center, National Science Foundation, Ohio Supercomputer Center, Pittsburgh Supercomputer Center, and American Heart Association. Over $1,000,000 in research funding has been experienced. Those research activities also resulted in 25 journal publications and 40 conference presentations.

Dr. Sztandera received his Ph.D. degree from the Department of Electrical Engineering and Computer Science, University of Toledo, Ohio, U.S.A., with a dissertation on "Fuzzy Sets in Self-Generating Neural Network Architectures." He earned his M.Sc. degree from the Department of Computer Science and Engineering, University of Missouri, Missouri, U.S.A., with a thesis on "Spatial Relations Among Fuzzy Subsets of an Image," and a Diploma in English from University of Cambridge, Cambridge, England.

Dr. Sztandera is a member of professional organizations in the U.S. and Canada: the North American Fuzzy Information Processing Society, Association for Computing Machinery, and Canadian Society for Fuzzy Information and Neural Systems. His scientific and scholarly research contributions to the fuzzy set theory are internationally recognized. He proposed, designed, and implemented fuzzy neural trees. For this and other contributions to the fuzzy sets and systems theory, he was included in the Encyclopedia of Computer Science and Technology, 1999 Edition. Dr. Sztandera is also listed in the Marquis Who's Who in Science and Engineering, Who's Who in the World, Who's Who in America, and Who's Who in the East.

Keyword Index

P

R

S

T

U

V